AF862979

Intelligent Control and Learning Systems

Volume 8

Series Editor

Dong Shen, School of Mathematics, Renmin University of China, Beijing, Beijing, China

The Springer book series Intelligent Control and Learning Systems addresses the emerging advances in intelligent control and learning systems from both mathematical theory and engineering application perspectives. It is a series of monographs and contributed volumes focusing on the in-depth exploration of learning theory in control such as iterative learning, machine learning, deep learning, and others sharing the learning concept, and their corresponding intelligent system frameworks in engineering applications. This series is featured by the comprehensive understanding and practical application of learning mechanisms. This book series involves applications in industrial engineering, control engineering, and material engineering, etc.

The Intelligent Control and Learning System book series promotes the exchange of emerging theory and technology of intelligent control and learning systems between academia and industry. It aims to provide a timely reflection of the advances in intelligent control and learning systems. This book series is distinguished by the combination of the system theory and emerging topics such as machine learning, artificial intelligence, and big data. As a collection, this book series provides valuable resources to a wide audience in academia, the engineering research community, industry and anyone else looking to expand their knowledge in intelligent control and learning systems.

Qiongxia Yu · Ting Lei · Fengchen Tian ·
Zhongsheng Hou · Xuhui Bu

Predictive Learning Control for Unknown Nonaffine Nonlinear Systems

Theory and Applications

Springer

Qiongxia Yu
Henan Key Laboratory of Intelligent Detection and Control of Coal Mine Equipment
School of Electrical Engineering and Automation
Henan Polytechnic University
Jiaozuo, Henan, China

Fengchen Tian
Henan Key Laboratory of Intelligent Detection and Control of Coal Mine Equipment
School of Electrical Engineering and Automation
Henan Polytechnic University
Jiaozuo, Henan, China

Xuhui Bu
Henan Key Laboratory of Intelligent Detection and Control of Coal Mine Equipment
School of Electrical Engineering and Automation
Henan Polytechnic University
Jiaozuo, Henan, China

Ting Lei
School of Electrical and Information Engineering
Zhengzhou University of Light Industry
Zhengzhou, Henan, China

Zhongsheng Hou
School of Automation
Academy of Systems Science and Control
Qingdao University
Qingdao, Shandong, China

ISSN 2662-5458 ISSN 2662-5466 (electronic)
Intelligent Control and Learning Systems
ISBN 978-981-19-8856-1 ISBN 978-981-19-8857-8 (eBook)
https://doi.org/10.1007/978-981-19-8857-8

© The Editor(s) (if applicable) and The Author(s), under exclusive license to Springer Nature Singapore Pte Ltd. 2023
This work is subject to copyright. All rights are solely and exclusively licensed by the Publisher, whether the whole or part of the material is concerned, specifically the rights of translation, reprinting, reuse of illustrations, recitation, broadcasting, reproduction on microfilms or in any other physical way, and transmission or information storage and retrieval, electronic adaptation, computer software, or by similar or dissimilar methodology now known or hereafter developed.
The use of general descriptive names, registered names, trademarks, service marks, etc. in this publication does not imply, even in the absence of a specific statement, that such names are exempt from the relevant protective laws and regulations and therefore free for general use.
The publisher, the authors, and the editors are safe to assume that the advice and information in this book are believed to be true and accurate at the date of publication. Neither the publisher nor the authors or the editors give a warranty, expressed or implied, with respect to the material contained herein or for any errors or omissions that may have been made. The publisher remains neutral with regard to jurisdictional claims in published maps and institutional affiliations.

This Springer imprint is published by the registered company Springer Nature Singapore Pte Ltd.
The registered company address is: 152 Beach Road, #21-01/04 Gateway East, Singapore 189721, Singapore

Preface

This monograph investigates both theory and applications of predictive learning control (PLC) for unknown and complex nonaffine nonlinear systems that operate in a repetitive pattern. PLC combines predictive control in time domain with learning control. Predictive control in time domain can use more future information for controller design, meanwhile can deal with system constraints. However, it cannot learn from historical experience and the undesired transient behaviors in the initial operation stage cannot be eliminated no matter how many times the system repeats. Learning control can learn from historical operation processes, and it can also achieve perfect tracking for each time point over the whole time interval. By absorbing the advantages of these two methods, PLC can not only get an optimal and predictive control input, but also achieve an improved control performance and eventual perfect tracking through learning.

This is the first monograph that focuses on PLC for unknown nonaffine nonlinear systems. Readers of this monograph will learn the design, theoretical analysis, and practical application of PLC methods without using any mechanism model information of the system and learn how to cope with various practical problems such as system constraints, varying trial lengths, unknown and time-varying input delay, available and unavailable system states, and so on.

This monograph consists of ten chapters and is divided into two parts. Chapter 1 is an introduction to predictive control, learning control, and predictive learning control (PLC). Part I focuses on design and theoretical analysis of predictive iterative learning control (PILC) which is the main and hot topic of PLC and is divided into six chapters. From Chaps. 2 to 7, PILC for unknown nonaffine nonlinear systems, constrained PILC, PILC with varying trial lengths, PILC with unknown time delay, PILC with full available and unavailable states are designed, respectively. Part II focuses on applications of PILC and predictive repetitive control (PRC) which is another main topic of PLC to practical railway and road transportation systems and is divided into three chapters. In Chap. 8, PILC is applied to repeatable automatic high-speed train operation systems. PRC is applied to periodic medium-scale and large-scale urban traffic systems in Chaps. 9 and 10, respectively.

The first author would like to thank her doctoral supervisor, Prof. Zhongsheng Hou, Qingdao University, for all the guidance and help he has given the first author. Moreover, the first author also wants to thank her brothers and sisters in Prof. Hou's team for their support. In addition, many thanks to the master students, Zhihao Fan and Yiteng Hou of the first author for their careful proofreading of this monograph. Finally, all five authors would like to express sincere appreciation to their families for their understanding and love.

The authors gratefully acknowledge the support of the National Natural Science Foundation of China (Nos. 62003133, 62273133, 61833001); Natural Science Foundation of Henan Province of China (No. 202300410177); Fundamental Research Funds for the Universities of Henan Province (Nos. NSFRF200324, NSFRF210449); Key Scientific Research Projects of Universities in Henan Province (No.20B413002); Research and Practice Project of Higher Education Teaching reform in Henan Province (No.2021SJGLX1011); Innovative Scientists and Technicians Team of Henan Polytechnic University under Grant (No.T2019-2); Innovative Scientists and Technicians Team of Henan Provincial High Education (No. 20IRTSTHN019); Doctoral Research Fund of Zhengzhou University of Light Industry (No. 2021BSJJ016).

Jiaozuo, China
September 2022

Qiongxia Yu

Contents

Chapter 1
Introduction

1.1 Predictive Control

Predictive control was developed from complex industrial processes in the late 1970s (Richalet 1978; García et al. 1989; Xi 1993; Grüne and Pannek 2011), and it has possesses abundant research results (Mayne 2014; Zhu and Xia 2016; Dubljevic and Humaloja 2020; Köhler et al. 2020; Chen et al. 2022). The basic idea of predictive control is to predict future output sequence by using system model. The main method is to solve a future finite horizon optimization problem (under system constraints) with required performance indexes at each time point and then obtain a certain number of future optimal control inputs. By utilizing receding horizon strategy which is a distinguishing characteristic of optimization in predictive control, only the current optimal control input is applied to the system. At the next time point, the optimization horizon will be moved one-step ahead and the finite horizon optimization problem will be solved again.

A simple block diagram of predictive control method in time domain is shown in Fig. 1.1, where $u(k)$ and $y(k)$ are the system input and output at time point k. $y_M(k)$ is a model description of the controlled plant. $y_P(k+i), i = 1, \ldots, m$ is the predicted output for the future m step time points at the current time point k. Note that if the design of $y_P(k+i), i = 1, \ldots, m$ relies only on the model output $y_M(k)$ and does not use the actual output information $y(k+1)$, it is called the open-loop prediction, while if the model error information $\tilde{e}^{\mathfrak{tw}}(k+1) = y(k+1) - y_M(k+1)$ is also used for the design of $y_P(k+i), i = 1, \ldots, m$ (as shown in the dashed line), it is called the closed-loop prediction. $y_d^{\mathfrak{tw}}(k)$ is the desired output trajectory. The cost function $J^{\mathfrak{tw}}(k+i), i = 1, \ldots, m$ aims at obtaining an optimal control input sequence that makes the predicted output $y_P(k+i), i = 1, \ldots, m$ as close to the desired trajectory $y_d^{\mathfrak{tw}}(k+i), i = 1, \ldots, m$ as possible.

Although this control method can get optimal and predictive control inputs and can deal with system constraints, it has some limitations when it is applied to practical complex industrial processes.

© The Author(s), under exclusive license to Springer Nature Singapore Pte Ltd. 2023

Q. Yu et al., *Predictive Learning Control for Unknown Nonaffine Nonlinear Systems*, Intelligent Control and Learning Systems 8,
https://doi.org/10.1007/978-981-19-8857-8_1

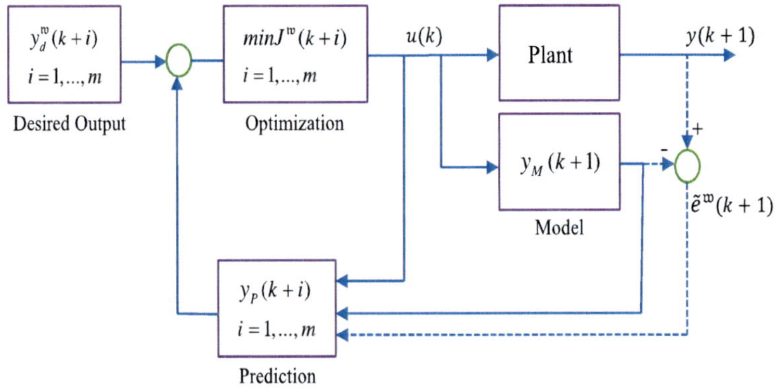

Fig. 1.1 Block diagram of predictive control in time domain

1. The observed transient behavior in the initial operation stage which is very common in time domain control methods will deteriorate the quality of operation control so as to affect the production quality of industrial products.
2. It cannot learn from historical experience in previous operations and hence cannot achieve gradually improved control performance.

In fact, historical experience in previous operations contains abundant operation information of the system. It is significant to make the predictive control have ability to learn from experience and eliminate the undesirable transient behavior, such that satisfactory control performance and broad economic and operational benefits for practical industrial system will be expected.

1.2 Learning Control

One of the most essential characteristics of living things is "learning". Learning is a basic and intelligent behavior of human beings and plays a very important role in the process of human evolution. In the field of control and system engineering, it is always a goal to make the control system have the ability of learning, so as to achieve an improved control performance gradually and iteratively.

To introduce learning into cybernetics, learning control is proposed and has two main research interests that are iterative learning control (ILC) (Uchiyama 1978; Arimoto et al. 1984) and repetitive control (RC) (Inoue et al. 1981; Omata et al. 1984). In fact, both ILC and RC have the common feature that they can learn dynamic behavior of the system from repetition. Up to now, these two control methods have been widely studied in both theory and applications (Rogers and Owens 1992; Hillerström and Walgama 1996; Longman 2000; Xu and Tan 2003; Zhang et al. 2003; Bristow et al. 2006; Wang et al. 2009; Wu et al. 2013; Owens 2016; Pandove and Singh 2018; Shen and Li 2019; Yu and Hou 2021; Astolfi et al. 2021; Bu et al. 2021; Zhang et al. 2022).

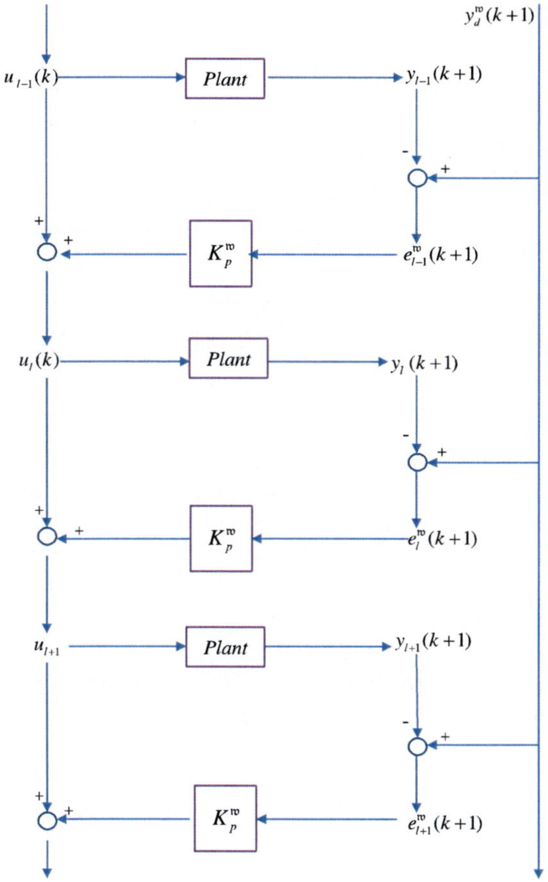

Fig. 1.2 Block diagram of classical P-type ILC

ILC is a typical method for systems that operates in a repetitive pattern over a finite time interval and track repeatable control tasks from iteration to iteration and has been applied to a variety of practical systems. For example, the high-speed train always runs on the specified line over the finite and specified time interval every day. Therefore, ILC is very suitable for high-speed train to achieve automatic train control.

A block diagram of classical proportional-type (P-type) ILC is presented in Fig. 1.2, where $u_l(k)$ and $y_l(k)$ are the system input and output at time k of the lth iteration. $k \in \{0, 1, \ldots, K^{\varpi}\}$ denotes the finite time instant. K^{ϖ} is the running time length at each iteration of the system. $e_l^{\varpi}(k) = y_d^{\varpi}(k) - y_l(k)$ is the tracking output error between the desired output $y_d^{\varpi}(k)$ and the actual output $y_l(k)$. K_p^{ϖ} is a proportional learning gain. According to Fig. 1.2, the iterative learning law of P-type ILC can be written as for $\forall k \in \{0, 1, \ldots, K^{\varpi}\}$

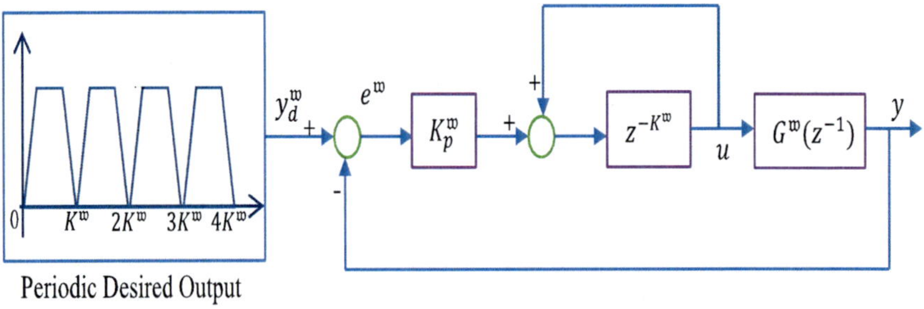

Fig. 1.3 Block diagram of classical P-type RC

$$u_{l+1}(k) = u_l(k) + K_p^{\mathfrak{w}} e_l^{\mathfrak{w}}(k+1) \tag{1.1}$$

In contrast to ILC, RC is an effective method for systems that repetitively and periodically operate on time domain and track periodic desired trajectories or reject periodic disturbances, and it also has been widely applied to practical systems such as traffic control systems that exhibit repeatability and periodicity in a daily, weekly, monthly or even yearly manner, proper control of power filter where the signals to be controlled are periodic in nature, and so on.

A block diagram of classical proportional-type (P-type) RC is presented in Fig. 1.3, where $K^{\mathfrak{w}}$ is the time length in a period of the periodic desired trajectory. z^{-1} is a backward shift operator. According to Fig. 1.3, the repetitive learning law of P-type RC is described as

$$u(k) = u(k - K^{\mathfrak{w}}) + K_p^{\mathfrak{w}} e^{\mathfrak{w}}(k - K^{\mathfrak{w}}) \tag{1.2}$$

1.3 Predictive Learning Control

To make the predictive control have the ability of learning, predictive learning control is developed and has two main branches that are predictive iterative learning control (PILC) (Amann et al. 1998; Lee and Lee 2000; Chu et al. 2016; Oh et al. 2018; Shi et al. 2014; Lu et al. 2019; Ma et al. 2020) and predictive repetitive control (PRC) (Lee et al. 2001; Gupta and Lee 2006; Wang et al. 2013; Lu et al. 2016; Wang et al. 2022).

With the deepening of theoretical research and promotion of application development, PILC has become a hot topic in the control of complex industrial processes (Amann et al. 1998; Lee and Lee 2000; Chu et al. 2016; Oh et al. 2018; Shi et al. 2014; Lu et al. 2019; Ma et al. 2020). A series of predictive ILC methods for known linear systems are proposed in Amann et al. (1998), Lee and Lee (2000), Chu et al. (2016), Oh et al. (2018). In Shi et al. (2014), a model-based predictive ILC in the framework of two-dimensional (2D) control system is designed. Lu et al. (2019) designs a multi-point model predictive ILC for a class of known nonlinear systems. In Ma et al. (2020), a control-affine feedforward neural network-based iterative learning model predictive control method is proposed for a class of affine nonlinear systems with known model structure. Compared with PILC, PRC mainly focuses on the control of periodic processes in time domain, and many PRC methods for known linear systems (Lee et al. 2001; Gupta and Lee 2006; Wang et al. 2013; Lu et al. 2016; Wang et al. 2022) are proposed until now.

It is worth pointing out that most of the existing works on PLC depend on the model information of the controlled system. In fact, accurate modeling for practical complex industrial processes is sometimes more difficult than the design of the control system itself. As a result, how to design a PLC for completely unknown nonaffine nonlinear systems without using any model information is of great significance.

1.4 Preview of This Monograph

This monograph is the first to investigate both theory and applications of predictive learning control (PLC) for unknown nonaffine nonlinear systems. To better demonstrate the design concept and analysis method, Part I focuses on predictive iterative learning control (PILC) which is the main and hot topic of PLC, and then, Part II gives applications of PILC together with another main topic of PLC, namely predictive repetitive control (PRC), on repeatable railway transportation systems and periodic road traffic systems, respectively.

Specifically, to avoid the difficult modeling problem for complex nonlinear systems, this monograph begins with the design and theoretical analysis of PILC without using any mechanism model information of the system. And then, a series of PILC methods are designed that can cope with system constraints, varying trial lengths, unknown and time-varying input delay, available and unavailable system states sequentially. Applications of PILC on automatic train operation systems and PRC on urban traffic systems are also studied. This monograph is intended for researchers, engineers, and graduate students who are interested in predictive control, learning control, intelligent transportation systems, and related fields.

This monograph is organized as follows. This chapter gives the introduction. Chaps. 2–7 show systematic procedures for design and analysis of a series of PILC methods for unknown nonaffine nonlinear systems with and without constraints, under varying trial lengths, unknown time delay, available and unavailable system states in sequence. The applications of PLC to practical systems are presented in Chaps. 8–10, including PILC for repeatable automatic high-speed train operation systems in Chap. 8 and PRC for periodic medium-scale and large-scale urban traffic systems in Chap. 9 and Chap. 10, respectively.

References

Amann N, Owens DH, Rogers E (1998) Predictive optimal iterative learning control. Int J Cont 69(2):203–226

Arimoto S, Kawamura S, Miyazaki F (1984) Bettering operation of robots by learning. J Robot Syst 1(2):123–140

Astolfi D, Marx S, van de Wouw N (2021) Repetitive control design based on forwarding for nonlinear minimum-phase systems. Automatica. https://doi.org/10.1016/j.automatica.109671

Bristow DA, Tharayil M, Alleyne AG (2006) A survey of iterative learning control. IEEE Cont Syst Magaz 26(3):96–114

Bu XH, Yu W, Yu QX, Hou ZS, Yang JQ (2021) Event-triggered model-free adaptive iterative learning control for a class of nonlinear systems over fading channels. IEEE Trans Cybernet 52(9):9597–9608

Chen SW, Wang T, Atanasov N, Kumar V, Morari M (2022) Large scale model predictive control with neural networks and primal active sets. Automatica 135:109947

Chu B, Owens DH, Freeman CT (2016) Iterative learning control with predictive trial information: convergence, robustness, and experimental verification. IEEE Trans Cont Syst Technol 24(3):1101–1108

Dubljevic S, Humaloja J-P (2020) Model predictive control for regular linear systems. Automatica 119:1–9

García CE, Prett DM, Morari M (1989) Model predictive control: theory and practice-A survey. Automatica 25(3):335–348

Grüne L, Pannek J (2011) Nonlinear model predictive control: theory and algorithms. Springer-Verlag, London

Gupta M, Lee JH (2006) Period-robust repetitive model predictive control. J Proc Cont 16:545–555

Hillerström G, Walgama K (1996) Repetitive control theory and applications-a survey. IFAC Proc 29(1):1446–1451

Inoue T, Nakano M, Iwai S (1981) High accuracy control of a proton synchrotron magnet power supply. IFAC Proceed 14(2):3137–3142

Köhler J, Müller MA, Allgöwer F (2020) A nonlinear model predictive control framework using reference generic terminal ingredients. IEEE Trans Autom Control 65(8):3576–3583

Lee KS, Lee JH (2000) Convergence of constrained model-based predictive control for batch processes. IEEE Trans Automat Cont 45(10):1928–1932

Lee JH, Natarajan S, Lee KS (2001) A model-based predictive control approach to repetitive control of continuous processes with periodic operations. J Proc Cont 11(2):195–207

Longman RW (2000) Iterative learning control and repetitive control for engineering practice. Int J Cont 73(10):930–954

Lu JY, Cao ZX, Gao FR (2016) Ellipsoid invariant setbased robust model predictive control for repetitive processes with constraints. IET Cont Theo Appl 10(9):1018–1026

Lu JY, Cao ZX, Gao FR (2019) Multi-point iterative learning model predictive control. IEEE Trans Ind Electron 66(8):6230–6240

Ma LL, Liu XJ, Kong XB, Lee KY (2020) Iterative learning model predictive control based on iterative data-driven modeling. IEEE Trans Neural Netw Learn Syst 32(8):3377–3390

Mayne DQ (2014) Model predictive control: recent developments and future promise. Automatica 50(12):2967–2986

Oh SK, Park BJ, Lee JM (2018) Point-to-point iterative learning model predictive control. Automatica 89:135–143

Omata T, Nakano M, Inoue T (1984) Application of repetitive control method to multivariable systems. Trans Soc Inst Cont Eng 20(9):795–800

Owens DH (2016) Iterative learning control: an optimization paradigm. Springer-Verlag, London

Pandove G, Singh M (2018) Robust repetitive control design for a three-phase four wire shunt active power filter. IEEE Trans Ind Inf 15(5):2810–2818

Richalet J (1978) Model predictive heuristic control: application to industrial process. Automatica 14(5):413–428

Rogers E, Owens DH (1992) Stability analysis for linear repetitive processes, the international symposium on mathematical models in automation and robotics, vol 175. Lecture Notes in Control and Information Sciences. Springer-Verlag, Berlin

Shen D, Li XF (2019) Iterative learning control for systems with iteration-varying trial lengths: synthesis and analysis. Springer, Singapore

Shi J, Zhou H, Cao Z, Jiang Q (2014) A design method for indirect iterative learning control based on two-dimensional generalized predictive control algorithm. J Proc Cont 24(10):1527–1537

Uchiyama M (1978) Formulation of high-speed motion pattern of a mechanical arm by trial. Trans Soc Inst Cont Eng 14(6):706–712

Wang Y, Gao F, Doyle FJ III (2009) Survey on iterative learning control, repetitive control, and run-to-run control. J Process Control 19(10):1589–1600

Wang LP, Freeman CT, Chai S, Rogers E (2013) Predictive-repetitive control with constraints: from design to implementation. J Proc Cont 23:956–967

Wang LP, Freeman CT, Rogers E (2022) Disturbance observer-based predictive repetitive control with constraints. Int J Cont 95(4):1060–1069

Wu M, Xu B, Cao W, She J (2013) Aperiodic disturbance rejection in repetitive-control systems. IEEE Trans Cont Syst Technol 22(3):1044–1051

Xi YG (1993) Predictive control. National Defense Industry Press, Beijing

Xu JX, Tan Y (2003) Linear and nonlinear iterative learning control. Springer-Verlag, Germany, Berlin

Yu QX, Hou ZS (2021) Adaptive fuzzy iterative learning control for high-speed trains with both randomly varying operation lengths and system constraints. IEEE Trans Fuzzy Syst 29(8):2408–2418

Zhang K, Kang Y, Xiong J, Chen J (2003) Direct repetitive control of SPWM inverter for UPS purpose. IEEE Trans Power Electron 18(3):784–792

Zhang Z, Chu B, Liu Y, Li Z, Owens DH (2022) Multimuscle functional-electrical-stimulation-based wrist tremor suppression using repetitive control. IEEE/ASME Trans Mechatron. https://doi.org/10.1109/TMECH.3150301

Zhu B, Xia XH (2016) Adaptive model predictive control for unconstrained discrete-time linear systems with parametric uncertainties. IEEE Trans Automat Control 61(10):3171–3176

Part I
Theory

Chapter 2
Predictive Iterative Learning Control for Unknown Systems

2.1 Introduction

Many practical systems always perform a given task repeatedly in a finite and fixed time interval, such as ultrasonic motor (Lu et al. 2020), bionic robotic fish (Wang et al. 2020), and hybrid energy storage system (Zhang et al. 2021). Iterative learning control (ILC) (Arimoto et al. 1984; Bristow et al. 2006; Wang et al. 2009; Freeman et al. 2015; Shen 2018; Liu et al. 2022) is an ideal control method for repeatable operation systems that operate in a finite time interval. This control method can learn from historical input and output information and therefore can achieve perfect tracking for each operating time over the whole finite operating time interval.

In recent years, many works on predictive iterative learning control (PILC) (Amann et al. 1998; Chu et al. 2016; Zhang and Gao 2018; Wang et al. 2021; Oh et al. 2018; Ma and Liu 2019; Liu et al. 2020; Rosolia et al. 2022; Qiu et al. 2020; Lu et al. 2019; Zhang et al. 2018; Shi et al. 2014) that combines time domain model predictive control (MPC) (Richalet 1978; García et al. 1989; Xi 1993; Grüne and Pannek 2011; Li et al. 2022) with ILC are developed. It is worth pointing out that most of the existing works on PILC rely on the mathematical model information of the system, but in fact, accurate model of the system is often difficult to be obtained due to the increasingly complex system devices and operation environments.

Motivated by this consideration, this chapter proposes a PILC for a class of unknown nonaffine nonlinear single input single output (SISO) systems.

The main contributions of this chapter are as follows.

- For the considered unknown nonaffine nonlinear system, only the measured input/output data rather than any system model information is used for the design of the proposed PILC.
- The learning gain in the proposed PILC is adaptively adjustable and the monotonic convergence property is guaranteed.

© The Author(s), under exclusive license to Springer Nature Singapore Pte Ltd. 2023

Q. Yu et al., *Predictive Learning Control for Unknown Nonaffine Nonlinear Systems*, Intelligent Control and Learning Systems 8,
https://doi.org/10.1007/978-981-19-8857-8_2

This chapter is organized as follows. Section 2.2 presents the problem formulation. Section 2.3 shows the designed predictive ILC method with theoretical analysis. In Sect. 2.4, simulation results are provided to show the effectiveness of the proposed method. Some conclusions are given in Sect. 2.5.

2.2 Problem Formulation

Consider the following repeatable SISO nonaffine nonlinear discrete-time system (Hou and Jin 2010; Chi et al. 2018; Hou and Jin 2013)

$$\begin{aligned} y_l(k+1) = f\big(&y_l(k), y_l(k-1), \ldots, y_l(k-n_y^\dagger), \\ &u_l(k), u_l(k-1), \ldots, u_l(k-n_u^\dagger)\big) \end{aligned} \tag{2.1}$$

where the subscript $l \in N^+$ denotes the index of the operation or iteration number and $k \in \{0, 1, \ldots, K^\dagger\}$ represents the time index. $y_l(k) \in \Re$ and $u_l(k) \in \Re$ are the system output and input at time k of the lth iteration, respectively. $n_u^\dagger \in N^+$ and $n_y^\dagger \in N^+$ are the unknown order of the system, and $f(\ldots) \in \Re$ is an unknown nonlinear function.

Assumption 2.1 The initial system output $y_l(0)$ is random but bounded for $\forall l = 0, 1, 2 \ldots$.

Assumption 2.2 The partial derivatives of $f(\ldots)$ with respect to $u_l(k)$ are continuous, for all $k \in \{0, 1, \ldots, K^\dagger\}$ and $l = 0, 1, 2 \ldots$ with finite exceptions.

Assumption 2.3 System (2.1) is generalized *Lipschitz* for all $k \in \{0, 1, \ldots, K^\dagger\}$ and $l = 0, 1, 2 \ldots$ with finite exceptions i.e.,

$$|y_{l_1}(k+1) - y_{l_2}(k+1)| \le b^\dagger |u_{l_1}(k) - u_{l_2}(k)| \tag{2.2}$$

for $u_{l_1}(k) \neq u_{l_2}(k)$ and any $l_1 \neq l_2$, $l_1, l_2 \geq 0$, where $y_{l_a}(k+1) = f(y_{l_a}(k), y_{l_a}(k-1), \ldots, y_{l_a}(k-n_y^\dagger), u_{l_a}(k), u_{l_a}(k-1), \ldots, u_{l_a}(k-n_u^\dagger))$, $a = 1, 2$, and $b^\dagger$ is a positive constant.

Assumption 2.4 $\forall l = 0, 1, 2, \ldots$, $\forall k \in \{0, 1, \ldots, K^\dagger\}$ and $\Delta u(k) \neq 0$, the symbol of $G_l^\dagger(k)$ defined in (2.3) remains unchanged, that is $G_l^\dagger(k) > \varepsilon^\dagger > 0$, or $G_l^\dagger(k) < -\varepsilon^\dagger$, where $\varepsilon^\dagger$ is a small positive constant. $\Delta u_l(k) = u_l(k) - u_{l-1}(k)$.

Remark 2.1 Assumption 2.1 allows the initial system outputs to be randomly varying. Assumption 2.2 is a typical constraint for control system design of general nonlinear systems. Assumption 2.3 imposes an upper bound restriction on the change rate of system outputs driven by the changes of control inputs. Assumption 2.4 is a common assumption (Zhang and Gao 2018; Wen et al. 2009) and is similar to the assumption on the control direction in model-based control methods.

The following lemma shows that the considered repeatable SISO nonaffine nonlinear discrete-time system satisfying Assumptions 2.1–2.3 can be transformed into an equivalent dynamical linearization model (Hou and Jin 2013); here, it is called iterative learning compact form dynamic linearization (ILCFDL) data model.

Lemma 2.1 *For nonlinear system (2.1) satisfying Assumptions 2.1–2.3, with* $|\Delta u_l(k)| \neq 0$ *for each fixed l, k, there must exist* $G_l^{\dagger}(k)$*, called the pseudo-partial derivative (PPD), such that (2.1) can be transformed into the following equivalent iterative learning compact form dynamic linearization (ILCFDL) data model:*

$$\Delta y_l(k+1) = G_l^{\dagger}(k)\Delta u_l(k) \tag{2.3}$$

where $\Delta u_l(k) = u_l(k) - u_{l-1}(k)$ *and* $\Delta y_l(k+1) = y_l(k+1) - y_{l-1}(k+1)$. Δ *denotes a difference operator with respect to the iterative index, and* $|G_l^{\dagger}(k)| \leq b^{\dagger}$.

Remark 2.2 For nonlinear system (2.1) satisfying Assumptions 2.1–2.3, if there exists an integer $l_0 \geq 1$ such that

$$\Delta u_j(t) \begin{cases} = 0, & j = 1, \ldots, l_0 - 1 \\ \neq 0, & j = l_0 \end{cases} \tag{2.4}$$

then for any integer $l \geq l_0$, a bounded integer $\sigma_l^{\dagger}$ can always be found such that

$$\Delta u_{l-j}(l) \begin{cases} = 0, & j = 0, \ldots, \sigma_l^{\dagger} - 2 \\ \neq 0, & j = \sigma_l^{\dagger} - 1 \end{cases} \tag{2.5}$$

Meanwhile, there exists a PPD $G_l^{\dagger}(k)$ that system (2.1) can be transformed into the following ILCFDL data model

$$y_l(k+1) - y_{l-\sigma_l^{\dagger}}(k+1) = G_l^{\dagger}(k)(u_l(k) - u_{l-\sigma_l^{\dagger}}(k)) \tag{2.6}$$

By virtue of the proof of Lemma 2.1, the above conclusion can be derived straightforwardly.

Remark 2.3 The constructed ILCFDL data model (2.3) is built along the iteration axis rather than time axis. The following proof of Lemma 2.1 implies that (2.3) is an equivalent description of the considered nonlinear system (2.1).

Proof
Differencing (2.1) along the iteration axis, one can obtain:

$$\begin{aligned}\Delta y_l(k+1) = {} & f(y_l(k), y_l(k-1), \ldots, y_l(k-n_y^{\dagger}), u_l(k), u_l(k-1), \ldots, \\ & u_l(k-n_u^{\dagger})) - f(y_{l-1}(k), y_{l-1}(k-1), \ldots, y_{l-1}(k-n_y^{\dagger}), \\ & u_{l-1}(k), u_{l-1}(k-1), \ldots, u_{l-1}(k-n_u^{\dagger}))\end{aligned}$$

$$\begin{aligned}
&= f(y_l(k), y_l(k-1), \ldots, y_l(k-n_y^\dagger), u_l(k), u_l(k-1), \ldots,\\
&\quad u_l(k-n_u^\dagger)) - f(y_l(k), y_l(k-1), \ldots, y_l(k-n_y^\dagger), u_{l-1}(k),\\
&\quad u_l(k-1), \ldots, u_l(k-n_u^\dagger)) + f(y_l(k), y_l(k-1), \ldots, y_l(k-n_y^\dagger),\\
&\quad u_{l-1}(k), u_l(k-1), \ldots, u_l(k-n_u^\dagger)) - f(y_{l-1}(k), y_{l-1}(k-1), \ldots,\\
&\quad y_{l-1}(k-n_y^\dagger), u_{l-1}(k), u_{l-1}(k-1), \ldots, u_{l-1}(k-n_u^\dagger))
\end{aligned} \tag{2.7}$$

Let $\xi_l^\dagger(k) = f(y_l(k), y_l(k-1), \ldots, y_l(k-n_y^\dagger), u_{l-1}(k), u_l(k-1), \ldots, u_l(k-n_u^\dagger)) - f(y_{l-1}(k), y_{l-1}(k-1), \ldots, y_{l-1}(k-n_y^\dagger), u_{l-1}(k), u_{l-1}(k-1), \ldots, u_{l-1}(k-n_u^\dagger))$. By virtue of Assumption 2.2 and differential mean value theorem, (2.7) can be rewritten as

$$\Delta y_l(k+1) = \frac{\partial f^*}{\partial u_l(k)} \Delta u_l(k) + \xi_l^\dagger(k) \tag{2.8}$$

where $\frac{\partial f^*}{\partial u_l(k)}$ represents the value of the partial derivative of $f(\ldots)$ with respect to the $(n_y^\dagger + 2)$th variable at a point between $[y_l(k), y_l(k-1), \ldots, y_l(k-n_y^\dagger), u_{l-1}(k), u_l(k-1), \ldots, u_l(k-n_u^\dagger)]^T$ and $[y_l(k), y_l(k-1), \ldots, y_l(k-n_y^\dagger), u_l(k), u_l(k-1), \ldots, u_l(k-n_u^\dagger)]^T$. For each fixed iteration l and time k, we consider the following equation with $\eta_l^\dagger(k) \in \Re$

$$\xi_l^\dagger(k) = \eta_l^\dagger(k) \Delta u_l(k) \tag{2.9}$$

Since the condition $|\Delta u_l(k)| \neq 0$ holds, (2.9) must have at least one solution $\eta_l^{\dagger *}(k)$. Let $G_l^\dagger(k) = \partial f^* / \partial u_l(k) + \eta_l^{\dagger *}(k)$, one can obtain $\Delta y_l(k+1) = G_l^\dagger(k) \Delta u_l(k)$. Then under Assumption 2.3, $|G_l^\dagger(k)| \leq b^\dagger$ can be derived. □

2.3 Predictive ILC Design

Rewrite (2.3) as

$$y_l(k+1) = y_{l-1}(k+1) + G_l^\dagger(k) \Delta u_l(k) \tag{2.10}$$

Define the tracking error trajectory

$$e_l^\dagger(k+1) = y_d^\dagger(k+1) - y_l(k+1) \tag{2.11}$$

where $y_d^\dagger(k+1)$ is the desired output trajectory, $y_l(k+1)$ is the actual output trajectory.

Subtracting $y_d^\dagger(k+1)$ from both sides of (2.10), then the tracking error for two consecutive iterations can be formulated as

$$e_l^\dagger(k+1) = e_{l-1}^\dagger(k+1) - G_l^\dagger(k)\Delta u_l(k) \tag{2.12}$$

According to (2.12), constructing the following model for m step iterative prediction

$$\begin{aligned}
e_{l+1|l}^\dagger(k+1) &= e_l^\dagger(k+1) - G_{l+1}^\dagger(k)\Delta u_{l+1}(k) \\
e_{l+2|l}^\dagger(k+1) &= e_l^\dagger(k+1) - G_{l+1}^\dagger(k)\Delta u_{l+1}(k) - G_{l+2}^\dagger(k)\Delta u_{l+2}(k) \\
&\vdots \\
e_{l+m|l}^\dagger(k+1) &= e_l^\dagger(k+1) - G_{l+1}^\dagger(k)\Delta u_{l+1}(k) - G_{l+2}^\dagger(k)\Delta u_{l+2}(k) - \cdots \\
&\quad - G_{l+m}^\dagger(k)\Delta u_{l+m}(k)
\end{aligned} \tag{2.13}$$

In (2.13), $e_{l+m|l}^\dagger(k+1)$ represents the prediction of the output error at iteration l, when the future control changes from $l+1$ to $l+m$.

Denote

$$\boldsymbol{G}_{l+1}^{\dagger m}(k) = [G_{l+1}^\dagger(k), G_{l+2}^\dagger(k), \ldots, G_{l+m}^\dagger(k)] \in \Re^{1\times m} \tag{2.14}$$

$$\Delta \boldsymbol{u}_{l+1}^m(k) = [\Delta u_{l+1}(k), \Delta u_{l+2}(k), \ldots, \Delta u_{l+m}(k)]^T \in \Re^{m\times 1} \tag{2.15}$$

Then $e_{l+m|l}^\dagger(k+1)$ can be expressed in the following vector form

$$e_{l+m|l}^\dagger(k+1) = e_l^\dagger(k+1) - \boldsymbol{G}_{l+1}^{\dagger m}(k)\Delta \boldsymbol{u}_{l+1}^m(k) \tag{2.16}$$

It is worth noting that $\boldsymbol{G}_{l+1}^{\dagger m}(k)$ in (2.16) is unknown, and it needs to be estimated and predicted if the subsequent control law is applied. Let $\hat{G}_{l+1}^\dagger(k)$ and $\hat{\boldsymbol{G}}_{l+1}^{\dagger m}(k)$ denote the estimation of $G_{l+1}^\dagger(k)$ and $\boldsymbol{G}_{l+1}^{\dagger m}(k)$. Then, the actual model for the prediction of the output error $e_{l+m|l}^\dagger(k+1)$ can be described as

$$e_{l+m|l}^\dagger(k+1) = e_l^\dagger(k+1) - \hat{\boldsymbol{G}}_{l+1}^{\dagger m}(k)\Delta \boldsymbol{u}_{l+1}^m(k) \tag{2.17}$$

where

$$\hat{\boldsymbol{G}}_{l+1}^{\dagger m}(k) = [\hat{G}_{l+1}^\dagger(k), \hat{G}_{l+2}^\dagger(k), \ldots, \hat{G}_{l+m}^\dagger(k)] \in \Re^{1\times m}, \tag{2.18}$$

and it is acquired by the designed iterative learning estimation and prediction algorithms in the following (2.23)–(2.30).

According to (2.17), if there are no constraints on input and output of the system, the optimal $\Delta \boldsymbol{u}_{l+1}^m(k)$ can be acquired by minimizing the following quadratic cost function

$$J_{l+1}^{\dagger}(k) = \min_{\Delta \boldsymbol{u}_{l+1}^m(k)} \frac{1}{2}\left\{q^{\dagger} e_{l+m|l}^{\dagger}(k+1)^2 + \Delta \boldsymbol{u}_{l+1}^{mT}(k) R^{\dagger} \Delta \boldsymbol{u}_{l+1}^m(k)\right\} \tag{2.19}$$

where $q^{\dagger} > 0$ is a positive constant. $R^{\dagger} \in \Re^{m\times m}$ is a coefficient matrix, for simplicity, define $R^{\dagger} = r^{\dagger} I_{m\times m}$ and $r^{\dagger} > 0$.

By using the optimality condition $\frac{\partial J_{l+1}^{\dagger}(k)}{\partial \Delta \boldsymbol{u}_{l+1}^m(k)} = \boldsymbol{0}$, the optimal control input sequences $\Delta \boldsymbol{u}_{l+1}^m(k)$ is obtained as

$$\Delta \boldsymbol{u}_{l+1}^m(k) = [\hat{\boldsymbol{G}}_{l+1}^{\dagger mT}(k) q^{\dagger} \hat{\boldsymbol{G}}_{l+1}^{\dagger m}(k) + \boldsymbol{R}^{\dagger}]^{-1} \hat{\boldsymbol{G}}_{l+1}^{\dagger mT}(k) q^{\dagger} \boldsymbol{e}_l^{\dagger}(k+1) \tag{2.20}$$

Note that the control algorithm (2.20) requires the calculation of matrix inversion. In practice, instead of (2.20), the following modified predictive ILC with low computational cost will be applied

$$\Delta \boldsymbol{u}_{l+1}^m(k) = \frac{\gamma^{\dagger} q^{\dagger} \hat{\boldsymbol{G}}_{l+1}^{\dagger mT}(k)}{r^{\dagger} + q^{\dagger} \|\hat{\boldsymbol{G}}_{l+1}^{\dagger m}(k)\|^2} e_l^{\dagger}(k+1) \tag{2.21}$$

where $\gamma^{\dagger} \in (0, 1]$ is a step factor introduced to make the control law more general.

Using the receding horizon control strategy at each iteration, only the first element $\Delta u_{l+1}(k)$ from the optimal $\Delta \boldsymbol{u}_{l+1}^m(k)$ is implemented, that is,

$$u_{l+1}(k) = u_l(k) + \boldsymbol{U}^T \Delta \boldsymbol{u}_{l+1}^m(k) \tag{2.22}$$

where $\boldsymbol{U} = \left[1\ \boldsymbol{0}_{1\times(m-1)}\right]^T$.

Now, we use the following iterative learning estimation and prediction algorithms to obtain $\hat{\boldsymbol{G}}_{l+1}^{\dagger m}(k)$ in (2.21).

Firstly, the modified projection algorithm along the iteration axis is used to estimate the unknown PPD $G_l^{\dagger}(k)$.

$$\hat{G}_l^{\dagger}(k) = \hat{G}_{l-1}^{\dagger}(k) + \frac{\eta^{\dagger}(\Delta y_{l-1}(k+1) - \hat{G}_{l-1}^{\dagger}(k)\Delta u_{l-1}(k))\Delta u_{l-1}(k)}{\mu^{\dagger} + |\Delta u_{l-1}(k)|^2} \tag{2.23}$$

$$\begin{aligned} \hat{G}_l^{\dagger}(k) = \hat{G}_0^{\dagger}(k), \text{ if } |\hat{G}_l^{\dagger}(k)| \le \varepsilon^{\dagger} \text{ or } |\Delta u_{l-1}(k)| \le \varepsilon^{\dagger} \\ \text{or } \operatorname{sign}(\hat{G}_l^{\dagger}(k)) \ne \operatorname{sign}(\hat{G}_0^{\dagger}(k)) \end{aligned} \tag{2.24}$$

where $\eta^\dagger \in (0, 1]$ is a step-size constant. $\mu^\dagger > 0$ is a weighting factor. $\hat{G}_0^\dagger(k)$ is the initial value of $\hat{G}_l^\dagger(k)$. The reset algorithm (2.24) is used to make the estimation algorithm (2.23) have a stronger ability to track the iteration time-varying PPD.

Note that based on the proposed iterative learning estimation algorithm (2.23)–(2.24), the estimated value $\hat{G}_l^\dagger(k)$ is available. By using these estimated values, the predicted values $\hat{G}_{l+i}^\dagger(k)$, $i = 1, 2, \ldots, m$ for future $l + i$ iterations will be obtained as follows.

At first, the one-step iterative learning prediction algorithm is constructed as:

$$\hat{G}_{l+1}^\dagger(k) = \Gamma_1^\dagger(k)\hat{G}_l^\dagger(k) + \Gamma_2^\dagger(k)\hat{G}_{l-1}^\dagger(k) + \cdots + \Gamma_{n_p^\dagger}^\dagger(k)\hat{G}_{l-n_p^\dagger+1}^\dagger(k) \tag{2.25}$$

where $\Gamma_j^\dagger(k) \in \Re$ $(j = 1, 2, \ldots, n_p^\dagger)$ are unknown parameters and $n_p^\dagger$ is a proper model order.

Define $\hat{\Gamma}_{j,l}^\dagger(k)$ as the estimation of the unknown parameters $\Gamma_j^\dagger(k)$, $(j=1, 2, \ldots, n_p^\dagger)$ at lth iteration, and they are designed in the following (2.30). Then, we can get the following available $(l + i)$ $(i = 1, 2, \ldots, m)$ step iterative learning prediction algorithm:

$$\hat{G}_{l+i}^\dagger(k) = \hat{\Gamma}_{1,l}^\dagger(k)\hat{G}_{l+i-1}^\dagger(k) + \hat{\Gamma}_{2,l}^\dagger(k)\hat{G}_{l+i-2}^\dagger(k) + \cdots + \hat{\Gamma}_{n_p^\dagger,l}^\dagger(k)\hat{G}_{l+i-n_p^\dagger}^\dagger(k) \tag{2.26}$$

where $i = 1, 2, \ldots m$.

Now define that:

$$\hat{\mathbf{\Gamma}}_l^\dagger(k) = [\hat{\Gamma}_{1,l}^\dagger(k),\ \hat{\Gamma}_{2,l}^\dagger(k), \ldots,\ \hat{\Gamma}_{n_p^\dagger,l}^\dagger(k)]^T \in \Re^{n_p^\dagger \times 1} \tag{2.27}$$

and

$$\hat{\mathbf{\Xi}}_{l-1}^\dagger(k) = \left[\hat{G}_{l-1}^\dagger(k), \hat{G}_{l-2}^\dagger(k), \ldots, \hat{G}_{l-n_p^\dagger}^\dagger(k)\right]^T \in \Re^{n_p^\dagger \times 1} \tag{2.28}$$

then (2.26) can be rewritten as

$$\hat{G}_{l+i}^\dagger(k) = \hat{\mathbf{\Xi}}_{l+i-1}^{\dagger T}(k)\,\hat{\mathbf{\Gamma}}_l^\dagger(k), \quad i = 1, 2, \ldots, m \tag{2.29}$$

Here, $\hat{\mathbf{\Gamma}}_l^\dagger(k)$ is computed by the following iterative learning algorithm:

$$\begin{aligned}\hat{\mathbf{\Gamma}}_l^\dagger(k) = \hat{\mathbf{\Gamma}}_{l-1}^\dagger(k) + \kappa^\dagger \hat{\mathbf{\Xi}}_{l-1}^\dagger(k)\left[\nu^\dagger + \hat{\mathbf{\Xi}}_{l-1}^{\dagger T}(k)\hat{\mathbf{\Xi}}_{l-1}^\dagger(k)\right]^{-1} \\ \times \left[\hat{G}_l^\dagger(k) - \hat{\mathbf{\Xi}}_{l-1}^{\dagger T}(k)\hat{\mathbf{\Gamma}}_{l-1}^\dagger(k)\right]\end{aligned} \tag{2.30}$$

where $\nu^\dagger$ and $\kappa^\dagger$ are known positive constants.

According to (2.23)–(2.30), $\hat{\boldsymbol{G}}_{l+1}^{\dagger m}(k)$ in (2.21) is available, such that the control law (2.21)–(2.22) can be implemented.

Theorem 2.1 *Considering the SISO nonlinear discrete-time system (2.1) satisfying Assumptions 2.1–2.4, the proposed predictive ILC (2.21)–(2.22), together with the iterative learning estimation and prediction algorithms (2.23)–(2.24) and (2.29)–(2.30) guarantees that*

(2.1.1) The estimation of the PPD $\hat{G}_l^{\dagger}(k)$ is bounded over the finite time interval $\{0, 1, \ldots, K^{\dagger}\}$ for all iterations l.

(2.1.2) The tracking error $e_l^{\dagger}(k)$ converges to zero monotonically along the iteration axis and pointwisely over the finite time interval $\{0, 1, 2, \ldots, K^{\dagger}\}$.

Proof

proof of (2.1.1):

Define $\tilde{G}_l^{\dagger}(k) = G_l^{\dagger}(k) - \hat{G}_l^{\dagger}(k)$ as the PPD estimation error. Subtracting $G_l^{\dagger}(k)$ from both sides of (2.23) gives

$$\begin{aligned}\tilde{G}_l^{\dagger}(k) &= \tilde{G}_{l-1}^{\dagger}(k) + G_l^{\dagger}(k) - G_{l-1}^{\dagger}(k) - \frac{\eta^{\dagger}\Delta u_{l-1}(k)^2}{\mu^{\dagger} + |\Delta u_{l-1}(k)|^2}\tilde{G}_{l-1}^{\dagger}(k) \\ &= G_l^{\dagger}(k) - G_{l-1}^{\dagger}(k) + \left[1 - \frac{\eta^{\dagger}\Delta u_{l-1}(k)^2}{\mu^{\dagger} + |\Delta u_{l-1}(k)|^2}\right]\tilde{G}_{l-1}^{\dagger}(k)\end{aligned} \tag{2.31}$$

It can be directly derived that for $0 < \eta^{\dagger} \leq 1$, $\mu^{\dagger} > 0$, there exists a constant $d_1^{\dagger}$ satisfying that

$$0 < \left|1 - \frac{\eta^{\dagger}\Delta u_{l-1}(k)^2}{\mu^{\dagger} + |\Delta u_{l-1}(k)|^2}\right| \leq d_1^{\dagger} < 1. \tag{2.32}$$

Since $\left|G_l^{\dagger}(k)\right| \leq b^{\dagger}$ according to Lemma 2.1, we can get $\left|G_l^{\dagger}(k) - G_{l-1}^{\dagger}(k)\right| \leq 2b^{\dagger}$. Then we have

$$\begin{aligned}\left|\tilde{G}_l^{\dagger}(k)\right| &\leq d_1^{\dagger}\left|\tilde{G}_{l-1}^{\dagger}(k)\right| + 2b^{\dagger} \leq d_1^{\dagger 2}\left|\tilde{G}_{l-2}^{\dagger}(k)\right| + 2d_1^{\dagger}b^{\dagger} + 2b^{\dagger} \leq \\ &\cdots \leq d_1^{\dagger(l-1)}\tilde{G}_1^{\dagger}(k) + \frac{2b^{\dagger}\left(1 - d_1^{\dagger(l-1)}\right)}{1 - d_1^{\dagger}}\end{aligned} \tag{2.33}$$

□

Therefore, $\tilde{G}_l^{\dagger}(k)$ is bounded. Based on the fact that $G_l^{\dagger}(k)$ is also bounded, it follows that $\hat{G}_l^{\dagger}(k)$ is bounded.

proof of (2.1.2):

Substituting the control laws (2.21) and (2.22) into (2.12) leads to

$$\begin{aligned} e_{l+1}^{\dagger}(k+1) &= e_l^{\dagger}(k+1) - G_{l+1}^{\dagger}(k)\,\Delta u_{l+1}(k) \\ &= e_l^{\dagger}(k+1) - \frac{\gamma^{\dagger} q^{\dagger} G_{l+1}^{\dagger}(k)\, \boldsymbol{U}^T \hat{\boldsymbol{G}}_{l+1}^{\dagger mT}(k)}{q^{\dagger}\|\hat{\boldsymbol{G}}_{l+1}^{\dagger m}(k)\|^2 + r^{\dagger}} e_l^{\dagger}(k+1) \\ &= \left(1 - \frac{\gamma^{\dagger} q^{\dagger} G_{l+1}^{\dagger}(k)\, \hat{G}_{l+1}^{\dagger}(k)}{q^{\dagger}\|\hat{\boldsymbol{G}}_{l+1}^{\dagger m}(k)\|^2 + r^{\dagger}}\right) e_l^{\dagger}(k+1) \end{aligned} \tag{2.34}$$

From Assumption 2.4 and the reset algorithm (2.24), we can get $G_{l+1}^{\dagger}(k)\,\hat{G}_{l+1}^{\dagger}(k) \geq 0$. If the parameters $\gamma^{\dagger}, q^{\dagger}, r^{\dagger}$ are suitably chosen, there exists two constants $0 < m_2^{\dagger} < 1$ and $0 < M_2^{\dagger} < 1$, such that

$$0 < \frac{\gamma^{\dagger} q^{\dagger} G_{l+1}^{\dagger}(k)\, \hat{G}_{l+1}^{\dagger}(k)}{q^{\dagger}\|\hat{\boldsymbol{G}}_{l+1}^{\dagger m}(k)\|^2 + r^{\dagger}} = m_2^{\dagger} < 1 \tag{2.35}$$

$$0 < 1 - \frac{\gamma^{\dagger} q^{\dagger} G_{l+1}^{\dagger}(k)\, \hat{G}_{l+1}^{\dagger}(k)}{q^{\dagger}\|\hat{\boldsymbol{G}}_{l+1}^{\dagger m}(k)\|^2 + r^{\dagger}} = M_2^{\dagger} < 1 \tag{2.36}$$

According to (2.34) and (2.36), one has

$$\left|e_{l+1}^{\dagger}(k+1)\right| \leq M_2^{\dagger}\left|e_l^{\dagger}(k+1)\right| \tag{2.37}$$

and

$$\left|e_{l+1}^{\dagger}(k+1)\right| \leq M_2^{\dagger}\left|e_l^{\dagger}(k+1)\right| \leq M_2^{\dagger 2}\left|e_{l-1}^{\dagger}(k+1)\right| \leq \cdots \leq M_2^{\dagger l}\left|e_1^{\dagger}(k+1)\right| \tag{2.38}$$

According to (2.38), we can get that $e_{l+1}^{\dagger}(k+1)$ converges to zero monotonically.

2.4 Simulation Validation

Consider the following nonlinear system

$$\begin{aligned} y_l(k+1) = &\frac{y_l(k)^2}{1+y_l(k)^2} + \frac{0.1y_l(k-1)^2}{1+y_l(k-1)^2+y_l(k-1)^3} \\ &+0.3a^{\dagger}(k)u_l(k) + 0.001rand(1) \end{aligned} \tag{2.39}$$

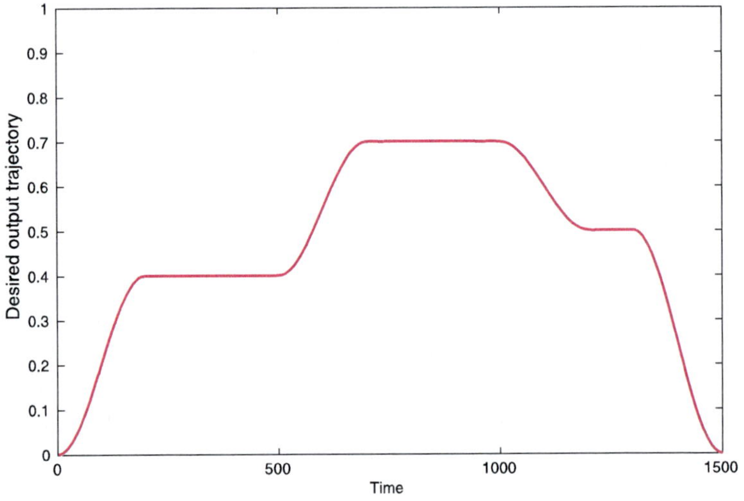

Fig. 2.1 Desired output trajectory to be tracked

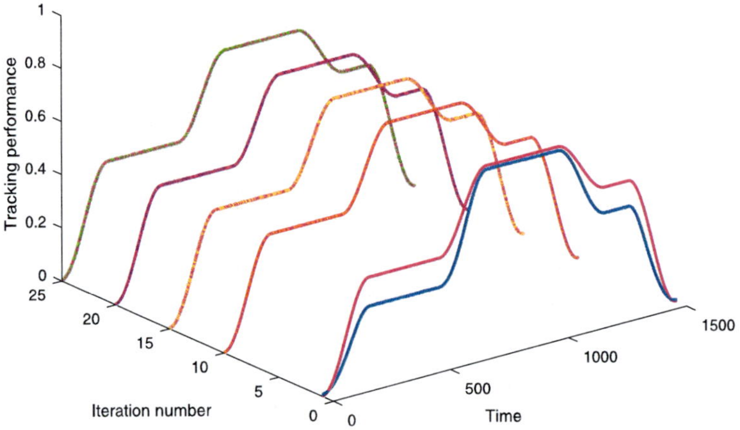

Fig. 2.2 Tracking performance of the proposed PILC at 1st, 10th, 15th, 20th, and 25th iterations

where $a^{\dagger}(k) = 1 + rand(1) * 0.01\sin(2\pi k/2000)$. $rand(1)$ takes a random value in the interval $(0, 1)$.

In the simulation, the proposed controller only uses input and output data generated by the above system model and does not use any specific model information.

The desired output trajectory to be tracked is shown in Fig. 2.1.

The proposed PILC (2.21)–(2.22) together with the iterative learning estimation and prediction laws (2.23)–(2.24) and (2.29)–(2.30) is simulated here. The control parameters are set as follows: $m = 2$, $\gamma^{\dagger} = 1$, $q^{\dagger} = 0.9$, $r^{\dagger} = 0.5$, $\eta^{\dagger} = 3$, $\mu^{\dagger} = 2$, $n_p^{\dagger} = 3$, $\kappa^{\dagger} = 0.01$, $\nu^{\dagger} = 5$, $\hat{G}_0^{\dagger} = 0.4$.

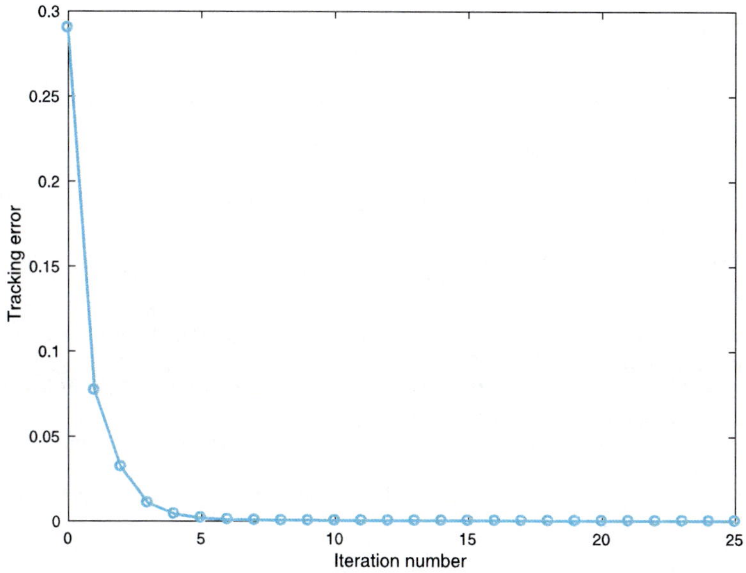

Fig. 2.3 Output tracking errors

The simulation results are shown in Figs. 2.2 and 2.3. Figure 2.2 shows the tracking performance of the proposed PILC along time axis at 1st, 10th, 15th, 20th, and 25th iterations. Figure 2.3 gives the root mean square (RMS) of output tracking errors over the whole time interval $\{0, 1, \ldots, 1500\}$ further. From Figs. 2.2 and 2.3, the tracking performance by applying the proposed PILC is improved gradually as the number of iterations increases.

2.5 Conclusion

This chapter proposes a PILC method that integrates ILC with predictive control for a class of unknown nonaffine nonlinear systems. Instead of using model information of the system, the proposed PILC only depends on the input/output data information. The monotonic and pointwise convergence property is guaranteed through theoretical analysis. Simulation results further demonstrate the effectiveness of the proposed PILC.

References

Amann N, Owens DH, Rogers E (1998) Predictive optimal iterative learning control. Int J Control 69(2):203–226

Arimoto S, Kawamura S, Miyazaki F (1984) Bettering operation of robots by learning. J Robot Syst 1:123–140

Bristow DA, Tharayil M, Alleyne AG (2006) A survey of iterative learning control. IEEE Control Syst Mag 26(3):96–114

Chi R, Hou Z, Jin S et al (2018) Computationally efficient data-driven higher order optimal iterative learning control. IEEE Trans Neural Netw Learn Syst 29(12):5971–5980

Chu B, Owens DH, Freeman CT (2016) Iterative learning control with predictive trial information: convergence, robustness, and experimental verification. IEEE Trans Control Syst Technol 24(3):1101–1108

Freeman CT, Rogers E, Burridge JH et al (2015) Iterative learning control for electrical stimulation and stroke rehabilitation. Springer, Heidelberg

García CE, Prett DM, Morari M (1989) Model predictive control: theory and practice—a survey. Automatica 25(3):335–348

Grüne L, Pannek J (2011) Nonlinear model predictive control: theory and algorithms. Springer, London

Hou Z, Jin S (2010) A novel data-driven control approach for a class of discrete-time nonlinear systems. IEEE Trans Control Syst Technol 19(6):1549–1558

Hou ZS, Jin ST (2013) Model free adaptive control: theory and applications. CRC Press, Florida

Li H, Song L, Jiang X et al (2022) Robust model predictive control for multi-phase batch processes with asynchronous switching. Int J Control Autom Syst 20(1):84–98

Liu X, Ma L, Kong X et al (2020) An efficient iterative learning predictive functional control for nonlinear batch processes. IEEE Trans Cybern 52(6):4147–4160

Liu J, Ruan X, Zheng Y (2022) Iterative learning control for discrete-time systems with full learnability. IEEE Trans Neural Netw Learn Syst 33(2):629–643

Lu JY, Cao ZX, Gao FR (2019) Multi-point iterative learning model predictive control. IEEE Trans Ind Electron 66(8):6230–6240

Lu S, Jingzhuo S, Ying Z (2020) Secant iterative learning control of ultrasonic motor. ISA Trans 103:343–354

Ma LL, Liu XJ (2019) Robust model predictive iterative learning control with iteration-varying reference trajectory. Acta Automatica Sinica 45(10):1933–1945

Oh SK, Park BJ, Lee JM (2018) Point-to-point iterative learning model predictive control. Automatica 89:135–143

Qiu WW, Xiong ZH, Zhang J, Hong YD, Li WZ (2020) Integrated predictive iterative learning control based on updating reference trajectory for point-to-point tracking. J Process Control 85:41–51

Richalet J (1978) Model predictive heuristic control: application to industrial process. Automatica 14(5):413–428

Rosolia U, Zhang XJ, Borrelli F (2022) Robust learning model-predictive control for linear systems performing iterative tasks. IEEE Trans Autom Control 67(2):856–869

Shen D (2018) Iterative learning control with incomplete information: a survey. IEEE/CAA J Automatica Sinica 5(5):885–901

Shi J, Zhou H, Cao Z, Jiang Q (2014) A design method for indirect iterative learning control based on two-dimensional generalized predictive control algorithm. J Process Control 24(10):1527–1537

Wang Y, Gao F, Doyle FJ III (2009) Survey on iterative learning control, repetitive control, and run-to-run control. J Process Control 19(10):1589–1600

Wang M, Zhang Y, Dong H et al (2020) Trajectory tracking control of a bionic robotic fish based on iterative learning. Sci China Inf Sci 63(7):1–9

Wang L, Yu J, Li P et al (2021) A 2D-FM model-based robust iterative learning model predictive control for batch processes. ISA Trans 110:271–282

Wen C, Zhou J, Wang W (2009) Decentralized adaptive backstepping stabilization of interconnected systems with dynamic input and output interactions. Automatica 45(1):55–67

Xi YG (1993) Predictive control. National Defense Industry Press, Beijing

Zhang R, Gao F (2018) Two-dimensional iterative learning model predictive control for batch processes: a new state space model compensation approach. IEEE Trans Syst Man Cybern: Syst 51(2):833–841

Zhang RD, Wu S, Tao JL (2018) A new design of predictive functional control strategy for batch processes in the two-dimensional framework. IEEE Trans Ind Inform 15(5):2905–2914

Zhang X, Wang B, Gamage D et al (2021) Model predictive and iterative learning control based hybrid control method for hybrid energy storage system. IEEE Trans Sustain Energy 12(4):2146–2158

Chapter 3
Constrained Predictive Iterative Learning Control

3.1 Introduction

Owing to the physical restrictions on actuators or for safety, practical systems always suffer from input and output/state constraints. For example, both the input force and actual operation speed of a high speed train operation system are constrained to ensure safe and comfortable operation. Motivated by this consideration, this chapter will extend the considered nonlinear systems in Chap. 2 to those with input and output/state constraints, and a new constrained PILC method is designed.

Until now, some works on constrained PILC have been proposed (Wang et al. 2019, 2013; Oh et al. 2018; Lu et al. 2016; Li et al. 2019; Wang et al. 2021; Ma and Liu 2019; Oh and Lee 2016; Li et al. 2016). A series of PILC methods for constrained linear systems are presented in Wang et al. (2019), Wang et al. (2013), Oh et al. (2018), Lu et al. (2016), Li et al. (2019), Wang et al. (2021), and the convergence property under system constraints is given in Oh et al. (2018). In Lu et al. (2016), Li et al. (2019), the constrained robust PILC methods for linear systems are designed, while in Wang et al. (2021), a predictive fault-tolerant controller with terminal constraints for linear systems is proposed. Ma and Liu (2019) designs a constrained robust PILC method for known nonlinear systems with theoretical analysis. A PILC method for constrained multivariable control for known nonlinear system can be found in Oh and Lee (2016). Li et al. (2016) proposes a real-time feedback-based PILC with the two-dimensional model for nonlinear systems with known model structure. Compared with these existing works on PILC, this chapter will design a constrained PILC method for unknown nonaffine nonlinear systems with unknown system parameters and system structure model under input and output constraints.

The main contributions of this chapter are as follows.

- A constrained PILC scheme is proposed for a class of unknown nonaffine nonlinear systems with both input and output constraints.
- Asymptotic convergence property of the proposed constrained PILC is guaranteed with rigorous theoretical analysis.

© The Author(s), under exclusive license to Springer Nature Singapore Pte Ltd. 2023

Q. Yu et al., *Predictive Learning Control for Unknown Nonaffine Nonlinear Systems*,
Intelligent Control and Learning Systems 8,
https://doi.org/10.1007/978-981-19-8857-8_3

This chapter is organized as follows. Section 3.2 presents the problem formulation. Section 3.3 shows the designed constrained predictive ILC method with theoretical analysis. In Sect. 3.4, simulation results are provided to show the effectiveness of the proposed method. Some conclusions are given in Sect. 3.5.

3.2 Problem Formulation

Consider the following repeatable SISO nonaffine nonlinear discrete-time system (Hou and Jin 2010; Chi et al. 2018; Hou and Jin 2013)

$$y_l(k+1) = f\big(y_l(k), y_l(k-1), \ldots, y_l(k-n_y^\diamond), \\ u_l(k), u_l(k-1), \ldots, u_l(k-n_u^\diamond)\big) \tag{3.1}$$

where the subscript $l \in N^+$ denotes the index of the operation or iteration number and $k \in \{0, 1, \ldots, K^\diamond\}$ represents the time index. $y_l(k) \in \Re$ and $u_l(k) \in \Re$ are the system output and input at time k of the lth iteration, respectively. $n_u^\diamond \in N^+$ and $n_y^\diamond \in N^+$ are the unknown order of the system, and $f(\ldots) \in \Re$ is an unknown nonlinear function.

In this chapter, system (3.1) suffers from the following constraints.

(1) Input constraint:

$$u_{\min}(k) \le u_l(k) \le u_{\max}(k) \tag{3.2}$$

where $u_{\min}(k)$, $u_{\max}(k)$ are the constraints on the control input.

(2) Output constraint:

$$y_{\min}(k) \le y_l(k) \le y_{\max}(k) \tag{3.3}$$

where $y_{\min}(k)$, $y_{\max}(k)$ are the constraints on the system output.

Assumption 3.1 The initial system output $y_l(0)$ is random but bounded for $\forall l = 0, 1, 2 \ldots$.

Assumption 3.2 The partial derivatives of $f(\ldots)$ with respect to $u_l(k)$ are continuous, for all $k \in \{0, 1, \ldots, K^\diamond\}$ and $l = 0, 1, 2 \ldots$ with finite exceptions.

Assumption 3.3 System (3.1) is generalized *Lipschitz* for all $k \in \{0, 1, \ldots, K^\diamond\}$ and $l = 0, 1, 2 \ldots$ with finite exceptions i.e.,

$$|y_{l_1}(k+1) - y_{l_2}(k+1)| \le b^\diamond |u_{l_1}(k) - u_{l_2}(k)| \tag{3.4}$$

for $u_{l_1}(k) \neq u_{l_2}(k)$ and any $l_1 \neq l_2, l_1, l_2 \ge 0$, where $y_{l_a}(k+1) = f(y_{l_a}(k), y_{l_a}(k-1), \ldots, y_{l_a}(k-n_y^\diamond), u_{l_a}(k), u_{l_a}(k-1), \ldots, u_{l_a}(k-n_u^\diamond))$, $a = 1, 2$, and $b^\diamond$ is a positive constant.

Assumption 3.4 There exists a feasible input trajectory $u_d(k)$ such that $e_l^\diamond(u_d(k)) = 0$, where $e_l^\diamond(k)$ is the tracking error trajectory defined as $e_l^\diamond(k) = y_d^\diamond(k) - y_l(k)$. $y_d^\diamond(k)$ is the desired output trajectory.

Assumption 3.5 $\forall l = 0, 1, 2, \ldots, \forall k \in \{0, 1, \ldots, K^\diamond\}$ and $\Delta u_l(k) \neq 0$, the symbol of $G_l^\diamond(k)$ defined in (3.5) remains unchanged, that is $G_l^\diamond(k) > \varepsilon^\diamond > 0$, or $G_l^\diamond(k) < -\varepsilon^\diamond$, where $\varepsilon^\diamond$ is a small positive constant. $\Delta u_l(k) = u_l(k) - u_{l-1}(k)$.

Remark 3.1 Assumption 3.4 means that there exists a feasible control input $u_d(k)$ that can drive the output $y_l(k)$ to track the desired output $y_d^\diamond(k)$ over the whole time interval $k \in \{0, 1, \ldots, K^\diamond\}$.

The following lemma shows that the considered nonlinear system satisfying Assumptions 3.1–3.3 can be transformed into an equivalent dynamical linearisation model (Hou and Jin 2013), here is called iterative learning compact form dynamic linearization (ILCFDL) data model.

Lemma 3.1 *For nonlinear system (3.1) satisfying Assumptions 3.1–3.3, with $|\Delta u_l(k)| \neq 0$ for each fixed l, k, there must exist $G_l^\diamond(k)$, called the pseudo-partial derivative (PPD), such that (3.1) can be transformed into the following equivalent iterative learning compact form dynamic linearization (ILCFDL) data model:*

$$\Delta y_l(k+1) = G_l^\diamond(k)\Delta u_l(k) \tag{3.5}$$

where $\Delta u_l(k) = u_l(k) - u_{l-1}(k)$ and $\Delta y_l(k+1) = y_l(k+1) - y_{l-1}(k+1)$. Δ denotes a difference operator along the iteration axis, and $|G_l^\diamond| \le b^\diamond$.

Proof The proof can be derived straightforwardly according the proof of Lemma 2.1 in Chap.2 and hence is omitted here. □

3.3 Constrained Predictive ILC Design

Rewrite (3.5) as

$$y_l(k+1) = y_{l-1}(k+1) + G_l^\diamond(k)\Delta u_l(k) \tag{3.6}$$

Define the tracking error trajectory

$$e_l^\diamond(k+1) = y_d^\diamond(k+1) - y_l(k+1) \tag{3.7}$$

where $y_d^\diamond(k+1)$ is the desired output trajectory, $y_l(k+1)$ is the actual output trajectory.

Subtracting $y_d^{\diamond}(k+1)$ from both sides of (3.6) yields

$$e_l^{\diamond}(k+1) = e_{l-1}^{\diamond}(k+1) - G_l^{\diamond}(k)\Delta u_l(k) \tag{3.8}$$

According to (3.8), constructing the following model for m step iterative prediction

$$\begin{aligned} e_{l+1|l}^{\diamond}(k+1) &= e_l^{\diamond}(k+1) - G_{l+1}^{\diamond}(k)\Delta u_{l+1}(k) \\ e_{l+2|l}^{\diamond}(k+1) &= e_l^{\diamond}(k+1) - G_{l+1}^{\diamond}(k)\Delta u_{l+1}(k) - G_{l+2}^{\diamond}(k)\Delta u_{l+2}(k) \\ &\vdots \\ e_{l+m|l}^{\diamond}(k+1) &= e_l^{\diamond}(k+1) - G_{l+1}^{\diamond}(k)\Delta u_{l+1}(k) - G_{l+2}^{\diamond}(k)\Delta u_{l+2}(k) - \cdots \\ &\quad - G_{l+m}^{\diamond}(k)\Delta u_{l+m}(k) \end{aligned} \tag{3.9}$$

In (3.9), $e_{l+m|l}^{\diamond}(k+1)$ denotes the prediction of the output error at iteration l, when the future control changes from $l+1$ to $l+m$.

Denote

$$\boldsymbol{G}_{l+1}^{\diamond m}(k) = [G_{l+1}^{\diamond}(k), G_{l+2}^{\diamond}(k), \ldots, G_{l+m}^{\diamond}(k)] \in \Re^{1\times m} \tag{3.10}$$

$$\Delta \boldsymbol{u}_{l+1}^{m}(k) = [\Delta u_{l+1}(k), \Delta u_{l+2}(k), \ldots, \Delta u_{l+m}(k)]^T \in \Re^{m\times 1} \tag{3.11}$$

Then, $e_{l+m|l}^{\diamond}(k+1)$ can be expressed in the following vector form

$$e_{l+m|l}^{\diamond}(k+1) = e_l^{\diamond}(k+1) - \boldsymbol{G}_{l+1}^{\diamond m}(k)\Delta \boldsymbol{u}_{l+1}^{m}(k) \tag{3.12}$$

It is worth noting that $\boldsymbol{G}_{l+1}^{\diamond m}(k)$ in (3.12) is unknown, and it needs to be estimated and predicted if the subsequent control law is applied. Let $\hat{G}_{l+1}^{\diamond}(k)$ and $\hat{\boldsymbol{G}}_{l+1}^{\diamond m}(k)$ denote the estimation of $G_{l+1}^{\diamond}(k)$ and $\boldsymbol{G}_{l+1}^{\diamond m}(k)$. Hence, the actual model for the prediction of the output error $e_{l+m|l}^{\diamond}(k+1)$ can be described as

$$e_{l+m|l}^{\diamond}(k+1) = e_l^{\diamond}(k+1) - \hat{\boldsymbol{G}}_{l+1}^{\diamond m}(k)\Delta \boldsymbol{u}_{l+1}^{m}(k) \tag{3.13}$$

where

$$\hat{\boldsymbol{G}}_{l+1}^{\diamond m}(k) = [\hat{G}_{l+1}^{\diamond}(k), \hat{G}_{l+2}^{\diamond}(k), \ldots, \hat{G}_{l+m}^{\diamond}(k)] \in \Re^{1\times m} \tag{3.14}$$

Next, we use the modified projection algorithm and multi-level hierarchical predicted method along the iteration axis to obtain $\hat{\boldsymbol{G}}_{l+1}^{\diamond m}(k)$ defined in (3.13)–(3.14).

Firstly, the modified projection algorithm along the iteration axis is used to estimate the unknown PPD $G_l^{\diamond}(k)$.

$$\hat{G}_l^\diamond(k) = \hat{G}_{l-1}^\diamond(k) + \frac{\eta^\diamond(\Delta y_{l-1}(k+1) - \hat{G}_{l-1}^\diamond(k)\Delta u_{l-1}(k))\Delta u_{l-1}(k)}{\mu^\diamond + |\Delta u_{l-1}(k)|^2} \tag{3.15}$$

$$\begin{aligned}\hat{G}_l^\diamond(k) = \hat{G}_0^\diamond(k),\ &\text{if } |\hat{G}_l^\diamond(k)| \le \varepsilon^\diamond \text{ or } |\Delta u_{l-1}(k)| \le \varepsilon^\diamond \\ &\text{or } \ \operatorname{sign}(\hat{G}_l^\diamond(k)) \ne \operatorname{sign}(\hat{G}_0^\diamond(k))\end{aligned} \tag{3.16}$$

where $\eta^\diamond \in (0, 1]$ is a step-size constant. $\mu^\diamond > 0$ is a weighting factor. $\hat{G}_0^\diamond(k)$ is the initial value of $\hat{G}_l^\diamond(k)$. The reset algorithm (3.16) is used to make the estimation algorithm (3.15) have a stronger ability to track the iteration time-varying PPD.

Note that based on the proposed iterative learning estimation algorithm (3.15)–(3.16), the estimated value $\hat{G}_l^\diamond(k)$ is available. By using these estimated values, the predicted values $\hat{G}_{l+i}^\diamond(k), i = 1, 2, \ldots, m$ for future $l + i$ iterations will be obtained as follows.

At first, the one-step iterative learning prediction algorithm is constructed as:

$$\hat{G}_{l+1}^\diamond(k) = \Gamma_1^\diamond(k)\hat{G}_l^\diamond(k) + \Gamma_2^\diamond(k)\hat{G}_{l-1}^\diamond(k) + \cdots + \Gamma_{n_p}^\diamond(k)\hat{G}_{l-n_p^\diamond+1}^\diamond(k) \tag{3.17}$$

where $\Gamma_j^\diamond(k) \in \Re$ $(j = 1, 2, \ldots, n_p^\diamond)$ are unknown parameters and $n_p^\diamond$ is a proper model order.

Define $\hat{\Gamma}_{j,l}^\diamond(k)$ as the estimation of the unknown parameters $\Gamma_j^\diamond(k)$, $(j = 1, 2, \ldots, n_p^\diamond)$ at l th iteration and they are designed in the following (3.22). Then, we can get the following available $(l + i)$ $(i = 1, 2, \ldots, m)$ step iterative learning prediction algorithm:

$$\hat{G}_{l+i}^\diamond(k) = \hat{\Gamma}_{1,l}^\diamond(k)\hat{G}_{l+i-1}^\diamond(k) + \hat{\Gamma}_{2,l}^\diamond(k)\hat{G}_{l+i-2}^\diamond(k) + \cdots + \hat{\Gamma}_{n_p^\diamond,l}^\diamond(k)\hat{G}_{l+i-n_p^\diamond}^\diamond(k) \tag{3.18}$$

where $i = 1, 2, \ldots, m$.

Now define that:

$$\hat{\boldsymbol{\Gamma}}_l^\diamond(k) = [\hat{\Gamma}_{1,l}^\diamond(k),\ \hat{\Gamma}_{2,l}^\diamond(k), \ldots,\ \hat{\Gamma}_{n_p^\diamond,l}^\diamond(k)]^T \in \Re^{n_p^\diamond \times 1} \tag{3.19}$$

and

$$\hat{\boldsymbol{\Xi}}_{l-1}^\diamond(k) = \left[\hat{G}_{l-1}^\diamond(k), \hat{G}_{l-2}^\diamond(k), \ldots, \hat{G}_{l-n_p^\diamond}^\diamond(k)\right]^T \in \Re^{n_p^\diamond \times 1} \tag{3.20}$$

then (3.18) can be rewritten as

$$\hat{G}_{l+i}^\diamond(k) = \hat{\boldsymbol{\Xi}}_{l+i-1}^{\diamond T}(k)\,\hat{\boldsymbol{\Gamma}}_l^\diamond(k), \quad i = 1, 2, \ldots, m \tag{3.21}$$

Here $\mathbf{\Gamma}_l^{\diamond}(k)$ is computed by the following iterative learning algorithm.

$$\hat{\mathbf{\Gamma}}_l^{\diamond}(k) = \hat{\mathbf{\Gamma}}_{l-1}^{\diamond}(k) + \kappa^{\diamond}\hat{\mathbf{\Xi}}_{l-1}^{\diamond}(k)\Big[\nu^{\diamond} + \hat{\mathbf{\Xi}}_{l-1}^{\diamond T}(k)\hat{\mathbf{\Xi}}_{l-1}^{\diamond}(k)\Big]^{-1} \times \Big[\hat{G}_l^{\diamond}(k) - \hat{\mathbf{\Xi}}_{l-1}^{\diamond T}(k)\hat{\mathbf{\Gamma}}_{l-1}^{\diamond}(k)\Big] \tag{3.22}$$

where $\nu^{\diamond}$ and $\kappa^{\diamond}$ are known positive constants.

Now based on (3.15)–(3.22), the predicted value $\hat{\boldsymbol{G}}_{l+1}^{\diamond m}(k)$ in (3.13) is obtained.

According to (3.13), as in Chap. 2, if there are no constraints on input and output of the system, $\Delta \boldsymbol{u}_{l+1}^{m}(k)$ can be calculated directly by minimizing the following quadratic cost function

$$J_{l+1}^{\diamond}(k) = \min_{\Delta \boldsymbol{u}_{l+1}^{m}(k)} \frac{1}{2}\Big\{q^{\diamond} e_{l+m|l}^{\diamond}(k+1)^2 + \Delta \boldsymbol{u}_{l+1}^{mT}(k) R^{\diamond} \Delta \boldsymbol{u}_{l+1}^{m}(k)\Big\} \tag{3.23}$$

where $q^{\diamond} > 0$ is a positive constant. $R^{\diamond} \in \Re^{m\times m}$ is a coefficient matrix, for simplicity, define $R^{\diamond} = r^{\diamond} I_{m\times m}$ and $r^{\diamond} > 0$.

Different from Chap. 2, in this chapter, the following constraints are considered for the optimization problem (3.23).

$$\begin{aligned} u_{\min}(k) &\le u_{l+1}(k) \le u_{\max}(k) \\ \Delta u_{\min}(k) &\le \Delta u_{l+1}(k) \le \Delta u_{\max}(k) \\ y_{\min}(k+1) - \tau_l^{\diamond}(k+1) &\le y_{l+m|l}(k+1) \le y_{\max}(k+1) + \tau_l^{\diamond}(k+1) \end{aligned} \tag{3.24}$$

where $u_{\min}(k)$, $u_{\max}(k)$ and $y_{\min}(k+1)$, $y_{\max}(k+1)$ are defined in (3.2)–(3.3). $\Delta u_{\min}(k)$ and $\Delta u_{\max}(k)$ are the lower and upper bounds of the change rate of control input along the iteration axis. A slack variable $\tau_l^{\diamond}(k+1) \ge 0$ is introduced to avoid infeasibility.

Remark 3.2 $\tau_l^{\diamond}(k+1)$ is a slack variable introduced to make the constraints always be feasible (Lee and Lee 2000; Zafiriou and Chiou 1993). It has been shown in Wang et al. (2013) that harsh input and output constraints may make the system become unstable. In practice, if the system constraints can be satisfied, $\tau_l^{\diamond}(k+1)$ can be set as zero. If not, an appropriate value of $\tau_l^{\diamond}(k+1)$ should be chosen.

Now under input and output constraints, the optimization problem can be described as:

$$J_{l+1}^{\diamond}(k) = \min_{\Delta \boldsymbol{u}_{l+1}^{m}(k), \tau_l^{\diamond}(k+1)} \frac{1}{2}\big\{\bar{q}^{\diamond} e_{l+m|l}^{\diamond}(k+1)^2 + \Delta \boldsymbol{u}_{l+1}^{mT}(k)\bar{R}^{\diamond}\Delta \boldsymbol{u}_{l+1}^{m}(k) + \bar{s}^{\diamond}\tau_l^{\diamond}(k+1)^2\big\} \quad \textit{subject to } (3.24) \tag{3.25}$$

where $\bar{R}^{\diamond} = \bar{r}^{\diamond} I_{m\times m} \in \Re^{m\times m}$ is positive definite and symmetric. $\bar{q}^{\diamond} > 0, \bar{r}^{\diamond} > 0$ and $\bar{s}^{\diamond} > 0$ are positive constants.

To facilitate the subsequent analysis of the convergence property, the following lemma is given.

Lemma 3.2 (Lee and Lee 2000) *Using Schwarz inequality, for positive definite constants $\bar{q}^\diamond$, $\bar{r}^\diamond$, $\bar{s}^\diamond$, $a^\diamond$, $b^\diamond$, $c^\diamond$, $d^\diamond$ and $e^\diamond$, we have*

$$\frac{1}{2}[(a^\diamond + b^\diamond)^2\bar{q}^\diamond + c^{\diamond 2}\bar{r}^\diamond + (d^\diamond + e^\diamond)^2\bar{s}^\diamond] \leq \left[\sqrt{\frac{1}{2}(a^{\diamond 2}\bar{q}^\diamond + c^{\diamond 2}\bar{r}^\diamond + d^{\diamond 2}\bar{s}^\diamond)} + \sqrt{\frac{1}{2}b^{\diamond 2}\bar{q}^\diamond} + \sqrt{\frac{1}{2}e^{\diamond 2}\bar{s}^\diamond}\right]^2 \quad (3.26)$$

Theorem 3.1 *Considering the SISO nonlinear discrete-time system (3.1) satisfying Assumptions 3.1–3.5, the proposed constrained predictive ILC (3.25), together with the iterative learning estimation and prediction algorithms (3.15)–(3.16) and (3.21)–(3.22) guarantees that the tracking error $e_l^\diamond(k)$ converges to zero asymptotically along the iteration axis and pointwisely over the finite time interval $\{0, 1, 2, \ldots, K^\diamond\}$.*

Proof Considering the cost function (3.25)

$$J_{l+1}^\diamond(k) = \min\Big[\Phi_{l+1}^\diamond(k) \triangleq \frac{1}{2}\{\bar{q}^\diamond e_{l+m|l}^{\diamond 2}(k+1) + \Delta\boldsymbol{u}_{l+1}^{mT}(k)\bar{R}^\diamond\Delta\boldsymbol{u}_{l+1}^m(k) + \bar{s}^\diamond\tau_l^\diamond(k+1)^2\}\Big] \quad (3.27)$$

The minimization of (3.25) will be performed in relation to $\big(e_{l+m|l}^\diamond(k+1),\ \Delta\boldsymbol{u}_{l+1}^m(k),\ \tau_l^\diamond(k+1)\big) \in \Omega_{l,k}^\diamond$. $\Omega_{l,k}^\diamond$ is a convex set defined by the system constraints (3.24) and the predictor (3.13).

In order to demonstrate that $J_{l+2}^\diamond(k)$ is less than or equal to $J_{l+1}^\diamond(k)$, $\Delta u_{l+m+1}(k)$ will be restricted to zero at $(l+2)$th optimization. And hence, one has

$$\Delta\boldsymbol{u}_{l+2}^{mT}(k)\bar{R}^\diamond\Delta\boldsymbol{u}_{l+2}^m(k)|_{\Delta u_{l+m+1}(k)=0} = \Delta\boldsymbol{u}_{l+1}^{mT}(k)\bar{R}^\diamond\Delta\boldsymbol{u}_{l+1}^m(k) - \bar{r}_1^\diamond\Delta u_{l+1}^2(k) \quad (3.28)$$

where $\bar{r}_1^\diamond = U^T\bar{R}^\diamond U$ and U is defined in (2.22) of Chap.2.

Define $\tilde{G}_l^\diamond(k) = G_l^\diamond(k) - \hat{G}_l^\diamond(k)$. According to (3.8) and (3.13), with $\Delta u_{l+m+1}(k) = 0$, the following equation can be derived

$$\begin{aligned}
e_{l+m+1|l+1}^\diamond(k+1) &= e_{l+1}^\diamond(k+1) - \hat{\boldsymbol{G}}_{l+2}^{\diamond m}(k)\Delta\boldsymbol{u}_{l+2}^m(k)\\
&= e_l^\diamond(k+1) - G_{l+1}^\diamond(k)\Delta u_{l+1}(k) - \hat{\boldsymbol{G}}_{l+2}^{\diamond m}(k)\Delta\boldsymbol{u}_{l+2}^m(k)\\
&= e_l^\diamond(k+1) - \tilde{G}_{l+1}^\diamond(k)\Delta u_{l+1}(k)\\
&\quad -\hat{G}_{l+1}^\diamond(k)\Delta u_{l+1}(k) - \hat{\boldsymbol{G}}_{l+2}^{\diamond m}(k)\Delta\boldsymbol{u}_{l+2}^m(k)\\
&= e_l^\diamond(k+1) - \tilde{G}_{l+1}^\diamond(k)\Delta u_{l+1}(k)\\
&\quad -\hat{G}_{l+1}^\diamond(k)\Delta u_{l+1}(k) - \hat{\boldsymbol{G}}_{l+1}^{\diamond m}(k)\Delta\boldsymbol{u}_{l+1}^m(k) + \hat{G}_{l+1}^\diamond(k)\Delta u_{l+1}(k)\\
&= e_l^\diamond(k+1) - \hat{\boldsymbol{G}}_{l+1}^{\diamond m}(k)\Delta\boldsymbol{u}_{l+1}^m(k) - \tilde{G}_{l+1}^\diamond(k)\Delta u_{l+1}(k)\\
&= e_{l+m|l}^\diamond(k+1) - \tilde{G}_{l+1}^\diamond(k)\Delta u_{l+1}(k)
\end{aligned} \quad (3.29)$$

Based on (3.29), the output constraint in (3.24) leads to

$$\begin{aligned} &y_d^\diamond(k) - y_{\max}(k+1) - (\tau_l^\diamond(k+1) + \tilde{G}_{l+1}^\diamond(k)\Delta u_{l+1}(k)) \\ &\quad \le e_{l+m+1|l+1}^\diamond(k+1)|_{\Delta u_{l+m+1}(k)=0} \\ &\quad \le y_d^\diamond(k) - y_{\min}(k+1) + (\tau_l^\diamond(k+1) - \tilde{G}_{l+1}^\diamond(k)\Delta u_{l+1}(k)) \end{aligned} \tag{3.30}$$

Define

$$\eta_{l+1}^\diamond(k) = \begin{cases} -\tilde{G}_{l+1}^\diamond(k)\Delta u_{l+1}(k) & \tilde{G}_{l+1}^\diamond(k)\Delta u_{l+1}(k) > 0 \\ \tilde{G}_{l+1}^\diamond(k)\Delta u_{l+1}(k) & \tilde{G}_{l+1}^\diamond(k)\Delta u_{l+1}(k) < 0 \end{cases} \tag{3.31}$$

Then (3.30) leads to

$$\begin{aligned} &y_d^\diamond(k) - y_{\max}(k+1) - (\tau_l^\diamond(k+1) - \eta_{l+1}^\diamond(k)) \\ &\quad \le e_{l+m+1|l+1}^\diamond(k+1)|_{\Delta u_{l+m+1}(k)=0} \\ &\quad \le y_d^\diamond(k) - y_{\min}(k+1) + (\tau_l^\diamond(k+1) - \eta_{l+1}^\diamond(k)) \end{aligned} \tag{3.32}$$

Let $(e_{l+m|l}^{\diamond *}(k+1), \Delta \boldsymbol{u}_{l+1}^{m*}(k), \tau_l^{\diamond *}(k+1))$ be the optimal solution for (3.27), then we can get that

$$\begin{aligned} &(e_{l+m+1|l+1}^\diamond(k+1), \Delta \boldsymbol{u}_{l+2}^{m}(k), \tau_{l+1}^\diamond(k+1)) = (e_{l+m|l}^{\diamond *}(k+1) \\ &\quad -\tilde{G}_{l+1}^\diamond(k)\Delta u_{l+1}(k), S\{\Delta \boldsymbol{u}_{l+1}^{m*}(k)\}, \tau_l^{\diamond *}(k+1) - \eta_{l+1}^\diamond(k)) \end{aligned} \tag{3.33}$$

is a feasible solution for (3.27), where S is a forward-shift operator such that $S\{\Delta \boldsymbol{u}_{l+1}^{m*}(k)\} = [\Delta u_{l+2}^*(k), \Delta u_{l+3}^*(k), \ldots, \Delta u_{l+m}^*(k), 0]^T$.

Based on (3.28) and (3.33), we have

$$\begin{aligned} J_{l+2}^\diamond(k) &\le \frac{1}{2}\left\{\bar{q}^\diamond e_{l+m+1|l+1}^\diamond(k+1)^2 + \Delta \boldsymbol{u}_{l+2}^{mT}(k)\bar{R}^\diamond \Delta \boldsymbol{u}_{l+2}^{m}(k) + \bar{s}^\diamond \tau_{l+1}^\diamond(k+1)^2\right\} \\ &\le \frac{1}{2}\left\{\begin{array}{l} \bar{q}^\diamond(e_{l+m|l}^{\diamond *}(k+1) - \tilde{G}_{l+1}^\diamond(k)\Delta u_{l+1}(k))^2 \\ +\Delta \boldsymbol{u}_{l+1}^{m*T}(k)\bar{R}^\diamond \Delta \boldsymbol{u}_{l+1}^{m*}(k) - \bar{r}_1^\diamond \Delta u_{l+1}(k)^2 \\ +\bar{s}^\diamond(\tau_l^{\diamond *}(k+1) - \eta_{l+1}^\diamond(k))^2 \end{array}\right\} \end{aligned} \tag{3.34}$$

By virtue of Lemma 3.2, (3.34) leads to

$$\begin{aligned} J_{l+2}^\diamond(k) \le &\left[\sqrt{J_{l+1}^\diamond(k)} + \sqrt{\frac{1}{2}\bar{q}^\diamond(\tilde{G}_{l+1}^\diamond(k)\Delta u_{l+1}(k))^2} \right. \\ &\left. + \sqrt{\frac{1}{2}\bar{s}^\diamond \eta_{l+1}^\diamond(k)^2}\right]^2 - \frac{1}{2}\bar{r}_1^\diamond \Delta u_{l+1}(k)^2 \end{aligned} \tag{3.35}$$

According to the proof (2.1.1) of Theorem 2.1 in Chap. 2, let $|\tilde{G}_l^{\diamond}(k)| \leq d_2^{\diamond}$ where $d_2^{\diamond} > 0$ is a positive constant, then (3.35) leads to

$$J_{l+2}^{\diamond}(k) \leq \left[\sqrt{J_{l+1}^{\diamond}(k)} + \sqrt{\frac{1}{2}\bar{q}^{\diamond} d_2^{\diamond 2}|\Delta u_{l+1}(k)|^2} + \sqrt{\frac{1}{2}\bar{s}^{\diamond} d_2^{\diamond 2}|\Delta u_{l+1}(k)|^2}\right]^2 - \frac{1}{2}\bar{r}_1^{\diamond}\Delta u_{l+1}(k)^2 \tag{3.36}$$

Note that there exists $J^{\diamond\max} > 0$ such that $J_l^{\diamond}(k) \leq J^{\diamond\max}, \forall l, k$. Let $\alpha^{\diamond} = \sqrt{\frac{1}{2}\bar{q}^{\diamond} d_2^{\diamond 2}} + \sqrt{\frac{1}{2}\bar{s}^{\diamond} d_2^{\diamond 2}}$, $\gamma^{\diamond} = 2\alpha^{\diamond}\sqrt{J^{\diamond\max}}$, $\beta^{\diamond} = \frac{1}{2}\bar{r}_1^{\diamond}$, then (3.36) leads to

$$\begin{aligned} J_{l+2}^{\diamond}(k) &\leq \left[\sqrt{J_{l+1}^{\diamond}(k)} + \alpha^{\diamond}|\Delta u_{l+1}(k)|\right]^2 - \beta^{\diamond}|\Delta u_{l+1}(k)|^2 \\ &\leq J_{l+1}^{\diamond}(k) + \gamma^{\diamond}|\Delta u_{l+1}(k)| - (\beta^{\diamond} - \alpha^{\diamond 2})|\Delta u_{l+1}(k)|^2 \end{aligned} \tag{3.37}$$

Suppose there exists a positive constant $\sigma^{\diamond}$ such that

$$\gamma^{\diamond} \leq (\beta^{\diamond} - \alpha^{\diamond 2} - \sigma^{\diamond})|\Delta u_{l+1}(k)| \tag{3.38}$$

then (3.37) leads to

$$J_{l+2}^{\diamond}(k) \leq J_{l+1}^{\diamond}(k) - \sigma^{\diamond}|\Delta u_{l+1}(k)|^2 \tag{3.39}$$

and

$$\lim_{l\to\infty} J_l^{\diamond}(k) \leq J_0^{\diamond}(k) - \sigma^{\diamond} \lim_{l\to\infty}\sum_{j=0}^{l-1}|\Delta u_j(k)|^2 \tag{3.40}$$

Note that the feasibility of the optimization problem (3.25) implies that $J_0^{\diamond}(k)$ is finite. Based on (3.39), (3.40), the boundedness of $J_0^{\diamond}(k)$ will guarantee the boundedness of $J_l^{\diamond}(k)$, and then the infinite series $\sum_{j=0}^{\infty}|\Delta u_j(k)|^2$ converges. Therefore, $\lim_{l\to\infty}\Delta u_l(k) = 0$.

Next, the convergence of $\Delta \boldsymbol{u}_l^{m*}(k)$ to $\mathbf{0}$ is displayed. By means of (3.27), we have

$$\begin{aligned} \Phi_{l+2}^{\diamond}(k) - J_{l+2}^{\diamond}(k) \geq{}& \frac{1}{2}\left\{\Delta \boldsymbol{u}_{l+2}^{mT}(k)\bar{R}^{\diamond}\Delta \boldsymbol{u}_{l+2}^{m}(k) - \Delta \boldsymbol{u}_{l+2}^{m*T}(k)\bar{R}^{\diamond}\Delta \boldsymbol{u}_{l+2}^{m*}(k)\right\} \\ &+ \frac{1}{2}\bar{s}^{\diamond}\left\{\tau_{l+1}^{\diamond}(k+1)^2 - \tau_{l+1}^{\diamond *}(k+1)^2\right\} \end{aligned} \tag{3.41}$$

then there exists a positive matrix $\boldsymbol{P}^{\diamond}$ such that (3.41) leads to

$$\Phi_{l+2}^{\diamond}(k)-J_{l+2}^{\diamond}(k)\geq\begin{bmatrix}\Delta\boldsymbol{u}_{l+2}^{m}(k)-\Delta\boldsymbol{u}_{l+2}^{m*}(k)\\ \tau_{l+1}^{\diamond}(k+1)-\tau_{l+1}^{\diamond*}(k+1)\end{bmatrix}^{T}\boldsymbol{P}^{\diamond}\begin{bmatrix}\Delta\boldsymbol{u}_{l+2}^{m}(k)-\Delta\boldsymbol{u}_{l+2}^{m*}(k)\\ \tau_{l+1}^{\diamond}(k+1)-\tau_{l+1}^{\diamond*}(k+1)\end{bmatrix}\tag{3.42}$$

According to (3.31), (3.33) and the fact that $\lim_{l\to\infty}\Delta u_l(k)=0$, we can get $\Delta\boldsymbol{u}_{l+2}^{m}(k)=S\{\Delta\boldsymbol{u}_{l+1}^{m*}(k)\}$, $\tau_{l+1}^{\diamond}(k+1)=\tau_l^{\diamond*}(k+1)$, and $\Phi_{l+2}^{\diamond}(k)=J_{l+1}^{\diamond}(k)-\frac{1}{2}\bar{r}_1^{\diamond}\Delta u_{l+1}^{*}(k)^2$ for sufficiently large l.

Then, (3.42) can be rewritten as

$$\begin{aligned}&J_{l+1}^{\diamond}(k)-J_{l+2}^{\diamond}(k)-\frac{1}{2}\bar{r}_1^{\diamond}\Delta u_{l+1}^{*}(k)^2\\&\geq\begin{bmatrix}S\{\Delta\boldsymbol{u}_{l+1}^{m*}(k)\}-\Delta\boldsymbol{u}_{l+2}^{m*}(k)\\ \tau_l^{\diamond*}(k+1)-\tau_{l+1}^{\diamond*}(k+1)\end{bmatrix}^{T}\boldsymbol{P}^{\diamond}\begin{bmatrix}S\{\Delta\boldsymbol{u}_{l+1}^{m*}(k)\}-\Delta\boldsymbol{u}_{l+2}^{m*}(k)\\ \tau_l^{\diamond*}(k+1)-\tau_{l+1}^{\diamond*}(k+1)\end{bmatrix}\end{aligned}\tag{3.43}$$

Note that $J_l^{\diamond}(k)$ is convergent and $\lim_{l\to\infty}\Delta u_l^{*}(k)=0$ according to (3.39)–(3.40). Then based on (3.43), we have $\lim_{l\to\infty}\Delta\boldsymbol{u}_{l+2}^{m*}(k)=\lim_{l\to\infty}S\{\Delta\boldsymbol{u}_{l+1}^{m*}(k)\}$, $\lim_{l\to\infty}\tau_l^{\diamond*}(k+1)=\lim_{l\to\infty}\tau_{l+1}^{\diamond*}(k+1)$. As a result, one can obtain $\lim_{l\to\infty}\Delta\boldsymbol{u}_l^{m*}(k)=\mathbf{0}, \forall\, k\in\{0,1,\ldots,K^{\diamond}\}$.

Now, we show that $\lim_{l\to\infty}e_l^{\diamond}(k+1)=0, \forall\, k\in\{0,1,\ldots,K^{\diamond}\}$.

Define

$$\boldsymbol{u}_d^{m}(k)=[u_d(k),u_d(k),\ldots,u_d(k)]^T\tag{3.44}$$

$$\boldsymbol{u}_{l+1}^{m}(k)=[u_{l+1}(k),u_{l+2}(k),\ldots,u_{l+m}(k)]^T\tag{3.45}$$

$$\Delta\boldsymbol{u}_{d,l+1}^{m}(k)=[\Delta u_{d,l+1}(k),\Delta u_{d,l+2}(k),\ldots,\Delta u_{d,l+m}(k)]^T\tag{3.46}$$

According to Assumption 3.4, $(e_{l+m|l}^{\diamond}(k+1),\Delta\boldsymbol{u}_{l+1}^{m}(k),\tau_l^{\diamond}(k+1))=(0,\Delta\boldsymbol{u}_{d,l+1}^{m}(k),0)$ is a feasible solution for $l\to\infty$, where $\Delta\boldsymbol{u}_{d,l+1}^{m}(k)=\boldsymbol{u}_d^{m}(k)-\boldsymbol{u}_{l+1}^{m}(k)$. Moreover, the optimal solution $(e_{l+m|l}^{\diamond*}(t+1),\Delta\boldsymbol{u}_{l+1}^{m*}(k),\tau_l^{\diamond*}(k+1))$ is also a feasible solution. As a result, the directional derivative of $\Phi_{l+1}^{\diamond}(k)$ from the optimal solution $(e_{l+m|l}^{\diamond*}(k+1),\Delta\boldsymbol{u}_{l+1}^{m*}(k),\tau_l^{\diamond*}(k+1))$ to the feasible solution $(0,\Delta\boldsymbol{u}_{d,l+1}^{m}(k),0)$ should be nonnegative. Then, we can obtain that

$$\Delta\boldsymbol{u}_{l+1}^{m*T}(k)\bar{R}^{\diamond}\Delta\boldsymbol{u}_{d,l+1}^{m}(k)\geq\bar{q}^{\diamond}e_{l+m|l}^{\diamond*}(k+1)^2+\Delta\boldsymbol{u}_{l+1}^{m*T}(k)\bar{R}^{\diamond}\Delta\boldsymbol{u}_{l+1}^{m*}(k)\geq 0\tag{3.47}$$

Since we have got $\lim_{l\to\infty}\Delta\boldsymbol{u}_l^{m*}(k)=\mathbf{0}$, then (3.47) implies $\lim_{l\to\infty}e_{l+m|l}^{\diamond*}(k+1)=0$.

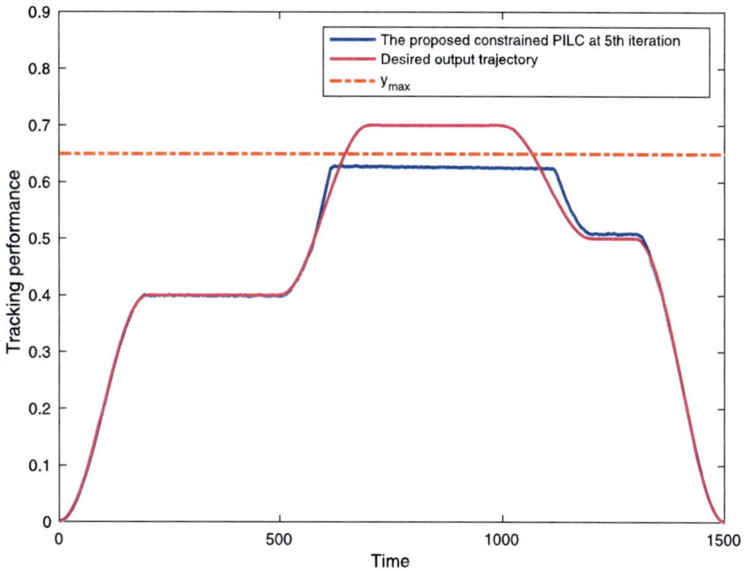

Fig. 3.1 Tracking performance of the proposed constrained PILC at 5th iteration

According to (3.13), the convergence of $\Delta \boldsymbol{u}_l^{m*}(k)$ and $e_{l+m|l}^{\diamond *}(k+1)$ to zero implies that the tracking error $e_l^{\diamond}(k+1)$ converges to 0 asymptotically, namely $\lim\limits_{l\to\infty} e_l^{\diamond}(k+1)=0$. □

3.4 Simulation Validation

The nonlinear system (2.39) in Chap. 2 and the desired output trajectory shown in Fig. 2.1 are also simulated here.

In the simulation, the following constraints on control input and system output are applied:

$$\begin{aligned} &0 \leq u_{1,l}(k) \leq 0.55 \\ &-0.1 \leq \Delta u_l(k) \leq 0.1 \\ &0 \leq y_{l+2|l}(k+1) \leq 0.65 \end{aligned} \tag{3.48}$$

Under (3.48), the proposed constrained PILC (3.25), together with the estimation and prediction algorithms (3.15)–(3.16) and (3.18)–(3.22) are implemented. The control parameters are set as follows: $m=2$, $\bar{q}^{\diamond}=0.9$, $\bar{R}^{\diamond}=0.3I_{2\times 2}$, $\eta^{\diamond}=3$, $\mu^{\diamond}=2$, $n_p^{\diamond}=3$, $\kappa^{\diamond}=0.01$, $\nu^{\diamond}=5$, $\hat{G}_0^{\diamond}=0.4$. In addition, interior-point algorithm is used to solve the constrained optimization problem (3.25).

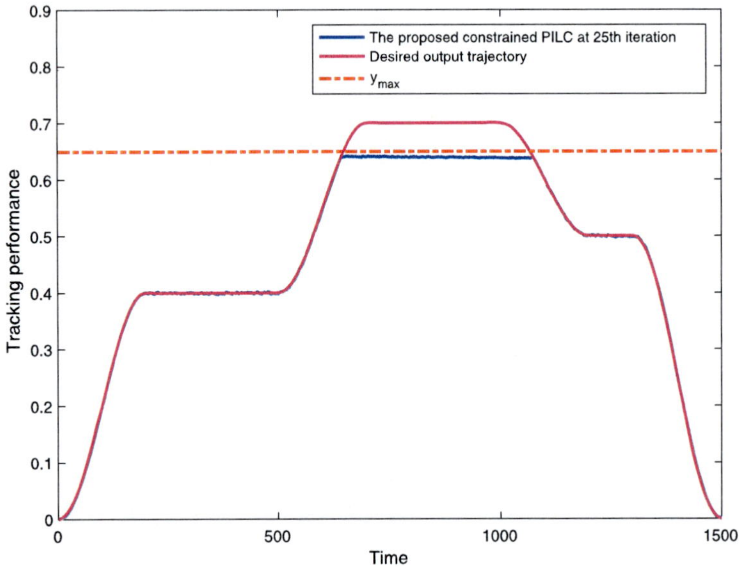

Fig. 3.2 Tracking performance of the proposed constrained PILC at 25th iteration

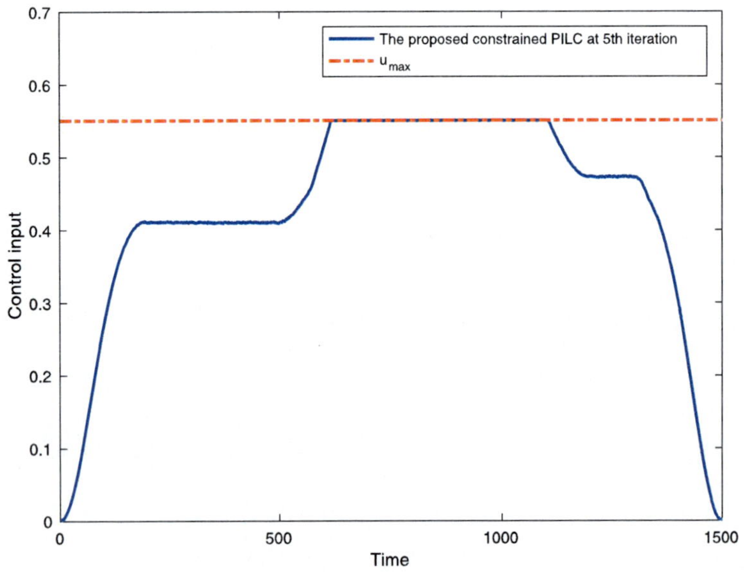

Fig. 3.3 Control input of the proposed constrained PILC at 5th iteration

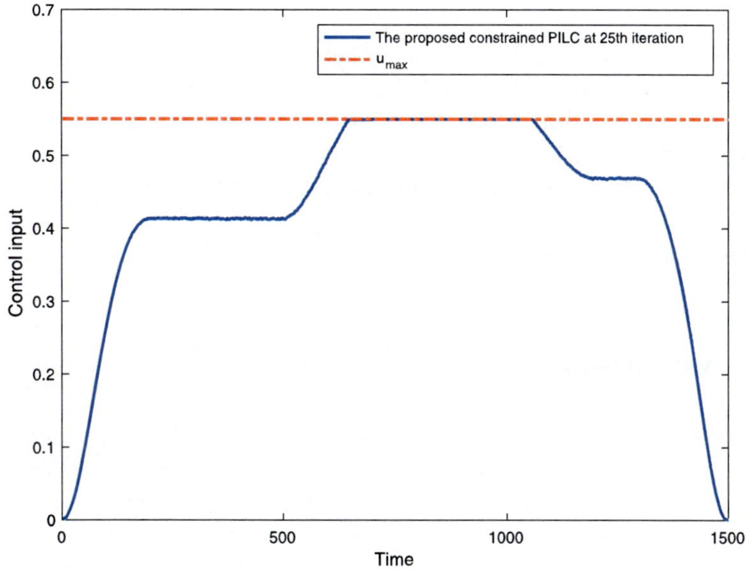

Fig. 3.4 Control input of the proposed constrained PILC at 25th iteration

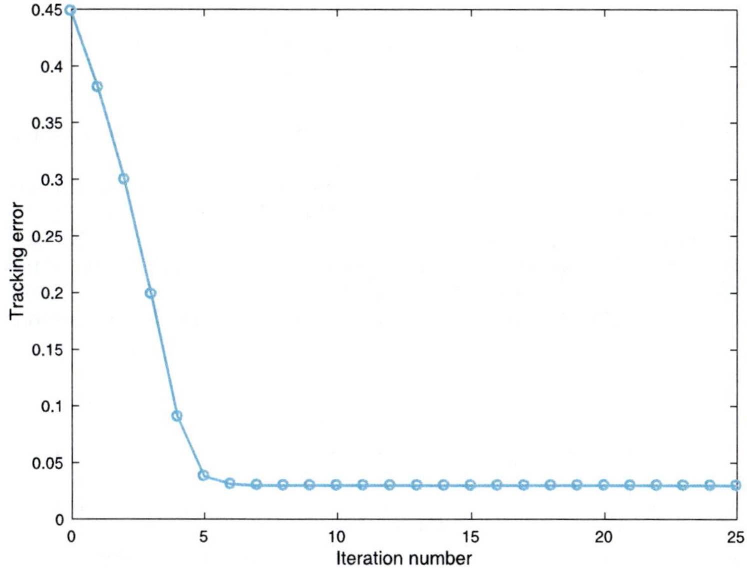

Fig. 3.5 Output tracking errors

Simulation results are shown in Figs. 3.1, 3.2, 3.3 and 3.4. Figures 3.1 and 3.2 give the tracking performance of the proposed constrained PILC at 5th and 25th iterations, respectively. Figures 3.3 and 3.4 demonstrate the control input profile of the proposed constrained PILC at these two iterations further. Figure 3.5 displays the output tracking errors along the iteration axis. As shown in Figs. 3.1, 3.2, 3.3 and 3.4, although both input and output of the system suffer from actuator restrictions, gradually improved control performance and reduced tracking control error are still obtained by utilizing the proposed constrained PILC method.

3.5 Conclusion

In this chapter, a constrained PILC for tracking control of unknown nonaffine nonlinear systems is proposed. Even though the considered unknown system is under both input and output constraints, asymptotic and pointwise convergence property of the proposed constrained PILC is guaranteed through theoretical analysis. Simulation results further verify both the ability to handle constraints and control performance of the proposed method.

References

Chi R, Hou Z, Jin S et al (2018) Computationally efficient data-driven higher order optimal iterative learning control. IEEE Trans Neural Netw Learn Syst 29(12):5971–5980

Hou Z, Jin S (2010) A novel data-driven control approach for a class of discrete-time nonlinear systems. IEEE Trans Cont Syst Technol 19(6):1549–1558

Hou ZS, Jin ST (2013) Model free adaptive control: theory and applications. CRC Press, Florida

Lee KS, Lee JH (2000) Convergence of constrained model-based predictive control for batch processes. IEEE Trans Autom Cont 45(10):1928–1932

Li D, Xi Y, Lu J et al (2016) Synthesis of real-time-feedback-based 2D iterative learning control-model predictive control for constrained batch processes with unknown input nonlinearity. Ind Eng Chem Res 55(51):13074–13084

Li D, He S, Xi Y et al (2019) Synthesis of ILC-MPC controller with data-driven approach for constrained batch processes. IEEE Trans Ind Electron 67(4):3116–3125

Lu J, Cao Z, Gao F (2016) Ellipsoid invariant setbased robust model predictive control for repetitive processes with constraints. IET Cont Theo Appl 10(9):1018–1026

Ma LL, Liu XJ (2019) Robust model predictive iterative learning control with iteration-varying reference trajectory. Acta Automat Sinica 45(10):1933–1945

Oh SK, Lee JM (2016) Iterative learning model predictive control for constrained multivariable control of batch processes. Comput Chem Eng 93:284–292

Oh SK, Park BJ, Lee JM (2018) Point-to-point iterative learning model predictive control. Automatica 89:135–143

Wang L, Freeman CT, Chai S et al (2013) Predictive-repetitive control with constraints: from design to implementation. J Proc Control 23(7):956–967

Wang L, Sun L, Luo W (2019) Robust constrained iterative learning predictive fault-tolerant control of uncertain batch processes. Sci China Inf Sci 62(11):1–3

Wang L, Song J, Zhang R et al (2021) Constrained model predictive fault-tolerant control for multi-time-delayed batch processes with disturbances: a Lyapunov-Razumikhin function method. J Franklin Inst 358(18):9483–9509

Zafiriou E, Chiou H W (1993) Output constraint softening for SISO model predictive control. American Control Conference, IEEE, pp 372–376

Chapter 4
Predictive Iterative Learning Control for Systems with Varying Trial Lengths

4.1 Introduction

In recent years, the existing PILC works (Amann et al. 1998; Chu et al. 2016; Zhang and Gao 2018; Wang et al. 2021; Oh et al. 2018; Ma and Liu 2019; Liu et al. 2020; Rosolia et al. 2022; Qiu et al. 2020; Lu et al. 2019; Zhang et al. 2018; Shi et al. 2014) are based on a common assumption that the time interval of each operation of the system must be the same. However, the actual operation of the system cannot strictly guarantee that the time interval of each operation is exactly the same. For example, a high-speed train carries out the task of transporting passengers to the destination every day, but sometimes, the arrival time will be earlier than the operating schedule, or will be later. Motivated by this consideration, in this chapter, the controlled systems considered in Chap.2 are extended to those with varying trial lengths, and the corresponding PILC method is designed.

At present, some works have been done to deal with the problem of varying trial lengths (Wang et al. 2021; Wei and Li 2017; Yu and Hou 2020; Bu et al. 2019; Shen et al. 2016; Zeng et al. 2019; Shen and Xu 2018; Shen et al. 2021). There are two typical technologies in these existing works to deal with this problem. One is to compensate the missing output error data with just zeros (Wang et al. 2021; Wei and Li 2017; Yu and Hou 2020; Bu et al. 2019; Shen et al. 2016), and the other is to compensate these missing data by the data generated at the last operating time of the current iteration (Zeng et al. 2019; Shen and Xu 2018; Shen et al. 2021). Compared with these existing compensation methods, this chapter will design a new data compensation-based PILC that more valid historical and predicted data can be used to compensate for the missing data attributed by the varying trial lengths.

The main contributions of this chapter are as follows.

- A new data compensation mechanism that exploits both historical actual operating data and predicted data is proposed to address the problem of varying trial lengths.

© The Author(s), under exclusive license to Springer Nature Singapore Pte Ltd. 2023
Q. Yu et al., *Predictive Learning Control for Unknown Nonaffine Nonlinear Systems*, Intelligent Control and Learning Systems 8,
https://doi.org/10.1007/978-981-19-8857-8_4

- Based on the proposed data compensation mechanism, a new iterative learning estimation and modeling algorithm and the corresponding data compensation-based PILC method are proposed. The convergence property can also be guaranteed theoretically.

This chapter is organized as follows. Section 4.2 presents the problem formulation. Section 4.3 shows the designed data compensation-based predictive ILC method with theoretical analysis. In Sect. 4.4, simulation results are provided to show the effectiveness of the proposed method. Some conclusions are given in Sect. 4.5.

4.2 Problem Formulation

Consider the following repeatable MIMO nonaffine nonlinear discrete-time system (Hou and Jin 2011, 2013):

$$\begin{aligned} \boldsymbol{y}_l(k+1) = \boldsymbol{f}\big(&\boldsymbol{y}_l(k), \boldsymbol{y}_l(k-1), \ldots, \boldsymbol{y}_l(k-n_y^{\wr}), \\ &\boldsymbol{u}_l(k), \boldsymbol{u}_l(k-1), \ldots, \boldsymbol{u}_l(k-n_u^{\wr})\big) \end{aligned} \tag{4.1}$$

where the subscript l denotes the number of iterations and $k \in \{0, 1, \ldots, K_l^{\wr}\}$ denotes the operation time. In this chapter, the actual trial length $K_l^{\wr}$ is dependent on the iteration number l, such that it is iteration varying. $\boldsymbol{y}_l(k) \in \Re^n$ and $\boldsymbol{u}_l(k) \in \Re^n$ are the output and input vectors of the system at time k of the lth iteration. $n_u^{\wr} \in N^+$ and $n_y^{\wr} \in N^+$ are the unknown order of the system. $\boldsymbol{f}(\cdots) \in \Re^n$ is an unknown vector-valued nonlinear function.

System (4.1) meets the following assumptions:

Assumption 4.1 $\forall k \in N^+$, the initial output of the system $\boldsymbol{y}_l(0)$ is random and bounded.

Assumption 4.2 Define $K_d^{\wr}$ as the desired and fixed trial length of each operation. For $\forall k \in \{0, 1, \ldots, K_d^{\wr}\}$ and $l = 0, 1, 2 \ldots$, the partial derivatives of $\boldsymbol{f}(\cdots)$ with respect to $\boldsymbol{u}_l(k)$ exist and are continuous.

Assumption 4.3 The system satisfies the generalized *Lipschitz* condition that for $\forall k \in \{0, 1, \ldots, K_d^{\wr}\}$ and $l = 0, 1, 2 \ldots$, we have $\|\boldsymbol{y}_{l_1}(k+1) - \boldsymbol{y}_{l_2}(k+1)\| \le b^{\wr}\|\boldsymbol{u}_{l_1}(k) - \boldsymbol{u}_{l_2}(k)\|$, where $\boldsymbol{u}_{l_1}(k) \neq \boldsymbol{u}_{l_2}(k)$ for any $l_1 \neq l_2, l_1, l_2 > 0$. $b^{\wr}$ is a positive constant.

Assumption 4.4 The pseudo-Jacobian matrix $\boldsymbol{G}_l^{\wr *}(k)$ defined in the following (4.2) is a diagonal dominant matrix and satisfies the following conditions: $|g_{ij,l}^{\wr *}(k)| \le b_1^{\wr}, b_2^{\wr} \le |g_{ii,l}^{\wr *}(k)| \le \alpha^{\wr} b_2^{\wr}, i = 1, \ldots, n, j = 1, \ldots, n, i \neq j, \alpha^{\wr} \ge 1, b_2^{\wr} \ge b_1^{\wr}(2\alpha^{\wr} + 1)(n-1)$ and all the element symbols in $\boldsymbol{G}_l^{\wr *}(k)$ remain unchanged.

The following lemma shows that the considered MIMO nonlinear discrete-time system satisfying Assumptions 4.1–4.3 can be transformed into an equivalent dynamical linearisation model (Hou and Jin 2013), here is called iterative learning compact form dynamic linearization (ILCFDL) data model.

Lemma 4.1 *If system (4.1) satisfies Assumptions 4.1–4.3, for* $k \in \{0, 1, \dots, K_d^{\wr}\}$ *and* $l = 0, 1, 2, \dots$ *and* $\Delta \boldsymbol{u}_l(k) \neq \boldsymbol{0}$*, then there must exist a pseudo-Jacobian matrix* $\boldsymbol{G}_l^{\wr *}(k)$*, such that the nonlinear system (4.1) can be equated to the following ILCFDL model:*

$$\Delta \boldsymbol{y}_l(k+1) = \boldsymbol{G}_l^{\wr *}(k) \Delta \boldsymbol{u}_l(k) \tag{4.2}$$

where $\Delta \boldsymbol{y}_l(k+1) = \boldsymbol{y}_l(k+1) - \boldsymbol{y}_{l-1}(k+1)$, $\Delta \boldsymbol{u}_l(k) = \boldsymbol{u}_l(k) - \boldsymbol{u}_{l-1}(k)$, *and*

$$\boldsymbol{G}_l^{\wr *}(k) = \begin{bmatrix} g_{11,l}^{\wr *}(k) & g_{12,l}^{\wr *}(k) & \dots & g_{1n,l}^{\wr *}(k) \\ g_{21,l}^{\wr *}(k) & g_{22,l}^{\wr *}(k) & \dots & g_{2n,l}^{\wr *}(k) \\ \vdots & \vdots & \ddots & \vdots \\ g_{n1,l}^{\wr *}(k) & g_{n2,l}^{\wr *}(k) & \dots & g_{nn,l}^{\wr *}(k) \end{bmatrix} \text{ is unknown but bounded.}$$

Proof From system (4.1), one has:

$$\begin{aligned} \Delta \boldsymbol{y}_l(k+1) &= \boldsymbol{f}(\boldsymbol{y}_l(k), \boldsymbol{y}_l(k-1), \dots, \boldsymbol{y}_l(k-n_y^{\wr}), \boldsymbol{u}_l(k), \boldsymbol{u}_l(k-1), \dots, \boldsymbol{u}_l(k \\ &\quad -n_u^{\wr})) - \boldsymbol{f}(\boldsymbol{y}_{l-1}(k), \boldsymbol{y}_{l-1}(k-1), \dots, \boldsymbol{y}_{l-1}(k-n_y^{\wr}), \boldsymbol{u}_{l-1}(k), \boldsymbol{u}_{l-1}(k \\ &\quad -1), \dots, \boldsymbol{u}_{l-1}(k-n_u^{\wr})) = \boldsymbol{f}(\boldsymbol{y}_l(k), \boldsymbol{y}_l(k-1), \dots, \boldsymbol{y}_l(k-n_y^{\wr}), \\ &\quad \boldsymbol{u}_l(k), \boldsymbol{u}_l(k-1), \dots, \boldsymbol{u}_l(k-n_u^{\wr})) - \boldsymbol{f}(\boldsymbol{y}_l(k), \dots, \boldsymbol{y}_l(k-1), \\ &\quad \dots, \boldsymbol{y}_l(k-n_y^{\wr}), \boldsymbol{u}_{l-1}(k), \boldsymbol{u}_l(k-1), \boldsymbol{u}_l(k-n_u^{\wr})) + \boldsymbol{f}(\boldsymbol{y}_l(k), \boldsymbol{y}_l(k-1), \\ &\quad \dots, \boldsymbol{y}_l(k-n_y^{\wr}), \boldsymbol{u}_{l-1}(k), \boldsymbol{u}_l(k-1), \dots, \boldsymbol{u}_l(k-n_u^{\wr})) - \boldsymbol{f}(\boldsymbol{y}_{l-1}(k), \\ &\quad \boldsymbol{y}_{l-1}(k-1), \dots, \boldsymbol{y}_{l-1}(k-n_y^{\wr}), \boldsymbol{u}_{l-1}(k), \boldsymbol{u}_{l-1}(k-1), \dots, \boldsymbol{u}_{l-1}(k \\ &\quad -n_u^{\wr})) \end{aligned} \tag{4.3}$$

Let $\boldsymbol{\zeta}_l^{\wr}(k) = \boldsymbol{f}(\boldsymbol{y}_l(k), \boldsymbol{y}_l(k-1), \dots, \boldsymbol{y}_l(k-n_y^{\wr}), \boldsymbol{u}_{l-1}(k), \boldsymbol{u}_l(k-1), \dots, \boldsymbol{u}_l(k - n_u^{\wr})) - \boldsymbol{f}(\boldsymbol{y}_{l-1}(k), \boldsymbol{y}_{l-1}(k-1), \dots, \boldsymbol{y}_{l-1}(k-n_y^{\wr}), \boldsymbol{u}_{l-1}(k), \boldsymbol{u}_{l-1}(k-1), \dots, \boldsymbol{u}_{l-1}(k - n_u^{\wr}))$. According to Assumption 4.2 and differential mean value theorem, (4.3) can be rewritten as

$$\Delta \boldsymbol{y}_l(k+1) = \frac{\partial \boldsymbol{f}^{\wr *}}{\partial \boldsymbol{u}_l(k)} \Delta \boldsymbol{u}_l(k) + \boldsymbol{\zeta}_l^{\wr}(k) \tag{4.4}$$

where

$$\frac{\partial \boldsymbol{f}^*}{\partial \boldsymbol{u}_l(k)} = \begin{bmatrix} \frac{\partial f_1^*}{\partial u_{1,l}(k)} & \frac{\partial f_1^*}{\partial u_{2,l}(t)} & \cdots & \frac{\partial f_1^*}{\partial u_{n,l}(k)} \\ \frac{\partial f_2^*}{\partial u_{1,l}(k)} & \frac{\partial f_2^*}{\partial u_{2,l}(k)} & \cdots & \frac{\partial f_2^*}{\partial u_{n,l}(k)} \\ \vdots & \vdots & \ddots & \vdots \\ \frac{\partial f_n^*}{\partial u_{1,l}(k)} & \frac{\partial f_n^*}{\partial u_{2,l}(k)} & \cdots & \frac{\partial f_n^*}{\partial u_{n,l}(k)} \end{bmatrix}$$

$\frac{\partial f_i^*}{\partial u_{j,l}(k)}(i = 1, \ldots, n)(j = 1, \ldots, n)$ denotes the value of the partial derivative of f_i with respect to the control input u at a point between $u_{j,l-1}(k)$ and $u_{j,l}(k)$. For each fixed iteration l and time k, consider the following equation containing the numerical matrix $\eta_l^{\wr}(k)$

$$\boldsymbol{\zeta}_l^{\wr}(k) = \eta_l^{\wr}(k)\Delta \boldsymbol{u}_l(k) \tag{4.5}$$

Since $\Delta \boldsymbol{u}_l(k) \neq \mathbf{0}$, there exists at least one solution $\eta_l^{\wr *}(k)$ to meet (4.5). Let $\boldsymbol{G}_l^{\wr *}(k) = \frac{\partial \boldsymbol{f}^*}{\partial \boldsymbol{u}_l(k)} + \eta_l^{\wr *}(k)$, then (4.4) can be rewritten as:

$$\Delta \boldsymbol{y}_l(k+1) = \boldsymbol{G}_l^{\wr *}(k)\Delta \boldsymbol{u}_l(k) \tag{4.6}$$

According to Assumption 4.3, it follows that $\boldsymbol{G}_l^{\wr *}(k)$ is bounded. □

4.3 Data Compensation-Based Predictive ILC Design

Define the tracking control error $\boldsymbol{e}_l^{\wr}(k+1)$ as

$$\boldsymbol{e}_l^{\wr}(k+1) = \boldsymbol{y}_d^{\wr}(k+1) - \boldsymbol{y}_l(k+1) \quad 0 \leq k \leq \min\{K_l^{\wr}, K_d^{\wr}\} \tag{4.7}$$

where $\boldsymbol{y}_d^{\wr}(k+1)$ is the desired output trajectory, $\boldsymbol{y}_l(k+1)$ is the actual output trajectory.

To better describe the problem of random varying trial lengths, a new tracking error $\boldsymbol{e}_l^{\wr *}(k+1)$ over the whole desired operation time interval $\{0, 1, \ldots, K_d^{\wr}\}$ is introduced

$$\begin{aligned} \boldsymbol{e}_l^{\wr *}(k+1) = {} & \Psi_l^{\wr}(k+1)\boldsymbol{e}_l^{\wr}(k+1) + (1 - \Psi_l^{\wr}(k+1))\Psi_{l-1}^{\wr}(k+1)\boldsymbol{e}_{l-1}^{\wr}(k+1) \\ & + (1 - \Psi_l^{\wr}(k+1))(1 - \Psi_{l-1}^{\wr}(k+1))\boldsymbol{e}_{l|l-1}^{\wr *}(k+1) \quad 0 \leq k \leq K_d^{\wr} \end{aligned} \tag{4.8}$$

where $\Psi_l^{\wr}(k+1)$ is a random variable satisfying Bernoulli distribution, taking binary values 0 and 1. $\Psi_l^{\wr}(k+1) = 1$ denotes that the system (4.1) can continue operating until the time $k+1$ at lth iteration, then we can get $\boldsymbol{e}_l^{\wr *}(k+1)=\boldsymbol{e}_l^{\wr}(k+1)$. On the contrary, $\Psi_l^{\wr}(k+1)=0$ implies that the lth operation of the system has already been finished before the time $k+1$. Similarly, $\Psi_{l-1}^{\wr}(k+1)=1$ will lead to $\boldsymbol{e}_l^{\wr *}(k+1)=\boldsymbol{e}_{l-1}^{\wr}(k+1)$. In addition, if $\Psi_l^{\wr}(k+1)=0$ and $\Psi_{l-1}^{\wr}(k+1)=0$ hold simultaneously, it is indicated that the actual data $\boldsymbol{e}_l^{\wr}(k+1)$ and $\boldsymbol{e}_{l-1}^{\wr}(k+1)$ are missing, such that the predicted value $\boldsymbol{e}_{l|l-1}^{\wr *}(k+1)$ will be used for data compensation, namely $\boldsymbol{e}_l^{\wr *}(k+1)=\boldsymbol{e}_{l|l-1}^{\wr *}(k+1)$ where $\boldsymbol{e}_{l|l-1}^{\wr *}(k+1)$ defined in (4.12) is the predicted tracking error for the lth iteration at the historical $l-1$th iteration.

By virtue of the above designed data compensation mechanism (4.8), the relationship between $\boldsymbol{e}_l^{\wr *}(k+1)$ and $\boldsymbol{e}_l^{\wr}(k+1)$ can be described as:
if $K_l^{\wr}<K_d^{\wr}$,

$$\boldsymbol{e}_l^{\wr *}(k+1)=\begin{cases}\boldsymbol{e}_l^{\wr}(k+1) & 0\le k\le K_l^{\wr}\\ (1-\Psi_l^{\wr}(k+1))\Psi_{l-1}^{\wr}(k+1)\boldsymbol{e}_{l-1}^{\wr}(k+1) & \\ +(1-\Psi_l^{\wr}(k+1))(1-\Psi_{l-1}^{\wr}(k+1))\boldsymbol{e}_{l|l-1}^{\wr *}(k+1) & K_l^{\wr}<k\le K_d^{\wr}\end{cases}\tag{4.9}$$

if $K_l^{\wr}\ge K_d^{\wr}$,

$$\boldsymbol{e}_l^{\wr *}(k+1)=\boldsymbol{e}_l^{\wr}(k+1),0\le k\le\ K_d^{\wr}\tag{4.10}$$

Rewrite (4.2) as $\boldsymbol{y}_l(k+1)=\boldsymbol{y}_{l-1}(k+1)+\boldsymbol{G}_l^{\wr *}(k)\Delta\boldsymbol{u}_l(k)$. Now based on (4.9)–(4.10), subtracting both sides of this equation by $\boldsymbol{y}_d^{\wr}(k+1)$ leads to:

$$\boldsymbol{e}_l^{\wr *}(k+1)=\boldsymbol{e}_{l-1}^{\wr *}(k+1)-\boldsymbol{G}_l^{\wr *}(k)\Delta\boldsymbol{u}_l(k)\tag{4.11}$$

According to (4.11), the tracking error for the future m step iterative prediction can be described as:

$$\begin{aligned}\boldsymbol{e}_{l+1|l}^{\wr *}(k+1)&=\boldsymbol{e}_l^{\wr *}(k+1)-\boldsymbol{G}_{l+1}^{\wr *}(k)\Delta\boldsymbol{u}_{l+1}(k)\\ \boldsymbol{e}_{l+2|l}^{\wr *}(k+1)&=\boldsymbol{e}_l^{\wr *}(k+1)-\boldsymbol{G}_{l+1}^{\wr *}(k)\Delta\boldsymbol{u}_{l+1}(k)-\boldsymbol{G}_{l+2}^{\wr *}(k)\Delta\boldsymbol{u}_{l+2}(k)\\ &\vdots\\ \boldsymbol{e}_{l+m|l}^{\wr *}(k+1)&=\boldsymbol{e}_l^{\wr *}(k+1)-\boldsymbol{G}_{l+1}^{\wr *}(k)\Delta\boldsymbol{u}_{l+1}(k)-\boldsymbol{G}_{l+2}^{\wr *}(k)\Delta\boldsymbol{u}_{l+2}(k)\\ &\quad-\ldots-\boldsymbol{G}_{l+m}^{\wr *}(k)\Delta\boldsymbol{u}_{l+m}(k)\end{aligned}\tag{4.12}$$

where $\boldsymbol{e}_{l+m|l}^{\wr *}(k+1)$ denotes the predicted tracking error of the future $l+m$th iteration at the current lth iteration.

Define $\boldsymbol{G}_{l+1}^{\wr *m}(k)\in\Re^{n\times mn}$ and $\Delta\boldsymbol{u}_{l+1}^{m}(k)\in\Re^{mn\times 1}$ as follows:

$$\boldsymbol{G}_{l+1}^{\wr *m}(k)=\left[\boldsymbol{G}_{l+1}^{\wr *}(k),\boldsymbol{G}_{l+2}^{\wr *}(k),\ldots,\boldsymbol{G}_{l+m}^{\wr *}(k)\right]\tag{4.13}$$

$$\Delta \boldsymbol{u}_{l+1}^{m}(k) = \left[\Delta \boldsymbol{u}_{l+1}^{T}(k), \Delta \boldsymbol{u}_{l+2}^{T}(k), \ldots, \Delta \boldsymbol{u}_{l+m}^{T}(k)\right]^{T} \tag{4.14}$$

then (4.12) can be rewritten as:

$$\boldsymbol{e}_{l+m|l}^{\wr *}(k+1) = \boldsymbol{e}_{l}^{\wr *}(k+1) - \boldsymbol{G}_{l+1}^{\wr *m}(k)\Delta \boldsymbol{u}_{l+1}^{m}(k) \tag{4.15}$$

Since $\boldsymbol{G}_{l}^{\wr *}(k)$ in (4.11) and $\boldsymbol{G}_{l+1}^{\wr *m}(k)$ in (4.15) are unknown, it is necessary to estimate $\boldsymbol{G}_{l}^{\wr *}(k)$ and predict $\boldsymbol{G}_{l+1}^{\wr *m}(k)$. Denote $\hat{\boldsymbol{G}}_{l}^{\wr *}(k)$ as the estimated value of $\boldsymbol{G}_{l}^{\wr *}(k)$ and $\hat{\boldsymbol{G}}_{l+1}^{\wr *m}(k)$ as the predicted value of $\boldsymbol{G}_{l+1}^{\wr *m}(k)$. Next, the corresponding estimation and prediction algorithms are designed sequentially.

Define the following modeling error:

$$\tilde{\boldsymbol{e}}_{l}^{\wr *}(k+1) = \boldsymbol{y}_{l}^{*}(k+1) - \hat{\boldsymbol{y}}_{l}(k+1) \tag{4.16}$$

where

$$\hat{\boldsymbol{y}}_{l}(k+1) = \boldsymbol{y}_{l-1}^{*}(k+1) + \hat{\boldsymbol{G}}_{l}^{\wr *}(k)\Delta \boldsymbol{u}_{l}(k) \tag{4.17}$$

is the estimation of $\boldsymbol{y}_{l}(k+1)$. $\hat{\boldsymbol{G}}_{l}^{\wr *}(k)$ is the estimation of $\boldsymbol{G}_{l}^{\wr *}(k)$. $\boldsymbol{y}_{l}^{*}(k+1)$ is a newly defined system output over the whole desired trial length $k\{0, 1, \ldots, K_{d}^{\wr}\}$ to compensate for the lost actual output data, that is,

$$\boldsymbol{y}_{l}^{*}(k+1) = \begin{cases} \boldsymbol{y}_{l}(k+1) & 0 \le k \le K_{l}^{\wr} \\ \boldsymbol{y}_{d}^{\wr}(k+1) & K_{l}^{\wr} < k \le K_{d}^{\wr} \end{cases} \tag{4.18}$$

Compared with the estimation algorithm (2.23) in Chap.2 and (3.15) in Chap.3, based on (4.16), the following new iterative learning estimation algorithm is constructed to obtain $\hat{\boldsymbol{G}}_{l}^{\wr *}(k)$:

$$\hat{\boldsymbol{G}}_{l}^{\wr *}(k) = \hat{\boldsymbol{G}}_{l-1}^{\wr *}(k) + \eta^{\wr}\tilde{\boldsymbol{e}}_{l-1}^{\wr *}(k+1)\Delta \boldsymbol{u}_{l-1}^{T}(k) \tag{4.19}$$

where $\eta^{\wr}$ is an adjustable learning parameter. Besides, the convergence of the estimation error can also be guaranteed in the following Theorem 4.1.

Moreover, to make the above estimation algorithm (4.19) have a stronger ability to track iteration time-varying parameters, the following reset algorithm is used

$$\hat{g}_{ii,l}^{\wr *}(k) = \hat{g}_{ii,0}^{\wr *}(k) \tag{4.20}$$

if $\left|\hat{g}_{ii,l}^{\wr *}(k)\right| < b_{2}^{\wr}$ or $\left|\hat{g}_{ii,l}^{\wr *}(k)\right| > \alpha^{\wr}b_{2}^{\wr}$ or $sign(\hat{g}_{ii,l}^{\wr *}(k)) \neq sign(\hat{g}_{ii,0}^{\wr *}(k))$.

$$\hat{g}_{ij,l}^{\wr *}(k) = \hat{g}_{ij,0}^{\wr *}(k) \tag{4.21}$$

if $\left|\hat{g}_{ij,l}^{\wr *}(k)\right| > b_1$ or $sign(\hat{g}_{ij,l}^{\wr *}(k)) \neq sign(\hat{g}_{ij,0}^{\wr *}(k))$.
where $\hat{g}_{ii,0}^{\wr *}(k)$ and $\hat{g}_{ij,0}^{\wr *}(k)$ are the values at the initial iteration of $\hat{g}_{ii,l}^{\wr *}(k)$ and $\hat{g}_{ij,l}^{\wr *}(k)$ respectively, namely $l = 0$.

Note that based on the proposed iterative learning estimation algorithm (4.19)–(4.21), the estimated value $\hat{\boldsymbol{G}}_l^{\wr *}(k)$ is available. By means of these estimated values, the predicted values $\hat{\boldsymbol{G}}_{l+i}^{\wr *}(k)$, $i = 1, 2, \ldots, m$ for future $l + i$ iterations will be acquired as follows.

First of all, the following one-step iterative learning prediction algorithm is constructed:

$$\hat{\boldsymbol{G}}_{l+1}^{\wr *}(k) = \boldsymbol{\Lambda}_1^{\wr}(k)\,\hat{\boldsymbol{G}}_l^{\wr *}(k) + \boldsymbol{\Lambda}_2^{\wr}(k)\,\hat{\boldsymbol{G}}_{l-1}^{\wr *}(k) + \ldots + \boldsymbol{\Lambda}_{n_p^{\wr}}^{\wr}(k)\,\hat{\boldsymbol{G}}_{l+1-n_p^{\wr}}^{\wr *}(k) \tag{4.22}$$

where $\boldsymbol{\Lambda}_q^{\wr}(k) \in \Re^{n \times n}(q^{\wr} = 1, 2, \ldots, n_p^{\wr})$ is an unknown parameter matrix. $n_p^{\wr}$ is an appropriate order.

Define $\hat{\boldsymbol{\Lambda}}_{q,l}^{\wr}(k)$ as the estimation of the unknown parameter matrix $\boldsymbol{\Lambda}_q^{\wr}(k)$, $(q = 1, 2, \ldots, n_p^{\wr})$ at lth iteration, which will be designed in the following (4.26). Then, the following available $(l + i)$ $(i = 1, 2, \ldots, m)$ step iterative learning prediction algorithm is constructed:

$$\hat{\boldsymbol{G}}_{l+i}^{\wr *}(k) = \hat{\boldsymbol{\Lambda}}_{1,l}^{\wr}(k)\hat{\boldsymbol{G}}_{l+i-1}^{\wr *}(k) + \hat{\boldsymbol{\Lambda}}_{2,l}^{\wr}(k)\hat{\boldsymbol{G}}_{l+i-2}^{\wr *}(k) + \ldots + \hat{\boldsymbol{\Lambda}}_{n_p^{\wr},l}^{\wr}(k)\hat{\boldsymbol{G}}_{l+i-n_p^{\wr}}^{\wr *}(k) \tag{4.23}$$

Now define that:

$$\begin{aligned} \hat{\boldsymbol{\Lambda}}_l^{\wr n_p^{\wr}}(k) &= \left[\hat{\boldsymbol{\Lambda}}_{1,l}^{\wr}(k)^T, \ldots, \hat{\boldsymbol{\Lambda}}_{n_p^{\wr},l}^{\wr}(k)^T\right]^T \in \Re^{nn_p^{\wr} \times n} \\ \hat{\boldsymbol{\Phi}}_{l-1}^{\wr}(k) &= \left[\hat{\boldsymbol{G}}_{l-1}^{\wr *}(k), \ldots, \hat{\boldsymbol{G}}_{l-n_p^{\wr}}^{\wr *}(k)\right]^T \in \Re^{nn_p^{\wr} \times n} \end{aligned} \tag{4.24}$$

then (4.23) can be rewritten as:

$$\hat{\boldsymbol{G}}_{l+i}^{\wr *}(k) = \hat{\boldsymbol{\Phi}}_{l+i-1}^{\wr T}(k)\hat{\boldsymbol{\Lambda}}_l^{\wr n_p^{\wr}}(k),\ i = 1, 2, \ldots, m \tag{4.25}$$

Here $\hat{\boldsymbol{\Lambda}}_l^{\wr n_p^{\wr}}(k)$ is computed by the following iterative learning algorithm:

$$\begin{aligned} \hat{\boldsymbol{\Lambda}}_l^{\wr n_p^{\wr}}(k) = \hat{\boldsymbol{\Lambda}}_{l-1}^{\wr n_p^{\wr}}(k) + \kappa^{\wr}\hat{\boldsymbol{\Phi}}_{l-1}^{\wr}(k)\left[\nu^{\wr} I + \hat{\boldsymbol{\Phi}}_{l-1}^{\wr T}(k)\hat{\boldsymbol{\Phi}}_{l-1}^{\wr}(k)\right]^{-1} \\ \left[\hat{\boldsymbol{G}}_l^{\wr *}(k) - \hat{\boldsymbol{\Phi}}_{l-1}^{\wr T}(k)\hat{\boldsymbol{\Lambda}}_{l-1}^{\wr n_p^{\wr}}(k)\right] \end{aligned} \tag{4.26}$$

where $\kappa^{\wr}$, $\nu^{\wr}$ are adjustable learning parameters.

Now based on the designed iterative learning prediction algorithm (4.23)–(4.26), the predicted values $\hat{\boldsymbol{G}}_{l+1}^{\wr*}(k), \hat{\boldsymbol{G}}_{l+2}^{\wr*}(k), \ldots, \hat{\boldsymbol{G}}_{l+m}^{\wr*}(k)$ can be obtained.

Denote

$$\hat{\boldsymbol{G}}_{l+1}^{\wr*m}(k) = \left[\hat{\boldsymbol{G}}_{l+1}^{\wr*}(k), \hat{\boldsymbol{G}}_{l+2}^{\wr*}(k), \ldots, \hat{\boldsymbol{G}}_{l+m}^{\wr*}(k)\right] \tag{4.27}$$

then instead of (4.15), we can use the following available data-based tracking error prediction model for controller design:

$$\boldsymbol{e}_{l+m|l}^{\wr*}(k+1) = \boldsymbol{e}_{l}^{\wr*}(k+1) - \hat{\boldsymbol{G}}_{l+1}^{\wr*m}(k)\Delta\boldsymbol{u}_{l+1}^{m}(k) \tag{4.28}$$

Based on (4.28), the optimal control input sequence $\Delta\boldsymbol{u}_{l+1}^{m}(k)$ can be obtained by computing the following quadratic cost function:

$$\begin{aligned} J_{l+1}^{\wr}(k) = \min_{\Delta\boldsymbol{u}_{l+1}^{m}(k)} & \left\{\boldsymbol{e}_{l+m|l}^{\wr*T}(k+1)\, \boldsymbol{Q}^{\wr}\boldsymbol{e}_{l+m|l}^{\wr*}(k+1)\right. \\ & \left. + \Delta\boldsymbol{u}_{l+1}^{mT}(k)\, \boldsymbol{R}^{\wr}\Delta\boldsymbol{u}_{l+1}^{m}(k)\right\} \end{aligned} \tag{4.29}$$

where $\boldsymbol{Q}^{\wr} \in \Re^{n\times n}$, $\boldsymbol{R}^{\wr} \in \Re^{nm\times nm}$ are two symmetric positive definite matrices, for simplicity, define $\boldsymbol{Q}^{\wr} = q^{\wr}\boldsymbol{I}_{n\times n}$, $\boldsymbol{R}^{\wr} = r^{\wr}\boldsymbol{I}_{nm\times nm}$, where $q^{\wr} > 0, r^{\wr} > 0$. According to the optimization condition $\frac{\partial J_{l+1}^{\wr}(k)}{\partial \Delta\boldsymbol{u}_{l+1}^{m}(k)} = \boldsymbol{0}$, we have:

$$\Delta\boldsymbol{u}_{l+1}^{m}(k) = \left[\hat{\boldsymbol{G}}_{l+1}^{\wr*mT}(k)\boldsymbol{Q}^{\wr}\hat{\boldsymbol{G}}_{l+1}^{\wr*m}(k) + \boldsymbol{R}^{\wr}\right]^{-1}\hat{\boldsymbol{G}}_{l+1}^{\wr*mT}(k)\boldsymbol{Q}^{\wr}\boldsymbol{e}_{l}^{\wr*}(k+1) \tag{4.30}$$

Since the control algorithm (4.30) contains matrix inverse operation, when the input and output dimensions of the system are large, the inverse operation is very time-consuming, which is not conducive to practical applications. In order to solve this problem, the following more practical control algorithm will be applied:

$$\Delta\boldsymbol{u}_{l+1}^{m}(k) = \frac{\gamma^{\wr}q^{\wr}\hat{\boldsymbol{G}}_{l+1}^{\wr*mT}(k)}{r^{\wr} + q^{\wr}\left\|\hat{\boldsymbol{G}}_{l+1}^{\wr*m}(k)\right\|^{2}}\boldsymbol{e}_{l}^{\wr*}(k+1) \tag{4.31}$$

where the step factor $\gamma^{\wr} \in (0, 1]$ is introduced to make the control law more general.

Using the receding optimization strategy, the first element in the designed control input sequence $\Delta\boldsymbol{u}_{l+1}^{m}(k)$ will be applied for the upcoming $(l+1)$th iteration, that is:

$$\boldsymbol{u}_{l+1}(k) = \boldsymbol{u}_{l}(k) + \boldsymbol{U}^{T}\Delta\boldsymbol{u}_{l+1}^{m}(k) \tag{4.32}$$

where $\boldsymbol{U} = \left[\boldsymbol{I}_{n\times n}\boldsymbol{0}_{n\times n(m-1)}\right]^{T}$.

To facilitate the subsequent analysis of the convergence property, the following lemma is given.

Lemma 4.2 *(Gerschgorin S 1931) Let* $\mathbf{A}^{\wr} = (a_{ij}^{\wr}) \in \mathbb{C}^{n\times n}$, *define the Gerschgorin disk:* $D_i^{\wr} = \{z^{\wr} \left| \left| z^{\wr} - a_{ii}^{\wr} \right| \leq \sum_{j=1, j\neq i}^{n} \left| a_{ij}^{\wr} \right| \right\}, z^{\wr} \in \mathbb{C}, 1 \leq i \leq n$. *Then, all the eigenroots of* $\mathbf{A}^{\wr}$ *matrix are in the disk and all the eigenroots of the matrix satisfy* $z_1^{\wr}, z_2^{\wr}, \ldots, z_n^{\wr} \in D_A^{\wr} = \bigcup_{i=1}^{n} D_i^{\wr}$.

Theorem 4.1 *If system (4.1) satisfies Assumptions 4.1–4.4, the proposed data compensation-based PILC method (4.9), (4.19)–(4.26), (4.31)–(4.32) satisfies the following condition:*

$$0 < \eta^{\wr} \frac{\partial \tilde{e}_{i,l}^{\wr *}(k+1)}{\partial \hat{\boldsymbol{G}}_{i,l}^{\wr *}(k)} \left(\frac{\partial \tilde{e}_{i,l}^{\wr *}(k+1)}{\partial \hat{\boldsymbol{G}}_{i,l}^{\wr *}(k)} \right)^T < 1, i = 1, 2, \ldots, n \tag{4.33}$$

where $\hat{\boldsymbol{G}}_{i,l}^{\wr}(k)$ *defined in (4.37) is the* i*th row of* $\hat{\boldsymbol{G}}_l^{\wr}(k)$, *and* $\tilde{e}_{i,l}^{\wr *}(k+1)$ *defined in (4.38) is the* i*th element of* $\tilde{\boldsymbol{e}}_{l+1}^{\wr}(k+1)$. *Then, we have:*
(4.1.1) The modeling error converges to zero when $l \to \infty$, *that is,*

$$\lim_{l\to\infty} \tilde{\boldsymbol{e}}_{l+1}^{\wr *}(k+1) = \mathbf{0}. \tag{4.34}$$

(4.1.2) The tracking control error can be converged to zero when $l \to \infty$, *that is,*

$$\lim_{l\to\infty} \left\| \boldsymbol{e}_{l+1}^{\wr *}(k+1) \right\|_{\upsilon} = 0. \tag{4.35}$$

Proof **proof of (4.1.1):**
Note that

$$\boldsymbol{y}_l^{\wr *}(k+1) = \left[y_{1,l}^{\wr *}(k+1), \ldots, y_{n,l}^{\wr *}(k+1) \right]^T \tag{4.36}$$

$$\hat{\boldsymbol{G}}_l^{\wr *}(k) = \left[\hat{\boldsymbol{G}}_{1,l+1}^{\wr * T}(k), \ldots, \hat{\boldsymbol{G}}_{n,l+1}^{\wr * T}(k) \right]^T \tag{4.37}$$

$$\tilde{\boldsymbol{e}}_l^{\wr *}(k+1) = [\tilde{e}_{1,l}^{\wr *}(k+1), \ldots, \tilde{e}_{n,l}^{\wr *}(k+1)] \tag{4.38}$$

According to (4.16) and (4.17), we can get:

$$\tilde{e}_{i,l+1}^{\wr *}(k+1) = y_{i,l+1}^{*}(k+1) - y_{i,l}^{*}(k+1) - \hat{\boldsymbol{G}}_{l+1}^{\wr *}(k) \Delta \boldsymbol{u}_{l+1}(k) \tag{4.39}$$

Based on the proposed compensation mechanism (4.18), there are four cases for the values of $y^*_{i,l+1}(k+1)$ and $y^*_{i,l}(k+1)$.

Case 1: $y^*_{i,l+1}(k+1)$ exists but $y^*_{i,l}(k+1)$ is missing, which means $y^*_{i,l+1}(k+1) = y_{i,l+1}(k+1)$ and $y^*_{i,l}(k+1) = y_d(k+1)$;

Case 2: Both $y^*_{i,l+1}(k+1)$ and $y^*_{i,l}(k+1)$ exist, which means $y^*_{i,l+1}(k+1) = y_{i,l+1}(k+1)$ and $y^*_{i,l}(k+1) = y_{i,l}(k+1)$;

Case 3: Both $y^*_{i,l+1}(k+1)$ and $y^*_{i,l}(k+1)$ are missing, which means $y^*_{i,l+1}(k+1) = y_d(k+1)$ and $y^*_{i,l}(k+1) = y_d(k+1)$;

Case 4: $y^*_{i,l+1}(k+1)$ is missing but $y^*_{i,l}(k+1)$ exists, which means $y^*_{i,l+1}(k+1) = y_d(k+1)$ and $y^*_{i,l}(k+1) = y_{i,l}(k+1)$.

Since the proof of the above four cases is similar, only the proof of Case 1 is shown here and the proof of the other three cases can also derive the convergence of $\tilde{e}^{\wr *}_{i,l+1}(k+1)$ according to (4.40)–(4.51).

For the above Case 1, (4.39) can be rewritten as:

$$\begin{aligned}\tilde{e}^{\wr *}_{i,l+1}(k+1) &= y^*_{i,l+1}(k+1) - y^*_{i,l}(k+1) - \hat{\boldsymbol{G}}^{\wr *}_{l+1}(k)\Delta \boldsymbol{u}_{l+1}(k) \\ &= y_{i,l+1}(k+1) - y_d(k+1) - \hat{\boldsymbol{G}}^{\wr *}_{l+1}(k)\Delta \boldsymbol{u}_{l+1}(k)\end{aligned} \tag{4.40}$$

Define the following composite energy function:

$$E^{\wr *}_{i,l+1}(k+1) = \tilde{e}^{\wr *}_{i,l+1}(k+1)^2, i = 1, 2, \dots, n \tag{4.41}$$

then its difference on the iterative axis can be expressed as:

$$\begin{aligned}\Delta E^{\wr *}_{i,l+1}(k+1) &= E^{\wr *}_{i,l+1}(k+1) - E^{\wr *}_{i,l}(k+1) \\ &= \tilde{e}^{\wr *}_{i,l+1}(k+1)^2 - \tilde{e}^{\wr *}_{i,l}(k+1)^2 \\ &= \Delta \tilde{e}^{\wr *}_{i,l+1}(k+1)^2 + 2\tilde{e}^{\wr *}_{i,l}(k+1)\Delta \tilde{e}^{\wr *}_{i,l+1}(k+1)\end{aligned} \tag{4.42}$$

where $\Delta \tilde{e}^{\wr *}_{i,l+1}(k+1) = \tilde{e}^{\wr *}_{i,l+1}(k+1) - \tilde{e}^{\wr *}_{i,l}(k+1)$.

Applying Taylor expansion to $\Delta \tilde{e}^{\wr *}_{i,l+1}(k+1)$ yields:

$$\Delta \tilde{e}^{\wr *}_{i,l+1}(k+1) = \frac{\partial \tilde{e}^{\wr *}_{i,l}(k+1)}{\partial \hat{\boldsymbol{G}}^{\wr *}_{i,l}(k)} \Delta \hat{\boldsymbol{G}}^{\wr * T}_{i,l+1}(k) \tag{4.43}$$

where $\Delta \hat{\boldsymbol{G}}^{\wr *}_{i,l+1}(k) = \hat{\boldsymbol{G}}^{\wr *}_{i,l+1}(k) - \hat{\boldsymbol{G}}^{\wr *}_{i,l}(k)$.

Taking (4.19) into (4.43) yields:

$$\Delta \tilde{e}^{\wr *}_{i,l+1}(k+1) = \eta^{\wr} \frac{\partial \tilde{e}^{\wr *}_{i,l}(k+1)}{\partial \hat{\boldsymbol{G}}^{\wr}_{i,l}(k)} \tilde{e}^{\wr *}_{i,l}(k+1)\Delta \boldsymbol{u}_l(k) \tag{4.44}$$

According to (4.40) and (4.44), we have:

$$\Delta\tilde{e}_{i,l+1}^{\wr *}(k+1) = -\eta^{\wr}\tilde{e}_{i,l}^{\wr *}(k+1)\frac{\partial\tilde{e}_{i,l}^{\wr *}(k+1)}{\partial\hat{\boldsymbol{G}}_{i,l}^{\wr *}(k)}\left(\frac{\partial\tilde{e}_{i,l}^{\wr *}(k+1)}{\partial\hat{\boldsymbol{G}}_{i,l}^{\wr *}(k)}\right)^{T} \tag{4.45}$$

Let

$$l_{i,l}^{\wr}(k+1) = \eta^{\wr}\frac{\partial\tilde{e}_{i,l}^{\wr *}(k+1)}{\partial\hat{\boldsymbol{G}}_{i,l}^{\wr *}(k)}\left(\frac{\partial\tilde{e}_{i,l}^{\wr *}(k+1)}{\partial\hat{\boldsymbol{G}}_{i,l}^{\wr *}(k)}\right)^{T} \tag{4.46}$$

Then, (4.45) can be rewritten as:

$$\Delta\tilde{e}_{i,l+1}^{\wr *}(k+1) = -\tilde{e}_{i,l}^{\wr *}(k+1)l_{i,l}^{\wr}(k+1) \tag{4.47}$$

Substituting (4.47) into (4.42) yields:

$$\begin{aligned}\Delta E_{i,l+1}^{\wr *}(k+1) &= \Delta\tilde{e}_{i,l+1}^{\wr *}(k+1)^2 + 2\tilde{e}_{i,l}^{\wr *}(k+1)\Delta\tilde{e}_{i,l+1}^{\wr *}(k+1) \\ &= \tilde{e}_{i,l}^{\wr *}(k+1)^2\left[l_{i,l}^{\wr}(k+1)^2 - 2l_{i,l}^{\wr}(k+1)\right]\end{aligned} \tag{4.48}$$

According to the condition (4.33), we know that $0 < l_{i,l}^{\wr}(k+1) < 1$, then from (4.42) and (4.48), we have:

$$\Delta E_{i,l+1}^{\wr *}(k+1) \le -l_{i,l}^{\wr}(k+1)\tilde{e}_{i,l}^{\wr *}(k+1)^2 \tag{4.49}$$

According to (4.42) we further have:

$$\tilde{e}_{i,l+1}^{\wr *}(k+1)^2 \le \left[1 - l_{i,l}^{\wr}(k+1)\right]\tilde{e}_{i,l}^{\wr *}(k+1)^2 \tag{4.50}$$

According to the condition (4.33), there exists a positive constant $\rho^{\wr}$ satisfying $0 < 1 - l_{i,l}^{\wr}(k+1) \le \rho^{\wr} < 1$, then (4.50) leads to:

$$\tilde{\boldsymbol{e}}_{l+1}^{\wr *}(k+1)^2 \le \rho^{\wr}\tilde{\boldsymbol{e}}_{l}^{\wr *}(k+1)^2 \tag{4.51}$$

and $\lim_{l\to\infty}\tilde{\boldsymbol{e}}_{l+1}^{\wr *}(k+1) = \mathbf{0}$.

proof of (4.1.2):

Substituting the proposed control law (4.31)–(4.32) into (4.11) yields:

$$\begin{aligned}\boldsymbol{e}_{l+1}^{\wr *}(k+1) &= \boldsymbol{e}_{l}^{\wr *}(k+1) - \boldsymbol{G}_{l+1}^{\wr *}(k)\Delta\boldsymbol{u}_{l+1}(k) \\ &= \boldsymbol{e}_{l}^{\wr *}(k+1) - \frac{\gamma^{\wr}q^{\wr}\boldsymbol{G}_{l+1}^{\wr *}(k)\boldsymbol{U}^T\hat{\boldsymbol{G}}_{l+1}^{\wr *mT}(k)}{r^{\wr} + q^{\wr}\left\|\hat{\boldsymbol{G}}_{l+1}^{\wr *m}(k)\right\|^2}\boldsymbol{e}_{l}^{\wr *}(k+1)\end{aligned}$$

$$
\begin{aligned}
&= \boldsymbol{e}_l^{\wr *}(k+1) - \frac{\gamma^{\wr} q^{\wr} \boldsymbol{G}_{l+1}^{\wr *}(k) \hat{\boldsymbol{G}}_{l+1}^{\wr * T}(k)}{r^{\wr} + q^{\wr} \left\| \hat{\boldsymbol{G}}_{l+1}^{\wr * m}(k) \right\|^2} \boldsymbol{e}_l^{\wr *}(k+1) \\
&= \left\{ \boldsymbol{I} - \frac{\gamma^{\wr} q^{\wr} \boldsymbol{G}_{l+1}^{\wr *}(k) \hat{\boldsymbol{G}}_{l+1}^{\wr * T}(k)}{r^{\wr} + q^{\wr} \left\| \hat{\boldsymbol{G}}_{l+1}^{\wr * m}(k) \right\|^2} \right\} \boldsymbol{e}_l^{\wr *}(k+1)
\end{aligned}
\tag{4.52}
$$

According to Lemma 4.2, it follows that:

$$
\begin{aligned}
D_j^{\wr} = \Bigg\{ z^{\wr} \Bigg| \Bigg| z^{\wr} &- \left(1 - \frac{\gamma^{\wr} q^{\wr} \sum_{i=1}^{n} g_{ji,l+1}^{\wr *}(k) \hat{g}_{ji,l+1}^{\wr *}(k)}{r^{\wr} + q^{\wr} \left\| \hat{\boldsymbol{G}}_{l+1}^{\wr * m}(k) \right\|^2} \right) \Bigg| \\
&\le \sum_{h=1, h \ne j}^{n} \left| \frac{\gamma^{\wr} q^{\wr} \sum_{i=1}^{n} g_{1,ji,l+1}^{\wr *}(k) \hat{g}_{hi,l+1}^{\wr *}(k)}{r^{\wr} + q^{\wr} \left\| \hat{\boldsymbol{G}}_{l+1}^{\wr * m}(k) \right\|^2} \right| \Bigg\}
\end{aligned}
\tag{4.53}
$$

where $z^{\wr}$ is the characteristic root of matrix $\boldsymbol{I} - \frac{\gamma^{\wr} q^{\wr} \boldsymbol{G}_{l+1}^{\wr *}(k) \hat{\boldsymbol{G}}_{l+1}^{\wr * T}(k)}{r^{\wr} + q^{\wr} \left\| \hat{\boldsymbol{G}}_{l+1}^{\wr * m}(k) \right\|^2}$, $D_w^{\wr}$, $w = 1, \ldots, m$ is the Gerschgorin disk.

Using the trigonometric inequality, (4.53) can be rewritten as:

$$
\begin{aligned}
D_w^{\wr} = \Bigg\{ z^{\wr} \Bigg| \left| z^{\wr} \right| &\le \left| 1 - \frac{\gamma^{\wr} q^{\wr} \sum_{i=1}^{n} g_{ji,l+1}^{\wr *}(k) \hat{g}_{ji,l+1}^{\wr *}(k)}{r^{\wr} + q^{\wr} \left\| \hat{\boldsymbol{G}}_{l+1}^{\wr * m}(k) \right\|^2} \right| \\
&+ \sum_{h=1, h \ne j}^{n} \left| \frac{\gamma^{\wr} q^{\wr} \sum_{i=1}^{n} g_{ji,l+1}^{\wr *}(k) \hat{g}_{hi,l+1}^{\wr *}(k)}{r^{\wr} + q^{\wr} \left\| \hat{\boldsymbol{G}}_{l+1}^{\wr * m}(k) \right\|^2} \right| \Bigg\}
\end{aligned}
\tag{4.54}
$$

On the other hand, the following two inequalities can be obtained from the reset algorithm (4.20)–(4.21) and Assumption 4.4:

$$
\begin{aligned}
1 - \frac{\gamma^{\wr} q^{\wr} \sum_{i=1}^{n} \left| g_{ji,l+1}^{\wr *}(k) \right| \left| \hat{g}_{ji,l+1}^{\wr *}(k) \right|}{r^{\wr} + q^{\wr} \left\| \hat{\boldsymbol{G}}_{l+1}^{\wr * m}(k) \right\|^2} &\le 1 - \frac{\gamma^{\wr} q^{\wr} \left| g_{jj,l+1}^{\wr *}(k) \right| \left| \hat{g}_{jj,l+1}^{\wr *}(k) \right|}{r^{\wr} + q^{\wr} \left\| \hat{\boldsymbol{G}}_{l+1}^{\wr * m}(k) \right\|^2} \\
&\le 1 - \frac{\gamma^{\wr} q^{\wr} b_2^{\wr 2}}{r^{\wr} + q^{\wr} \left\| \hat{\boldsymbol{G}}_{l+1}^{\wr * m}(k) \right\|^2}
\end{aligned}
\tag{4.55}
$$

and

$$
\begin{aligned}
&\sum_{h=1, h \ne j}^{n} \frac{\gamma^{\wr} q^{\wr} \sum_{i=1}^{n} g_{ji,l+1}^{\wr *}(k) \, \hat{g}_{hi,l+1}^{\wr *}(k)}{r^{\wr} + q^{\wr} \left\| \hat{\boldsymbol{G}}_{l+1}^{\wr * m}(k) \right\|^2} \\
&\le \gamma^{\wr} q^{\wr} \sum_{h=1, h \ne j}^{n} \frac{\sum_{i=1}^{n} \left| g_{ji,l+1}^{\wr *}(k) \right| \left| \hat{g}_{hi,l+1}^{\wr *}(k) \right|}{r^{\wr} + q^{\wr} \left\| \hat{\boldsymbol{G}}_{l+1}^{\wr * m}(k) \right\|^2}
\end{aligned}
$$

$$
\begin{aligned}
&\le \gamma^{\wr} q^{\wr} \frac{\sum_{h=1,h\neq j}^{n} \left| g_{jj,l+1}^{\wr *}(k) \right| \left| \hat{g}_{hj,l+1}^{\wr *}(k) \right|}{r^{\wr} + q^{\wr} \left\| \hat{\boldsymbol{G}}_{l+1}^{\wr *m}(k) \right\|^{2}} \\
&+\gamma^{\wr} q^{\wr} \sum_{h=1,h\neq j}^{n} \frac{\sum_{i=1,i\neq j}^{n} \left| g_{ji,l+1}^{\wr *}(k) \right| \left| \hat{g}_{hi,l+1}^{\wr *}(k) \right|}{r^{\wr} + q^{\wr} \left\| \hat{\boldsymbol{G}}_{l+1}^{\wr *m}(k) \right\|^{2}} \\
&\le \gamma^{\wr} q^{\wr} \frac{\sum_{h=1,h\neq j}^{n} \left| g_{jj,l+1}^{\wr *}(k) \right| \left| \hat{g}_{hj,l+1}^{\wr *}(k) \right|}{r^{\wr} + q^{\wr} \left\| \hat{\boldsymbol{G}}_{l+1}^{\wr *m}(k) \right\|^{2}} \\
&\quad + \gamma^{\wr} q^{\wr} \frac{\sum_{h=1,h\neq j}^{n} \left| g_{jh,l+1}^{\wr *}(k) \right| \left| \hat{g}_{hh,l+1}^{\wr *}(k) \right|}{r^{\wr} + q^{\wr} \left\| \hat{\boldsymbol{G}}_{l+1}^{\wr *m}(k) \right\|^{2}} \\
&+\gamma^{\wr} q^{\wr} \sum_{h=1,h\neq j}^{n} \frac{\sum_{i=1,i\neq j,l}^{n} \left| g_{ji,l+1}^{\wr *}(k) \right| \left| \hat{g}_{hi,l+1}^{\wr *}(k) \right|}{r^{\wr} + q^{\wr} \left\| \hat{\boldsymbol{G}}_{l+1}^{\wr *m}(k) \right\|^{2}} \\
&\le \gamma^{\wr} q^{\wr} \frac{2\alpha^{\wr} b_1{}^{\wr} b_2{}^{\wr} (n-1) + b_1^{\wr 2} (n-1)(n-2)}{r^{\wr} + q^{\wr} \left\| \hat{\boldsymbol{G}}_{l+1}^{\wr *m}(k) \right\|^{2}}
\end{aligned}
\tag{4.56}
$$

According to (4.55) and (4.56), we have:

$$
\begin{aligned}
&1 - \frac{\gamma^{\wr} q^{\wr} \sum_{i=1}^{n} \left| g_{ji,l+1}^{\wr *}(k) \right| \left| \hat{g}_{ji,l+1}^{\wr *}(k) \right|}{r^{\wr} + q^{\wr} \left\| \hat{\boldsymbol{G}}_{l+1}^{\wr *m}(k) \right\|^{2}} + \sum_{h=1,h\neq j}^{n} \frac{\gamma^{\wr} q^{\wr} \sum_{i=1}^{n} g_{ji,l+1}^{\wr *}(k)\, \hat{g}_{hi,l+1}^{\wr *}(k)}{r^{\wr} + q^{\wr} \left\| \hat{\boldsymbol{G}}_{l+1}^{\wr *m}(k) \right\|^{2}} \\
&\le 1 - \frac{\gamma^{\wr} q^{\wr} b_2^{\wr 2}}{r^{\wr} + q^{\wr} \left\| \hat{\boldsymbol{G}}_{l+1}^{\wr *m}(k) \right\|^{2}} + \gamma^{\wr} q^{\wr} \frac{2\alpha^{\wr} b_1{}^{\wr} b_2{}^{\wr} (n-1) + b_1^{\wr 2} (n-1)(n-2)}{r^{\wr} + q^{\wr} \left\| \hat{\boldsymbol{G}}_{l+1}^{\wr *m}(k) \right\|^{2}} \\
&\le 1 - \gamma^{\wr} q^{\wr} \frac{b_2^{\wr 2} - 2\alpha^{\wr} b_1{}^{\wr} b_2{}^{\wr} (n-1) - b_1^{\wr 2} (n-1)(n-2)}{r^{\wr} + q^{\wr} \left\| \hat{\boldsymbol{G}}_{l+1}^{\wr *m}(k) \right\|^{2}}
\end{aligned}
\tag{4.57}
$$

According to Assumption 4.4, (4.57) leads to:

$$
\begin{aligned}
&1 - \frac{\gamma^{\wr} q^{\wr} \sum_{i=1}^{n} \left| g_{ji,l+1}^{\wr *}(k) \right| \left| \hat{g}_{ji,l+1}^{\wr *}(k) \right|}{r^{\wr} + q^{\wr} \left\| \hat{\boldsymbol{G}}_{l+1}^{\wr *m}(k) \right\|^{2}} + \sum_{h=1,h\neq j}^{n} \frac{\gamma^{\wr} q^{\wr} \sum_{i=1}^{n} g_{ji,l+1}^{\wr *}(k)\, \hat{g}_{hi,l+1}^{\wr *}(k)}{r^{\wr} + q^{\wr} \left\| \hat{\boldsymbol{G}}_{l+1}^{\wr *m}(k) \right\|^{2}} \\
&\le 1 - \gamma^{\wr} q^{\wr} \frac{b_2^{\wr} \left(b_2^{\wr} - 2\alpha^{\wr} b_1^{\wr} (n-1) \right) - b_1^{\wr 2} (n-1)(n-2)}{r^{\wr} + q^{\wr} \left\| \hat{\boldsymbol{G}}_{l+1}^{\wr *m}(k) \right\|^{2}} \\
&\le 1 - \gamma^{\wr} q^{\wr} \frac{b_1^{\wr} b_2^{\wr} (n-1) - b_1^{\wr 2} (n-1)(n-2)}{r^{\wr} + q^{\wr} \left\| \hat{\boldsymbol{G}}_{l+1}^{\wr *m}(k) \right\|^{2}}
\end{aligned}
$$

$$
\begin{aligned}
&\le 1-\gamma^{\wr}q^{\wr}\frac{b_1^{\wr}b_2^{\wr}(n-1)-b_1^{\wr 2}(n-1)^2}{r^{\wr}+q^{\wr}\left\|\hat{\boldsymbol{G}}_{l+1}^{\wr *m}(k)\right\|^2}\\
&\le 1-\gamma^{\wr}q^{\wr}\frac{b_1^{\wr}(n-1)\left(b_2^{\wr}-b_1^{\wr}(n-1)\right)}{r^{\wr}+q^{\wr}\left\|\hat{\boldsymbol{G}}_{l+1}^{\wr *m}(k)\right\|^2}\\
&\le 1-\gamma^{\wr}q^{\wr}\frac{2\alpha^{\wr}b_1^{\wr 2}(n-1)^2}{r^{\wr}+q^{\wr}\left\|\hat{\boldsymbol{G}}_{l+1}^{\wr *m}(k)\right\|^2}
\end{aligned}
\tag{4.58}
$$

By the reset algorithm (4.20)–(4.21) and Assumption 4.4, we can obtain that $g_{ji,l+1}^{\wr *}(k)\,\hat{g}_{ji,l+1}^{\wr}(k)>0$. Hence, there exists a $r_{\min}^{\wr}>0$ such that when $r^{\wr}>r_{\min}^{\wr}$, one has:

$$
\begin{aligned}
\frac{\sum_{i=1}^{n}g_{ji,l+1}^{\wr *}(k)\,\hat{g}_{ji,l+1}^{\wr *}(k)}{r^{\wr}+q^{\wr}\left\|\hat{\boldsymbol{G}}_{l+1}^{\wr *m}(k)\right\|^2}
&=\frac{\sum_{i=1}^{n}\left|g_{ji,l+1}^{\wr *}(k)\right|\left|\hat{g}_{ji,l+1}^{\wr *}(k)\right|}{r^{\wr}+q^{\wr}\left\|\hat{\boldsymbol{G}}_{l+1}^{\wr *m}(k)\right\|^2}\\
\le\frac{\alpha^{\wr 2}b_2^{\wr 2}+b_1^{\wr 2}(n-1)}{r^{\wr}+q^{\wr}\left\|\hat{\boldsymbol{G}}_{l+1}^{\wr *m}(k)\right\|^2}
&\le\frac{\alpha^{\wr 2}b_2^{\wr 2}+b_1^{\wr 2}(n-1)}{r_{\min}^{\wr}+q^{\wr}\left\|\hat{\boldsymbol{G}}_{l+1}^{\wr *m}(k)\right\|^2}<1
\end{aligned}
\tag{4.59}
$$

When $r^{\wr}>r_{\min}^{\wr}>0$ and $0<\gamma^{\wr}q^{\wr}\le 1$, we can get that:

$$
\begin{aligned}
0<M^{\wr}&\le\frac{2\alpha^{\wr}b_1^{\wr 2}(n-1)^2}{r^{\wr}+q^{\wr}\left\|\hat{\boldsymbol{G}}_{1,l+1}^{\wr *m}(k)\right\|^2}<\frac{b_2^{\wr 2}}{r^{\wr}+q^{\wr}\left\|\hat{\boldsymbol{G}}_{1,l+1}^{\wr *m}(k)\right\|^2}\\
&\le\frac{\alpha^{\wr 2}b_2^{\wr 2}+b_1^{\wr 2}(n-1)}{r^{\wr}+q^{\wr}\left\|\hat{\boldsymbol{G}}_{1,l+1}^{\wr *m}(k)\right\|^2}<\frac{\alpha^{\wr 2}b_2^{\wr 2}+b_1^{\wr 2}(n-1)}{r_{\min}^{\wr}+q^{\wr}\left\|\hat{\boldsymbol{G}}_{1,l+1}^{\wr *m}(k)\right\|^2}<1
\end{aligned}
\tag{4.60}
$$

where $M^{\wr}>0$ is a positive constant.

According to (4.58) and (4.60), one can get:

$$
\begin{aligned}
&\left|1-\frac{\gamma^{\wr}q^{\wr}\sum_{i=1}^{n}g_{ji,l+1}^{\wr *}(k)\,\hat{g}_{ji,l+1}^{\wr *}(k)}{r^{\wr}+q^{\wr}\left\|\hat{\boldsymbol{G}}_{l+1}^{\wr *m}(k)\right\|^2}\right|\\
&\qquad+\sum_{h=1,h\neq j}^{n}\frac{\gamma^{\wr}q^{\wr}\sum_{i=1}^{n}g_{ji,l+1}^{\wr *}(k)\,\hat{g}_{hi,l+1}^{\wr *}(k)}{r^{\wr}+q^{\wr}\left\|\hat{\boldsymbol{G}}_{l+1}^{\wr *m}(k)\right\|^2}\\
&\quad=1-\frac{\gamma^{\wr}q^{\wr}\sum_{i=1}^{n}\left|g_{ji,l+1}^{\wr *}(k)\right|\left|\hat{g}_{ji,l+1}^{\wr *}(k)\right|}{r^{\wr}+q^{\wr}\left\|\hat{\boldsymbol{G}}_{l+1}^{\wr *m}(k)\right\|^2}\\
&\qquad+\sum_{h=1,h\neq j}^{n}\frac{\gamma^{\wr}q^{\wr}\sum_{i=1}^{n}g_{ji,l+1}^{\wr *}(k)\,\hat{g}_{hi,l+1}^{\wr *}(k)}{r^{\wr}+q^{\wr}\left\|\hat{\boldsymbol{G}}_{l+1}^{\wr *m}(k)\right\|^2}
\end{aligned}
$$

$$< 1 - \frac{\gamma^{\wr} q^{\wr} 2\alpha^{\wr} b_1^{\wr 2}(n-1)^2}{r^{\wr} + q^{\wr} \left\| \hat{\boldsymbol{G}}_{l+1}^{\wr * m}(k) \right\|^2} < 1 - \gamma^{\wr} q^{\wr} M^{\wr} < 1 \tag{4.61}$$

Then based on (4.54) and (4.61), one can obtain:

$$s\left(\boldsymbol{I} - \frac{\gamma^{\wr} q^{\wr} \boldsymbol{G}_{l+1}^{\wr *}(k)\, \hat{\boldsymbol{G}}_{l+1}^{\wr * T}(k)}{r^{\wr} + q^{\wr} \left\| \hat{\boldsymbol{G}}_{l+1}^{\wr * m}(k) \right\|^2} \right) < 1 - \gamma^{\wr} q^{\wr} M^{\wr} < 1 \tag{4.62}$$

where $s\left(\mathbf{A}^{\wr}\right)$ is the spectral radius of the matrix $\mathbf{A}^{\wr}$, that is $s\left(\mathbf{A}^{\wr}\right) = \max_{s \in \{1,2,\ldots,m\}} \left| z_s^{\wr} \right|$. $z_s^{\wr}$ $(s = 1, 2, \ldots, m)$ is the eigenvalue of the matrix $\mathbf{A}^{\wr}$. According to the knowledge of spectral radius of the matrix, it follows that there exists any small $\varepsilon^{\wr} > 0$ and $0 < d_1^{\wr} < 1$ such that:

$$\begin{aligned} 0 \le \left\| \boldsymbol{I} - \frac{\gamma^{\wr} q^{\wr} \boldsymbol{G}_{l+1}^{\wr *}(k)\, \hat{\boldsymbol{G}}_{l+1}^{\wr * T}(k)}{r^{\wr} + q^{\wr} \left\| \hat{\boldsymbol{G}}_{l+1}^{\wr * m}(k) \right\|^2} \right\|_{\upsilon} &\le s\left(\boldsymbol{I} - \frac{\gamma^{\wr} q^{\wr} \boldsymbol{G}_{l+1}^{\wr *}(k)\, \hat{\boldsymbol{G}}_{l+1}^{\wr * T}(k)}{r^{\wr} + q^{\wr} \left\| \hat{\boldsymbol{G}}_{l+1}^{\wr * m}(k) \right\|^2} \right) + \varepsilon^{\wr} \\ &\le 1 - \gamma^{\wr} q^{\wr} M^{\wr} + \varepsilon^{\wr} \le d_1^{\wr} < 1 \end{aligned} \tag{4.63}$$

where $\left\| \mathbf{A}^{\wr} \right\|_{\upsilon}$ is the compatible norm of matrix $\mathbf{A}^{\wr}$.

According to (4.52), we can get:

$$\begin{aligned} \left\| \boldsymbol{e}_{l+1}^{\wr *}(k+1) \right\|_{\upsilon} &\le \left\| \boldsymbol{I} - \frac{\gamma^{\wr} q^{\wr} \boldsymbol{G}_{l+1}^{\wr *}(k)\, \hat{\boldsymbol{G}}_{l+1}^{\wr * T}(k)}{r^{\wr} + q^{\wr} \left\| \hat{\boldsymbol{G}}_{l+1}^{\wr * m}(k) \right\|^2} \right\|_{\upsilon} \left\| \boldsymbol{e}_{l}^{\wr *}(k+1) \right\|_{\upsilon} \\ &\le d_1^{\wr} \left\| \boldsymbol{e}_{l}^{\wr *}(k+1) \right\|_{\upsilon} \le d_1^{\wr 2} \left\| \boldsymbol{e}_{l-1}^{\wr *}(k+1) \right\|_{\upsilon} \\ &\le \ldots \le d_1^{\wr (l+1)} \left\| \boldsymbol{e}_{0}^{\wr *}(k+1) \right\|_{\upsilon} \end{aligned} \tag{4.64}$$

Since $0 \le d_1^{\wr} < 1$, (4.64) leads to:

$$\lim_{l \to \infty} \left\| \boldsymbol{e}_{l+1}^{\wr *}(k+1) \right\|_{\upsilon} = 0 \tag{4.65}$$

That is, the tracking error $\boldsymbol{e}_{l}^{\wr *}(k+1)$ can be converged to zero when l tends to infinity. □

4.4 Simulation Validation

Consider the following linear motor model (Lee et al. 2000)

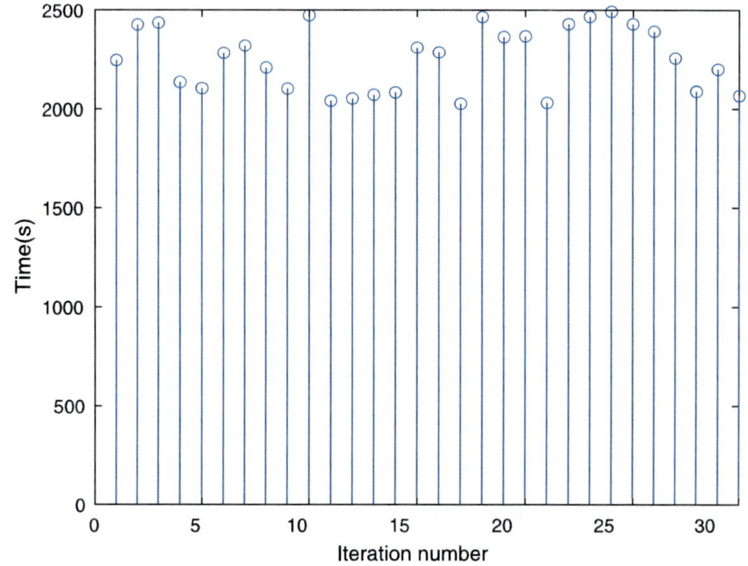

Fig. 4.1 Varying trial lengths

$$\begin{cases} \dot{x}(k) = v(k) \\ \dot{v}(k) = \frac{u(k)-f_{fric}(k)-f_{rip}(k)}{m^{\wr}} \\ f_{fric}(k) = [f_c + (f_s - f_c)e^{-(v/v_s)^2} + f_v v]sgn(v) \\ f_{rip}(k) = A_r^{\wr} sin(\omega_0^{\wr} \int_0^k v\ (\tau)d\tau) \end{cases} \tag{4.66}$$

where $x(k)$, $v(k)$ and $u(k)$ are the position, speed and control input thrust of linear motor, respectively. $m^{\wr}$ is the mass of motor. f_{fric} is the frictional force, where f_s, f_c, v_s and f_v are the static friction, minimum value of coulomb friction, lubrication parameter and coefficient of viscous friction, respectively. f_{rip} with two parameters $A_r^{\wr}$ and $\omega^{\wr}$ is the ripple force.

The simulation parameters of the considered motor model (4.66) are set as follows: $m^{\wr} = 0.60\,\text{kg}$, $f_c = 10\,\text{N}$, $f_s = 20\,\text{N}$, $v_s = 0.01\,\text{m/s}$, $f_v = 10\,\text{N}\,s m^{-1}$, $A_r^{\wr} = 8.6\,\text{N}$, $\omega_0^{\wr} = 314\,\text{rad/s}$, and the sampling period is 1s.

The above linear motor model is only used to produce input and output data and is not used in the design of the controller.

Here, we take the speed of the motor $v(k)$ as the system output. The following desired trajectory is repeatedly tracked by the motor for 30 iterations in total.

$$y_d^{\wr}(k) = 3.5\sin^2(\pi k/500), (0 \leq k \leq 2500) \tag{4.67}$$

In the simulation, the desired trial length for each iteration is 2500, but the actual trial length is randomly varying around the desired one, shown in Fig. 4.1.

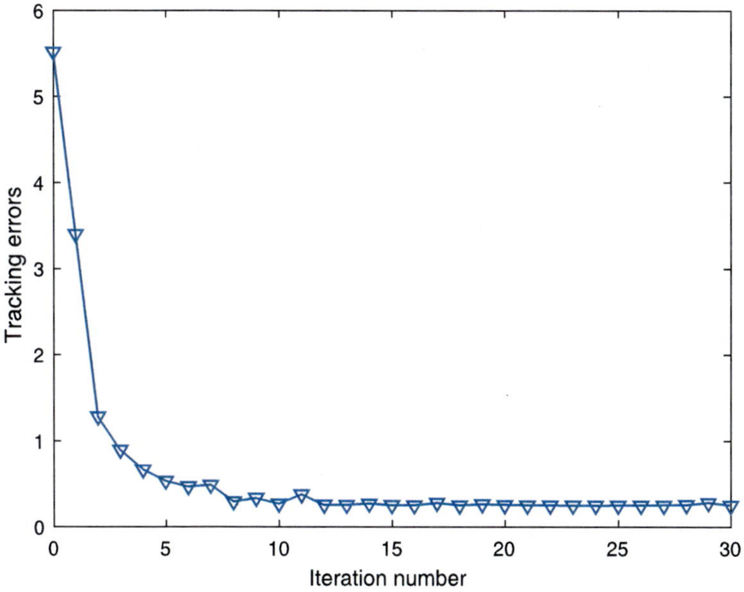

Fig. 4.2 Output tracking errors

The control parameters of the proposed method (4.19)–(4.26), (4.31)–(4.32) are set as follows: $r^{\wr} = 0.28$, $q^{\wr} = 1.05$, $\alpha^{\wr} = 10^3$, $b_1^{\wr} = 10^{-7}$, $b_2^{\wr} = 0.9 \times 10^{-3}$, $\eta^{\wr} = 0.5$, $\gamma^{\wr} = 0.001$, $\kappa^{\wr} = 0.01$, $\nu^{\wr} = 5$, $n_p^{\wr} = 2$, $\hat{G}_0^{\wr} = 0.5$, $\hat{G}_1^{\wr} = 0.5$, $\Lambda_0^{\wr} = 0.2$, $\Lambda_1^{\wr} = 0.2$.

The simulation results are shown in Figs. 4.2, 4.3, 4.4 and 4.5. The output tracking errors defined in (4.7) are given in Fig. 4.2. Furthermore, the actual control input and output profiles at 15th and 30th iterations are demonstrated in Fig. 4.3 and Fig. 4.4, respectively. As shown in Figs. 4.2, 4.3 and 4.4, even if the actual trial length of each operation is randomly varying around the desired 2500, the proposed data compensation-based PILC can effectively compensate the missing operation data for a guaranteed control performance. In addition, the modeling errors defined in (4.16) is represented in Fig. 4.5. From Fig. 4.5, we can see that the proposed algorithm is well qualified for model identification, although the system is under randomly varying trial lengths.

4.5 Conclusion

In this chapter, we propose a PILC with data compensation mechanism for unknown MIMO nonaffine nonlinear systems with varying trial lengths. By searching for actual historical operation data and predicted data at historical operations, a new

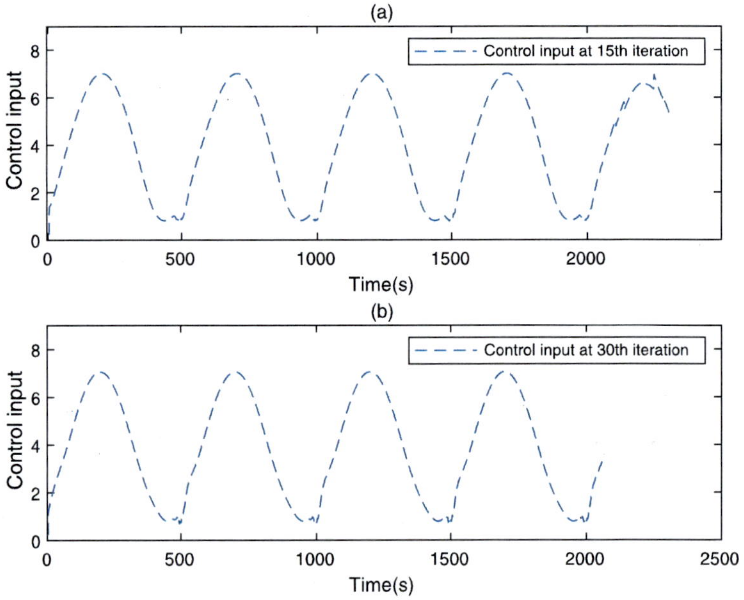

Fig. 4.3 Actual control input at different iterations

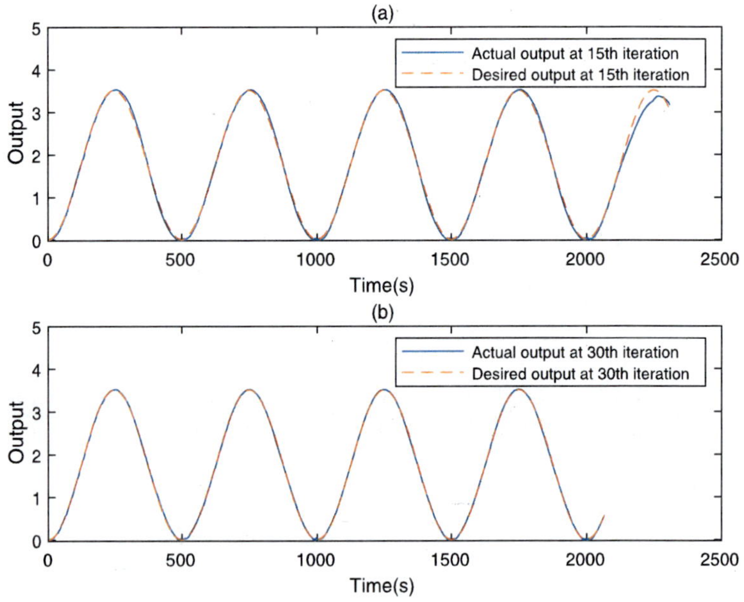

Fig. 4.4 Actual system output at different iterations

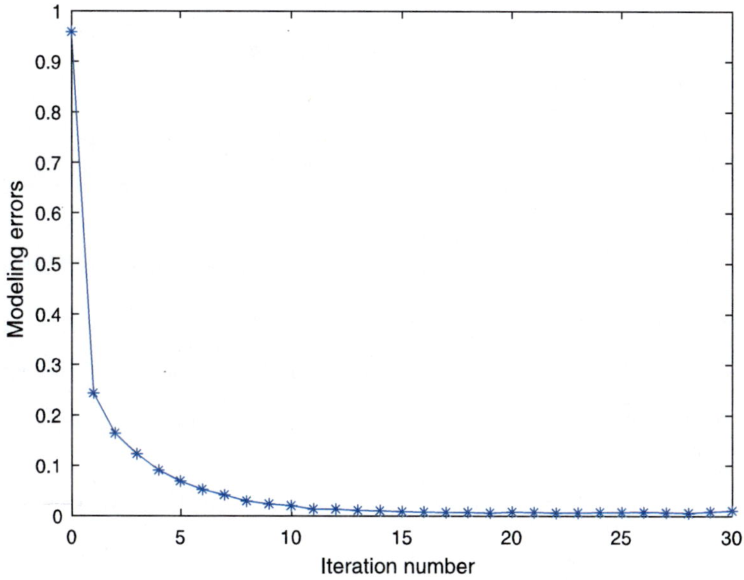

Fig. 4.5 Modeling errors

data compensation mechanism is proposed to remedy the lost operation data arising from randomly varying trial lengths. asymptotic and pointwise convergence property of the proposed PILC is guaranteed through theoretical analysis. Simulation results further verify both the ability to handle varying trial lengths and control performance of the proposed method.

References

Amann N, Owens DH, Rogers E (1998) Predictive optimal iterative learning control. Int J Control 69(2):203–226

Bu X, Wang S, Hou Z et al (2019) Model free adaptive iterative learning control for a class of nonlinear systems with randomly varying iteration lengths. J Franklin Inst 356(5):2491–2504

Chu B, Owens DH, Freeman CT (2016) Iterative learning control with predictive trial information: convergence, robustness, and experimental verification. IEEE Trans Control Syst Technol 24(3):1101–1108

Gerschgorin S (1931) Uber die abgrenzung der eigenwerte einer matrix (On the delimitation of the eigenvalues of a matrix). Bulletin de l Academie des Sciences de l URSS. Classe des sciences mathematiques et na 7(6):749–754

Hou Z, Jin S (2011) Data-driven model-free adaptive control for a class of MIMO nonlinear discrete-time systems. IEEE Trans Neural Netw 22(12):2173–2188

Hou ZS, Jin ST (2013) Model free adaptive control: theory and applications. CRC Press, Florida

Lee TH, Tan KK, Lim SY, Dou HF (2000) Iterative learning control of permanent magnet linear motor with relay automatic tuning. Mechatronics 10(1–2):169–190

Liu X, Ma L, Kong X et al (2020) An efficient iterative learning predictive functional control for nonlinear batch processes. IEEE Trans Cybern 52(6):4147–4160

Lu JY, Cao ZX, Gao FR (2019) Multi-point iterative learning model predictive control. IEEE Trans Ind Electron 66(8):6230–6240

Ma LL, Liu XJ (2019) Robust model predictive iterative learning control with iteration-varying reference trajectory. Acta Automatica Sinica 45(10):1933–1945

Oh SK, Park BJ, Lee JM (2018) Point-to-point iterative learning model predictive control. Automatica 89:135–143

Qiu WW, Xiong ZH, Zhang J, Hong YD, Li WZ (2020) Integrated predictive iterative learning control based on updating reference trajectory for point-to-point tracking. J Process Control 85:41–51

Rosolia U, Zhang XJ, Borrelli F (2022) Robust learning model-predictive control for linear systems performing iterative tasks. IEEE Trans Autom Control 67(2):856–869

Shen D, Xu JX (2018) Adaptive learning control for nonlinear systems with randomly varying iteration lengths. IEEE Trans Neural Netw Learn Syst 30(4):1119–1132

Shen D, Zhang W, Xu JX (2016) Iterative learning control for discrete nonlinear systems with randomly iteration varying lengths. Syst Control Lett 96:81–87

Shen M, Wu X, Park JH et al (2021) Iterative learning control of constrained systems with varying trial lengths under alignment condition. IEEE Trans Neural Netw Learn Syst. https://doi.org/10.1109/TNNLS.2021.3135504

Shi J, Zhou H, Cao Z, Jiang Q (2014) A design method for indirect iterative learning control based on two-dimensional generalized predictive control algorithm. J Process Control 24(10):1527–1537

Wang J, Zhou N, Wang S, Shen D, Li BQ (2021) Feedback-assisted PD-type quantized iterative learning control with randomly iteration varying lengths. Control Decis 36(10):2569–2576

Wang L, Yu J, Li P et al (2021) A 2D-FM model-based robust iterative learning model predictive control for batch processes. ISA Trans 110:271–282

Wei YS, Li XD (2017) Robust higher-order ILC for non-linear discrete-time systems with varying trail lengths and random initial state shifts. IET Control Theor Appl 11(15):2440–2447

Yu Q, Hou Z (2020) Adaptive fuzzy iterative learning control for high-speed trains with both randomly varying operation lengths and system constraints. IEEE Trans Fuzzy Syst 29(8):2408–2418

Zeng C, Shen D, Wang JR (2019) Adaptive learning tracking for robot manipulators with varying trial lengths. J Franklin Inst 356(12):5993–6014

Zhang R, Gao F (2018) Two-dimensional iterative learning model predictive control for batch processes: a new state space model compensation approach. IEEE Trans Syst Man Cybern: Syst 51(2):833–841

Zhang RD, Wu S, Tao JL (2018) A new design of predictive functional control strategy for batch processes in the two-dimensional framework. IEEE Trans Ind Inform 15(5):2905–2914

Chapter 5
Predictive Iterative Learning Control for Systems with Unknown Time Delay

5.1 Introduction

The phenomenon of time delay inevitably occurs in many real systems such as motor systems (Kali et al. 2020), incubation periods of infectious diseases (Bentout et al. 2022), circuit signal systems (Ravelo et al. 2022), and urban traffic management systems (Wan et al. 2018). In fact, time delay generated by the system will make the controller design very difficult, and hence, it is of practical importance to analyze and compensate the undesirable time delay, expecting a satisfactory control performance.

So far, some researches on ILC with time delay (Browne et al. 2020; Feng et al. 2017; Wang et al. 2021; Tao et al. 2019; Wei et al. 2017; Qiang et al. 2020) have been developed. An ILC method for the twin-roll strip casting process with known constant time delay is designed in Browne et al. (2020). An observer-based iterative learning scheme for a class of nonlinear systems with known constant time delay is proposed in Feng et al. (2017). For a class of linear systems with unknown constant time delay, Wang et al. (2021); Tao et al. (2019) can compensate the unknown time delay by using the maximum time delay information. For nonlinear systems with known model structure, unknown time delay can be compensated by combining appropriate Lyapunov–Krasovskii function with Young's inequality (Wei et al. 2017) or by previous state information (Qiang et al. 2020), but both methods have to meet the condition that the derivative of the time delay is less than 1. Compared with these existing works on the topic of time delay, this chapter will focus on unknown time delay for unknown nonlinear systems, meanwhile the restriction on the derivative of the time delay is removed.

The main contributions of this chapter are as follows.

- By utilizing the upper and lower bound information of the unknown time-varying delay, a new iterative learning estimation and modeling algorithm with time delay compensation, and the corresponding time delay compensation-based predictive ILC method are designed for unknown nonaffine nonlinear systems.

© The Author(s), under exclusive license to Springer Nature Singapore Pte Ltd. 2023
Q. Yu et al., *Predictive Learning Control for Unknown Nonaffine Nonlinear Systems*, Intelligent Control and Learning Systems 8,
https://doi.org/10.1007/978-981-19-8857-8_5

- Even if the considered unknown nonlinear systems is under unknown time-varying delay, the convergence property can still be guaranteed.

This chapter is organized as follows. Section 5.2 presents the problem formulation. In Sect. 5.3, the proposed time delay compensation-based predictive ILC is shown. In Sect. 5.4, simulation results are provided to show the effectiveness of the proposed method. Some conclusions are given in Sect. 5.5.

5.2 Problem Formulation

Consider the following repeatable MIMO nonaffine nonlinear system with unknown time delay (Hou and Jin 2011, 2013; Liu et al. 2020):

$$\begin{aligned} \boldsymbol{y}_l(k+1) = \boldsymbol{f}\big(&\boldsymbol{y}_l(k), \boldsymbol{y}_l(k-1), \ldots, \boldsymbol{y}_l(k-n_y^{\Diamond}), \boldsymbol{u}_l(k-h^{\Diamond}(k)), \\ &\boldsymbol{u}_l(k-h^{\Diamond}(k)-1), \ldots, \boldsymbol{u}_l(k-h^{\Diamond}(k)-n_u^{\Diamond})\big) \end{aligned} \tag{5.1}$$

where the subscript l denotes the number of iterations and $k \in \{0, 1, \ldots, K^{\Diamond}\}$ denotes the time index. $\boldsymbol{y}_l(k) \in \Re^n$ and $\boldsymbol{u}_l(k) \in \Re^n$ are the output and input of the system at time k of the lth iteration. $h^{\Diamond}(k) \in N^+$ is an unknown and time-varying input time delay, satisfying $\tau_1^{\Diamond}(k) \leq h^{\Diamond}(k) \leq \tau_2^{\Diamond}(k)$ where $\tau_1^{\Diamond}(k)$ and $\tau_2^{\Diamond}(k)$ are the known lower and upper bounds of the unknown time delay $h^{\Diamond}(k)$. $n_u^{\Diamond} \in N^+$ and $n_y^{\Diamond} \in N^+$ are the unknown order of the system. $\boldsymbol{f}(\cdots) \in \Re^n$ is an unknown nonlinear vector-valued function.

The considered system (5.1) meets the following assumptions:

Assumption 5.1 $\forall k \in N^+$, the initial output of the system $\boldsymbol{y}_l(0)$ is random and bounded.

Assumption 5.2 $\forall k \in \{0, 1, \ldots, K^{\Diamond}\}$ and $l = 0, 1, 2 \ldots$, the partial derivatives of $\boldsymbol{f}(\cdots)$ with respect to $\boldsymbol{u}_l(k)$ exist and are continuous.

Assumption 5.3 The system satisfies the generalized *Lipschitz* condition for $\forall k \in \{0, 1, \ldots, K^{\Diamond}\}$ and $l = 0, 1, 2 \ldots$, that is $\|\boldsymbol{y}_{l_1}(k+1) - \boldsymbol{y}_{l_2}(k+1)\| \leq b^{\Diamond}\|\boldsymbol{u}_{l_1}(k) - \boldsymbol{u}_{l_2}(k)\|$, where $\boldsymbol{u}_{l_1}(k) \neq \boldsymbol{u}_{l_2}(k)$, for any $l_1 \neq l_2, l_1, l_2 > 0$. $b^{\Diamond}$ is a positive constant.

Assumption 5.4 The pseudo-Jacobian matrix $\boldsymbol{G}_{a,l}^{\Diamond}(k)(a = 1, 2, 3)$ defined in the following (5.2) is a diagonal dominant matrix and satisfies the following conditions: $|g_{a,ij,l}^{\Diamond}(k)| \leq b_1^{\Diamond}, b_2^{\Diamond} \leq |g_{a,ii,l}^{\Diamond}(k)| \leq \alpha^{\Diamond} b_2^{\Diamond}, i = 1, \ldots, n, j = 1, \ldots, n, i \neq j$, $\alpha^{\Diamond} \geq 1, b_2^{\Diamond} \geq b_1^{\Diamond}(2\alpha^{\Diamond} + 1)(n-1)$ and all the element symbols in $\boldsymbol{G}_{a,l}^{\Diamond}(k)$ are unchanged at any time k.

The following lemma shows that the general MIMO nonlinear discrete-time system with unknown time delay satisfying Assumptions 5.1–5.3 can be transformed into an equivalent dynamical linearisation model Hou and Jin (2013), here is called time delay-based iterative learning compact form dynamic linearization (TD-ILCFDL) data model.

Lemma 5.1 *If the system (5.1) satisfies Assumptions 5.1–5.3, for* $k \in \{0, 1, \ldots, K^{\between}\}$ *and* $l = 0, 1, 2, \ldots$ *and* $\Delta \boldsymbol{u}_l(k) \neq \boldsymbol{0}$, *then there must exist three pseudo-Jacobian matrices* $\boldsymbol{G}^{\between}_{a,l}(k) \in \Re^{n\times n}(a = 1, 2, 3)$, *such that the nonlinear system (5.1) can be equated to the following time delay-based iterative learning compact form dynamic linearization (TD-ILCFDL) model:*

$$\begin{aligned}\Delta \boldsymbol{y}_l(k+1) = \boldsymbol{G}^{\between}_{1,l}(k)\Delta \boldsymbol{u}_l(k) + \boldsymbol{G}^{\between}_{2,l}(k)\Delta \boldsymbol{u}_l(k - \tau^{\between}_1(k)) \\ + \boldsymbol{G}^{\between}_{3,l}(k)\Delta \boldsymbol{u}_l(k - \tau^{\between}_2(k))\end{aligned} \tag{5.2}$$

where $\Delta \boldsymbol{y}_l(k+1) = \boldsymbol{y}_l(k+1) - \boldsymbol{y}_{l-1}(k+1)$, $\Delta \boldsymbol{u}_l(k) = \boldsymbol{u}_l(k) - \boldsymbol{u}_{l-1}(k)$,

$$\boldsymbol{G}^{\between}_{a,l}(k) = \begin{bmatrix} g^{\between}_{a,11,l}(k) & g^{\between}_{a,12,l}(k) & \ldots & g^{\between}_{a,1n,l}(k) \\ g^{\between}_{a,21,l}(k) & g^{\between}_{a,22,l}(k) & \ldots & g^{\between}_{a,2n,l}(k) \\ \vdots & \vdots & \ddots & \vdots \\ g^{\between}_{a,n1,l}(k) & g^{\between}_{a,n2,l}(k) & \ldots & g^{\between}_{a,nn,l}(k) \end{bmatrix}$$

, $a = 1, 2, 3$ *is unknown but bounded.*

Proof Based on system (5.1), one has:

$$\begin{aligned}
&\Delta \boldsymbol{y}_l(k+1) = \boldsymbol{f}\big(\boldsymbol{y}_l(k), \boldsymbol{y}_l(k-1), \ldots, \boldsymbol{y}_l(k-n^{\between}_y), \boldsymbol{u}_l(k-\tau^{\between}_1(k)), \ldots, \boldsymbol{u}_l(k-h^{\between}(k) \\
&), \boldsymbol{u}_l(k-1-h^{\between}(k)), \ldots, \boldsymbol{u}_l(k-\tau^{\between}_2(k)), \ldots, \boldsymbol{u}_l(k-n^{\between}_u-h^{\between}(k))\big) - \boldsymbol{f}\big(\boldsymbol{y}_{l-1}(k), \\
&\boldsymbol{y}_{l-1}(k-1), \ldots, \boldsymbol{y}_{l-1}(k-n^{\between}_y), \boldsymbol{u}_{l-1}(k-\tau^{\between}_1(k)), \ldots, \boldsymbol{u}_{l-1}(k-h^{\between}(k)), \boldsymbol{u}_{l-1}(k- \\
&1-h^{\between}(k)), \ldots, \boldsymbol{u}_{l-1}(k-\tau^{\between}_2(k)), \ldots, \boldsymbol{u}_{l-1}(k-n^{\between}_u-h^{\between}(k))\big) = \boldsymbol{f}\big(\boldsymbol{y}_l(k), \boldsymbol{y}_l(k-1) \\
&, \ldots, \boldsymbol{y}_l(k-n^{\between}_y), \boldsymbol{u}_l(k-h^{\between}(k)), \ldots, \boldsymbol{u}_l(k-n^{\between}_u-h^{\between}(k))\big) - \boldsymbol{f}\big(\boldsymbol{y}_{l-1}(k), \boldsymbol{y}_{l-1}(k- \\
&1), \ldots, \boldsymbol{y}_{l-1}(k-n^{\between}_y), \boldsymbol{u}_{l-1}(k-h^{\between}(k)), \boldsymbol{u}_{l-1}(k-1-h^{\between}(k)), \ldots, \boldsymbol{u}_{l-1}(k-n^{\between}_u- \\
&h^{\between}(k))\big) + \boldsymbol{f}\big(\boldsymbol{y}_l(k), \ldots, \boldsymbol{y}_l(k-n^{\between}_y), \boldsymbol{u}_l(k), \boldsymbol{u}_l(k-h^{\between}(k)), \boldsymbol{u}_l(k-1-h^{\between}(k)), \ldots, \\
&\boldsymbol{u}_l(k-n^{\between}_u-h^{\between}(k))\big) - \boldsymbol{f}\big(\boldsymbol{y}_l(k), \ldots, \boldsymbol{y}_l(k-n^{\between}_y), \boldsymbol{u}_{l-1}(k), \boldsymbol{u}_l(k-h^{\between}(k)), \boldsymbol{u}_l(k- \\
&1-h^{\between}(k)), \ldots, \boldsymbol{u}_l(k-n^{\between}_u-h^{\between}(k))\big) + \boldsymbol{f}\big(\boldsymbol{y}_l(k), \ldots, \boldsymbol{y}_l(k-n^{\between}_y), \boldsymbol{u}_l(k-\tau^{\between}_1(k)), \\
&\ldots, \boldsymbol{u}_l(k-h^{\between}(k)), \boldsymbol{u}_l(k-1-h^{\between}(k)), \ldots, \boldsymbol{u}_l(k-n^{\between}_u-h^{\between}(k))\big) - \boldsymbol{f}\big(\boldsymbol{y}_l(k), \ldots, \\
&\boldsymbol{y}_l(k-n^{\between}_y), \boldsymbol{u}_{l-1}(k-\tau^{\between}_1(k)), \ldots, \boldsymbol{u}_l(k-h^{\between}(k)), \boldsymbol{u}_l(k-1-h^{\between}(k)), \ldots, \boldsymbol{u}_l(k- \\
&n^{\between}_u-h^{\between}(k))\big) + \boldsymbol{f}\big(\boldsymbol{y}_l(k), \ldots, \boldsymbol{y}_l(k-n^{\between}_y), \boldsymbol{u}_l(k-h^{\between}(k)), \boldsymbol{u}_l(k-1-h^{\between}(k)), \ldots, \\
&\boldsymbol{u}_l(k-\tau^{\between}_2(k)), \ldots, \boldsymbol{u}_l(k-n^{\between}_u-h^{\between}(k))\big) - \boldsymbol{f}\big(\boldsymbol{y}_l(k), \ldots, \boldsymbol{y}_l(k-n^{\between}_y), \boldsymbol{u}_l(k- \\
&h^{\between}(k)), \boldsymbol{u}_l(k-1-h^{\between}(k)), \ldots, \boldsymbol{u}_{l-1}(k-\tau^{\between}_2(k)), \ldots, \boldsymbol{u}_l(k-n^{\between}_u-h^{\between}(k))\big) - \\
&\boldsymbol{f}\big(\boldsymbol{y}_l(k), \ldots, \boldsymbol{y}_l(k-n^{\between}_y), \boldsymbol{u}_l(k), \boldsymbol{u}_l(k-h^{\between}(k)), \boldsymbol{u}_l(k-1-h^{\between}(k)), \ldots, \boldsymbol{u}_l(k- \\
&n^{\between}_u-h^{\between}(k))\big) + \boldsymbol{f}\big(\boldsymbol{y}_l(k), \ldots, \boldsymbol{y}_l(k-n^{\between}_y), \boldsymbol{u}_{l-1}(k), \boldsymbol{u}_l(k-h^{\between}(k)), \boldsymbol{u}_l(k-1-
\end{aligned}$$

$$h^{\wp}(k)), \ldots, \boldsymbol{u}_l(k - n_u^{\wp} - h^{\wp}(k))\big) - \boldsymbol{f}\big(\boldsymbol{y}_l(k), \ldots, \boldsymbol{y}_l(k - n_y^{\wp}), \boldsymbol{u}_l(k - \tau_1^{\wp}(k)), \boldsymbol{u}_l(k - h^{\wp}(k)), \boldsymbol{u}_l(k - 1 - h^{\wp}(k)), \ldots, \boldsymbol{u}_l(k - n_u^{\wp} - h^{\wp}(k))\big) + \boldsymbol{f}\big(\boldsymbol{y}_l(k), \ldots, \boldsymbol{y}_l(k - n_y^{\wp}), \boldsymbol{u}_{l-1}(k - \tau_1^{\wp}(k))), \boldsymbol{u}_l(k - h^{\wp}(k)), \boldsymbol{u}_l(k - 1 - h^{\wp}(k)), \ldots, \boldsymbol{u}_l(k - n_u^{\wp} - h^{\wp}(k))\big) - \boldsymbol{f}\big(\boldsymbol{y}_l(k), \ldots, \boldsymbol{y}_l(k - n_y^{\wp}), \boldsymbol{u}_l(k - h^{\wp}(k)), \boldsymbol{u}_l(k - 1 - h^{\wp}(k)), \ldots, \boldsymbol{u}_l(k - \tau_2^{\wp}(k)), \ldots, \boldsymbol{u}_l(k - n_u^{\wp} - h^{\wp}(k))\big) + \boldsymbol{f}\big(\boldsymbol{y}_l(k), \ldots, \boldsymbol{y}_l(k - n_y^{\wp}), \boldsymbol{u}_l(k - h^{\wp}(k)), \boldsymbol{u}_l(k - 1 - h^{\wp}(k)), \ldots, \boldsymbol{u}_{l-1}(k - \tau_2^{\wp}(k)), \ldots, \boldsymbol{u}_l(k - n_u^{\wp} - h^{\wp}(k))\big) \tag{5.3}$$

Let $\boldsymbol{\zeta}_{1,l}^{\wp}(k) = \boldsymbol{f}\big(\boldsymbol{y}_l(k), \ldots, \boldsymbol{y}_l(k - n_y^{\wp}), \boldsymbol{u}_{l-1}(k), \boldsymbol{u}_l(k - h^{\wp}(k)), \boldsymbol{u}_l(k - 1 - h^{\wp}(k)), \ldots, \boldsymbol{u}_l(k - n_u^{\wp} - h^{\wp}(k))\big) - \boldsymbol{f}\big(\boldsymbol{y}_l(k), \ldots, \boldsymbol{y}_l(k - n_y^{\wp}), \boldsymbol{u}_l(k), \boldsymbol{u}_l(k - h^{\wp}(k)), \boldsymbol{u}_l(k - 1 - h^{\wp}(k)), \ldots, \boldsymbol{u}_l(k - n_u^{\wp} - h^{\wp}(k))\big)$, $\boldsymbol{\zeta}_{2,l}^{\wp}(k - \tau_1^{\wp}(k)) = \boldsymbol{f}\big(\boldsymbol{y}_l(k), \ldots, \boldsymbol{y}_l(k - n_y^{\wp}), \boldsymbol{u}_{l-1}(k - \tau_1^{\wp}(k))), \boldsymbol{u}_l(k - h^{\wp}(k)), \boldsymbol{u}_l(k - 1 - h^{\wp}(k)), \ldots, \boldsymbol{u}_l(k - n_u^{\wp} - h^{\wp}(k))\big) - \boldsymbol{f}\big(\boldsymbol{y}_l(k), \ldots, \boldsymbol{y}_l(k - n_y^{\wp}), \boldsymbol{u}_l(k - \tau_1^{\wp}(k)), \boldsymbol{u}_l(k - h^{\wp}(k)), \boldsymbol{u}_l(k - 1 - h^{\wp}(k)), \ldots, \boldsymbol{u}_l(k - n_u^{\wp} - h^{\wp}(k))\big)$, $\boldsymbol{\zeta}_{3,l}^{\wp}(k - \tau_2^{\wp}(k)) = \boldsymbol{f}\big(\boldsymbol{y}_l(k), \ldots, \boldsymbol{y}_l(k - n_y^{\wp}), \boldsymbol{u}_l(k - h^{\wp}(k)), \boldsymbol{u}_l(k - 1 - h^{\wp}(k)), \ldots, \boldsymbol{u}_{l-1}(k - \tau_2^{\wp}(k)), \ldots, \boldsymbol{u}_l(k - n_u^{\wp} - h^{\wp}(k))\big) - \boldsymbol{f}\big(\boldsymbol{y}_l(k), \ldots, \boldsymbol{y}_l(k - n_y^{\wp}), \boldsymbol{u}_l(k - h^{\wp}(k)), \boldsymbol{u}_l(k - 1 - h^{\wp}(k)), \ldots, \boldsymbol{u}_l(k - \tau_2^{\wp}(k)), \ldots, \boldsymbol{u}_l(k - n_u^{\wp} - h^{\wp}(k))\big) + \boldsymbol{f}\big(\boldsymbol{y}_l(k), \boldsymbol{y}_l(k - 1), \ldots, \boldsymbol{y}_l(k - n_y^{\wp}), \boldsymbol{u}_l(k - h^{\wp}(k)), \ldots, \boldsymbol{u}_l(k - n_u^{\wp} - h^{\wp}(k))\big) - \boldsymbol{f}\big(\boldsymbol{y}_{l-1}(k), \boldsymbol{y}_{l-1}(k - 1), \ldots, \boldsymbol{y}_{l-1}(k - n_y^{\wp}), \boldsymbol{u}_{l-1}(k - h^{\wp}(k)), \ldots, \boldsymbol{u}_{l-1}(k - n_u^{\wp} - h^{\wp}(k))\big)$.

Then according to Assumption 5.2 and differential mean value theorem, (5.3) can be rewritten as

$$\Delta \boldsymbol{y}_l(k + 1) = \frac{\partial \boldsymbol{f}^*}{\partial \boldsymbol{u}_l(k)} \Delta \boldsymbol{u}_l(k) + \frac{\partial \boldsymbol{f}^*}{\partial \boldsymbol{u}_l(k - \tau_1^{\wp}(k))} \Delta \boldsymbol{u}_l(k - \tau_1^{\wp}(k)) + \frac{\partial \boldsymbol{f}^*}{\partial \boldsymbol{u}_l(k - \tau_2^{\wp}(k))} \Delta \boldsymbol{u}_l(k - \tau_2^{\wp}(k)) + \boldsymbol{\zeta}_{1,l}^{\wp}(k) + \boldsymbol{\zeta}_{2,l}^{\wp}(k - \tau_1^{\wp}(k)) + \boldsymbol{\zeta}_{3,l}^{\wp}(k - \tau_2^{\wp}(k)) \tag{5.4}$$

where

$$\frac{\partial \boldsymbol{f}^*}{\partial \boldsymbol{u}_l(k)} = \begin{bmatrix} \frac{\partial f_1^*}{\partial u_{1,l}(k)} & \frac{\partial f_1^*}{\partial u_{2,l}(k)} & \cdots & \frac{\partial f_1^*}{\partial u_{n,l}(k)} \\ \frac{\partial f_2^*}{\partial u_{1,l}(k)} & \frac{\partial f_2^*}{\partial u_{2,l}(k)} & \cdots & \frac{\partial f_2^*}{\partial u_{n,l}(k)} \\ \vdots & \vdots & \ddots & \vdots \\ \frac{\partial f_n^*}{\partial u_{1,l}(k)} & \frac{\partial f_n^*}{\partial u_{2,l}(k)} & \cdots & \frac{\partial f_n^*}{\partial u_{n,l}(k)} \end{bmatrix}$$

$$\frac{\partial \boldsymbol{f}^*}{\partial \boldsymbol{u}_l(k-\tau_1^{\lozenge}(k))}=\begin{bmatrix} \frac{\partial f_1^*}{\partial u_{1,l}(k-\tau_1^{\lozenge}(k))} & \frac{\partial f_1^*}{\partial u_{2,l}(k-\tau_1^{\lozenge}(k))} & \cdots & \frac{\partial f_1^*}{\partial u_{n,l}(k-\tau_1^{\lozenge}(k))} \\ \frac{\partial f_2^*}{\partial u_{1,l}(k-\tau_1^{\lozenge}(k))} & \frac{\partial f_2^*}{\partial u_{2,l}(k-\tau_1^{\lozenge}(k))} & \cdots & \frac{\partial f_2^*}{\partial u_{n,l}(k-\tau_1^{\lozenge}(k))} \\ \vdots & \vdots & \ddots & \vdots \\ \frac{\partial f_n^*}{\partial u_{1,l}(k-\tau_1^{\lozenge}(k))} & \frac{\partial f_n^*}{\partial u_{2,l}(k-\tau_1^{\lozenge}(k))} & \cdots & \frac{\partial f_n^*}{\partial u_{n,l}(k-\tau_1^{\lozenge}(k))} \end{bmatrix}$$

$$\frac{\partial \boldsymbol{f}^*}{\partial \boldsymbol{u}_l(k-\tau_2^{\lozenge}(k))}=\begin{bmatrix} \frac{\partial f_1^*}{\partial u_{1,l}(k-\tau_2^{\lozenge}(k))} & \frac{\partial f_1^*}{\partial u_{2,l}(k-\tau_2^{\lozenge}(k))} & \cdots & \frac{\partial f_1^*}{\partial u_{n,l}(k-\tau_2^{\lozenge}(k))} \\ \frac{\partial f_2^*}{\partial u_{1,l}(k-\tau_2^{\lozenge}(k))} & \frac{\partial f_2^*}{\partial u_{2,l}(k-\tau_2^{\lozenge}(k))} & \cdots & \frac{\partial f_2^*}{\partial u_{n,l}(k-\tau_2^{\lozenge}(k))} \\ \vdots & \vdots & \ddots & \vdots \\ \frac{\partial f_n^*}{\partial u_{1,l}(k-\tau_2^{\lozenge}(k))} & \frac{\partial f_n^*}{\partial u_{2,l}(k-\tau_2^{\lozenge}(k))} & \cdots & \frac{\partial f_n^*}{\partial u_{n,l}(k-\tau_2^{\lozenge}(k))} \end{bmatrix}$$

$\frac{\partial f_i^*}{\partial u_{j,l}(k)}(i=1,\ldots,n,\ j=1,\ldots,n)$ denotes the value of the partial derivative of f_i with respect to $u_{j,l}$ at a point between $u_{j,l-1}(k)$ and $u_{j,l}(k)$. For each fixed iteration l and time k, consider the following equation containing the numerical matrix $\phi_{1,l}^{\lozenge}(k)$, namely $\boldsymbol{\zeta}_{1,l}^{\lozenge}(k)=\phi_{1,l}^{\lozenge}(k)\Delta\boldsymbol{u}_l(k)$. Since $\Delta\boldsymbol{u}_l(k)\neq\boldsymbol{0}$, there exists at least one solution $\phi_{1,l}^{\lozenge*}(k)$ to let $\boldsymbol{\zeta}_{1,l}^{\lozenge}(k)=\phi_{1,l}^{\lozenge*}(k)\Delta\boldsymbol{u}_l(k)$. Similarly, it can be obtained that $\boldsymbol{\zeta}_{2,l}^{\lozenge}(k-\tau_1^{\lozenge}(k))=\phi_{2,l}^{\lozenge}(k)\Delta\boldsymbol{u}_l(k-\tau_1^{\lozenge}(k))$ and $\boldsymbol{\zeta}_{3,l}^{\lozenge}(k-\tau_2^{\lozenge}(k))=\phi_{3,l}^{\lozenge}(k)\Delta\boldsymbol{u}_l(k-\tau_2^{\lozenge}(k))$ also exist at least one solution $\phi_{2,l}^{\lozenge}(k)=\phi_{2,l}^{\lozenge*}(k)$ and $\phi_{3,l}^{\lozenge}(k)=\phi_{3,l}^{\lozenge*}(k)$ to make the above two equations work. Let $\boldsymbol{G}_{1,l}^{\lozenge}(k)=\frac{\partial \boldsymbol{f}^*}{\partial \boldsymbol{u}_l(k)}+\phi_{1,l}^{\lozenge*}(k)$, $\boldsymbol{G}_{2,l}^{\lozenge}(k)=\frac{\partial \boldsymbol{f}^*}{\partial \boldsymbol{u}_l(k-\tau_1^{\lozenge}(k))}+\phi_{2,l}^{\lozenge*}(k)$, $\boldsymbol{G}_{3,l}^{\lozenge}(k)=\frac{\partial \boldsymbol{f}^*}{\partial \boldsymbol{u}_l(k-\tau_2^{\lozenge}(k))}+\phi_{3,l}^{\lozenge*}(k)$, then (5.4) can be written as:

$$\begin{aligned}\Delta\boldsymbol{y}_l(k+1)=&\ \boldsymbol{G}_{1,l}^{\lozenge}(k)\Delta\boldsymbol{u}_l(k)+\boldsymbol{G}_{2,l}^{\lozenge}(k)\Delta\boldsymbol{u}_l(k-\tau_1^{\lozenge}(k))\\&+\boldsymbol{G}_{3,l}^{\lozenge}(k)\Delta\boldsymbol{u}_l(k-\tau_2^{\lozenge}(k))\end{aligned}\tag{5.5}$$

According to Assumption 5.3, it follows that $\boldsymbol{G}_{a,l}^{\lozenge}(k)$, $a=1,2,3$ is bounded.

5.3 Time Delay Compensation-Based Predictive ILC Design

Rewrite the established TD-ILCFDL model (5.2) as:

$$\begin{aligned}\boldsymbol{y}_l(k+1)=&\ \boldsymbol{y}_{l-1}(k+1)+\boldsymbol{G}_{1,l}^{\lozenge}(k)\Delta\boldsymbol{u}_l(k)\\&+\boldsymbol{G}_{2,l}^{\lozenge}(k)\Delta\boldsymbol{u}_l(k-\tau_1^{\lozenge}(k))+\boldsymbol{G}_{3,l}^{\lozenge}(k)\Delta\boldsymbol{u}_l(k-\tau_2^{\lozenge}(k))\end{aligned}\tag{5.6}$$

Define the tracking control error $\boldsymbol{e}_l^{\between}(k+1)$ as

$$\boldsymbol{e}_l^{\between}(k+1) = \boldsymbol{y}_d^{\between}(k+1) - \boldsymbol{y}_l(k+1) \tag{5.7}$$

where $\boldsymbol{y}_d^{\between}(k+1)$ is the desired output trajectory, $\boldsymbol{y}_l(k+1)$ is the actual output trajectory.

Then, subtracting both sides of (5.6) by $\boldsymbol{y}_d^{\between}(k+1)$ leads to:

$$\begin{aligned}\boldsymbol{e}_l^{\between}(k+1) = \boldsymbol{e}_{l-1}^{\between}(k+1) - \boldsymbol{G}_{1,l}^{\between}(k)\Delta\boldsymbol{u}_l(k)\\ -\boldsymbol{G}_{2,l}^{\between}(k)\Delta\boldsymbol{u}_l(k-\tau_1^{\between}(k)) - \boldsymbol{G}_{3,l}^{\between}(k)\Delta\boldsymbol{u}_l(k-\tau_2^{\between}(k))\end{aligned} \tag{5.8}$$

According to (5.8), the tracking error for the future m step iterative prediction can be described as:

$$\begin{aligned}
\boldsymbol{e}_{l+1|l}^{\between}(k+1) &= \boldsymbol{e}_l^{\between}(k+1) - \boldsymbol{G}_{1,l+1}^{\between}(k)\Delta\boldsymbol{u}_{l+1}(k) - \boldsymbol{G}_{2,l+1}^{\between}(k)\Delta\boldsymbol{u}_{l+1}(k-\tau_1^{\between}(k))\\
&\quad -\boldsymbol{G}_{3,l+1}^{\between}(k)\Delta\boldsymbol{u}_{l+1}(k-\tau_2^{\between}(k))\\
\boldsymbol{e}_{l+2|l}^{\between}(k+1) &= \boldsymbol{e}_l^{\between}(k+1) - \boldsymbol{G}_{1,l+1}^{\between}(k)\Delta\boldsymbol{u}_{l+1}(k) - \boldsymbol{G}_{2,l+1}^{\between}(k)\Delta\boldsymbol{u}_{l+1}(k-\tau_1^{\between}(k))\\
&\quad -\boldsymbol{G}_{3,l+1}^{\between}(k)\Delta\boldsymbol{u}_{l+1}(k-\tau_2^{\between}(k)) - \boldsymbol{G}_{1,l+2}^{\between}(k)\Delta\boldsymbol{u}_{l+2}(k)\\
&\quad -\boldsymbol{G}_{2,l+2}^{\between}(k)\Delta\boldsymbol{u}_{l+2}(k-\tau_1^{\between}(k)) - \boldsymbol{G}_{3,l+2}^{\between}(k)\Delta\boldsymbol{u}_{l+2}(k-\tau_2^{\between}(k))\\
&\ \vdots\\
\boldsymbol{e}_{l+m|l}^{\between}(k+1) &= \boldsymbol{e}_l^{\between}(k+1) - \boldsymbol{G}_{1,l+1}^{\between}(k)\Delta\boldsymbol{u}_{l+1}(k) - \boldsymbol{G}_{2,l+1}^{\between}(k)\Delta\boldsymbol{u}_{l+1}(k-\tau_1^{\between}(k))\\
&\quad -\boldsymbol{G}_{3,l+1}^{\between}(k)\Delta\boldsymbol{u}_{l+1}(k-\tau_2^{\between}(k)) - \boldsymbol{G}_{1,l+2}^{\between}(k)\Delta\boldsymbol{u}_{l+2}(k)\\
&\quad -\boldsymbol{G}_{2,l+2}^{\between}(k)\Delta\boldsymbol{u}_{l+2}(k-\tau_1^{\between}(k)) - \boldsymbol{G}_{3,l+2}^{\between}(k)\Delta\boldsymbol{u}_{l+2}(k-\tau_2^{\between}(k))\\
&\quad -\ldots - \boldsymbol{G}_{1,l+m}^{\between}(k)\Delta\boldsymbol{u}_{l+m}(k) - \boldsymbol{G}_{2,l+m}^{\between}(k)\Delta\boldsymbol{u}_{l+m}(k-\tau_1^{\between}(k))\\
&\quad -\boldsymbol{G}_{3,l+m}^{\between}(k)\Delta\boldsymbol{u}_{l+m}(k-\tau_2^{\between}(k))
\end{aligned} \tag{5.9}$$

where $\boldsymbol{e}_{l+m|l}^{\between}(k+1)$ denotes the predicted tracking error of the future $l+m$ th iteration at the current lth iteration.

Define $\boldsymbol{G}_{a,l+1}^{\between m}(k) \in \Re^{n\times mn} (a = 1, 2, 3)$, $\Delta\boldsymbol{u}_{l+1}^m(k) \in \Re^{mn\times 1}$, $\Delta\boldsymbol{u}_{l+1}^m(k-\tau_1^{\between}(k)) \in \Re^{mn\times 1}$, $\Delta\boldsymbol{u}_{l+1}^m(k-\tau_2^{\between}(k)) \in \Re^{mn\times 1}$ as follows:

$$\boldsymbol{G}_{a,l+1}^{\between m}(k) = \left[\boldsymbol{G}_{a,l+1}^{\between}(k), \boldsymbol{G}_{a,l+2}^{\between}(k), \ldots, \boldsymbol{G}_{a,l+m}^{\between}(k)\right] \tag{5.10}$$

$$\Delta\boldsymbol{u}_{l+1}^m(k) = \left[\Delta\boldsymbol{u}_{l+1}^T(k), \Delta\boldsymbol{u}_{l+2}^T(k), \ldots, \Delta\boldsymbol{u}_{l+m}^T(k)\right]^T \tag{5.11}$$

$$\Delta \boldsymbol{u}_{l+1}^{m}(k-\tau_1^{\phi}(k)) = \big[\Delta \boldsymbol{u}_{l+1}^{T}(k-\tau_1^{\phi}(k)),\\ \Delta \boldsymbol{u}_{l+2}^{T}(k-\tau_1^{\phi}(k))\ldots,\Delta \boldsymbol{u}_{l+m}^{T}(k-\tau_1^{\phi}(k))\big]^T \tag{5.12}$$

$$\Delta \boldsymbol{u}_{l+1}^{m}(k-\tau_2^{\phi}(k)) = \big[\Delta \boldsymbol{u}_{l+1}^{T}(k-\tau_2^{\phi}(k)),\\ \Delta \boldsymbol{u}_{l+2}^{T}(k-\tau_2^{\phi}(k))\ldots,\Delta \boldsymbol{u}_{l+m}^{T}(k-\tau_2^{\phi}(k))\big]^T \tag{5.13}$$

Then, (5.9) can be rewritten as:

$$\boldsymbol{e}_{l+m|l}^{\phi}(k+1) = \boldsymbol{e}_l^{\phi}(k+1) - \boldsymbol{G}_{1,l+1}^{\phi m}(k)\Delta \boldsymbol{u}_{l+1}^{m}(k)\\ - \boldsymbol{G}_{2,l+1}^{\phi m}(k)\Delta \boldsymbol{u}_{l+1}^{m}(k-\tau_1^{\phi}(k)) - \boldsymbol{G}_{3,l+1}^{\phi m}(k)\Delta \boldsymbol{u}_{l+1}^{m}(k-\tau_2^{\phi}(k)) \tag{5.14}$$

Since $\boldsymbol{G}_{a,l}^{\phi}(k)(a=1,2,3)$ in (5.8) and $\boldsymbol{G}_{a,l+1}^{\phi m}(k)(a=1,2,3)$ in (5.14) are unknown, it is necessary to estimate $\boldsymbol{G}_{a,l}^{\phi}(k)$ and meanwhile predict $\boldsymbol{G}_{a,l+1}^{\phi m}(k)$.

Define the estimated value of $\boldsymbol{G}_{a,l}^{\phi}(k)$ as $\hat{\boldsymbol{G}}_{a,l}^{\phi}(k)$ and the predicted value of $\boldsymbol{G}_{a,l+1}^{\phi m}(k)$ as $\hat{\boldsymbol{G}}_{a,l+1}^{\phi m}(k)$. Next, the estimation algorithm and prediction algorithm are designed, respectively.

Define the following modeling error:

$$\tilde{\boldsymbol{e}}_l^{\phi}(k+1) = \boldsymbol{y}_l(k+1) - \hat{\boldsymbol{y}}_l(k+1) \tag{5.15}$$

where

$$\hat{\boldsymbol{y}}_l(k+1) = \boldsymbol{y}_{l-1}(k+1) + \hat{\boldsymbol{G}}_{1,l}^{\phi}(k)\Delta \boldsymbol{u}_l(k) + \hat{\boldsymbol{G}}_{2,l}^{\phi}(k)\Delta \boldsymbol{u}_l(k-\tau_1^{\phi}(k))\\ + \hat{\boldsymbol{G}}_{3,l}^{\phi}(k)\Delta \boldsymbol{u}_l(k-\tau_2^{\phi}(k)) \tag{5.16}$$

is the estimated value of $\boldsymbol{y}_l(k+1)$, and $\hat{\boldsymbol{G}}_{a,l}^{\phi}(k)$ is the estimated value of $\boldsymbol{G}_{a,l}^{\phi}(k)$, $(a=1,2,3)$.

Compared with the proposed estimation algorithm (4.19) in Chap.4, based on (5.15), the following new iterative learning estimation algorithm with time delay compensation is designed to estimate $\hat{\boldsymbol{G}}_{a,l}^{\phi}(k)$, $(a=1,2,3)$:

$$\hat{\boldsymbol{G}}_{1,l}^{\phi}(k) = \hat{\boldsymbol{G}}_{1,l-1}^{\phi}(k) + \eta_1^{\phi}\tilde{\boldsymbol{e}}_{l-1}^{\phi}(k+1)\Delta \boldsymbol{u}_{l-1}^{T}(k)\\ \hat{\boldsymbol{G}}_{2,l}^{\phi}(k) = \hat{\boldsymbol{G}}_{2,l-1}^{\phi}(k) + \eta_2^{\phi}\tilde{\boldsymbol{e}}_{l-1}^{\phi}(k+1)\Delta \boldsymbol{u}_{l-1}^{T}(k-\tau_1^{\phi}(k))\\ \hat{\boldsymbol{G}}_{3,l}^{\phi}(k) = \hat{\boldsymbol{G}}_{3,l-1}^{\phi}(k) + \eta_3^{\phi}\tilde{\boldsymbol{e}}_{l-1}^{\phi}(k+1)\Delta \boldsymbol{u}_{l-1}^{T}(k-\tau_2^{\phi}(k)) \tag{5.17}$$

where $\eta_1^{\phi} > 0$, $\eta_2^{\phi} > 0$, $\eta_3^{\phi} > 0$ are adjustable learning parameters. Besides, the convergence of the estimation error can also be guaranteed in the following Theorem 5.1.

The following reset algorithm is adopted to make the estimation algorithm (5.17) have a stronger ability to track iteration time-varying parameters:

$$\hat{g}^{\phi}_{a,ii,l}(k) = \hat{g}^{\phi}_{a,ii,0}(k) \tag{5.18}$$

if $\left|\hat{g}^{\phi}_{a,ii,l}(k)\right| < b_2^{\phi}$ or $\left|\hat{g}^{\phi}_{a,ii,l}(k)\right| > \alpha^{\phi} b_2^{\phi}$ or $sign(\hat{g}^{\phi}_{a,ii,l}(k)) \neq sign(\hat{g}^{\phi}_{a,ii,0}(k))$

$$\hat{g}^{\phi}_{a,ij,l}(k) = \hat{g}^{\phi}_{a,ij,0}(k) \tag{5.19}$$

if $\left|\hat{g}^{\phi}_{a,ij,l}(k)\right| > b_1^{\phi}$ or $sign(\hat{g}^{\phi}_{a,ij,l}(k)) \neq sign(\hat{g}^{\phi}_{a,ij,0}(k))$

where $\hat{g}^{\phi}_{a,ii,0}(k)$ and $\hat{g}^{\phi}_{a,ij,0}(k)$ are the values at the initial iteration of $\hat{g}^{\phi}_{a,ii,l}(k)$ and $\hat{g}^{\phi}_{a,ij,l}(k)$, respectively, namely $l = 0$.

Note that based on the proposed iterative learning estimation algorithm (5.17)–(5.19), the estimated values $\hat{\boldsymbol{G}}^{\phi}_{1,l}(k)$,$\hat{\boldsymbol{G}}^{\phi}_{2,l}(k)$ and $\hat{\boldsymbol{G}}^{\phi}_{3,l}(k)$ are available. By using these estimated values, the predicted values $\hat{\boldsymbol{G}}^{\phi}_{1,l+i}(k)$, $\hat{\boldsymbol{G}}^{\phi}_{2,l+i}(k)$, $\hat{\boldsymbol{G}}^{\phi}_{3,l+i}(k)$, $(i = 1, 2, \ldots, m)$ for future $l + i$ iterations will be obtained as follows.

Firstly, the one-step iterative learning prediction algorithm is constructed as:

$$\begin{aligned}
\hat{\boldsymbol{G}}^{\phi}_{1,l+1}(k) &= \boldsymbol{\aleph}^{\phi}_1(k)\,\hat{\boldsymbol{G}}^{\phi}_{1,l}(k) + \boldsymbol{\aleph}^{\phi}_2(k)\,\hat{\boldsymbol{G}}^{\phi}_{1,l-1}(k) + \\
&\quad \ldots + \boldsymbol{\aleph}^{\phi}_{n_p^{\phi}}(k)\,\hat{\boldsymbol{G}}^{\phi}_{1,l-n_p^{\phi}+1}(k) \\
\hat{\boldsymbol{G}}^{\phi}_{2,l+1}(k) &= \bar{\boldsymbol{\aleph}}^{\phi}_1(k)\hat{\boldsymbol{G}}^{\phi}_{2,l}(k) + \bar{\boldsymbol{\aleph}}^{\phi}_2(k)\hat{\boldsymbol{G}}^{\phi}_{2,l-1}(k) + \\
&\quad \ldots + \bar{\boldsymbol{\aleph}}^{\phi}_{n_p^{\phi}}(k)\hat{\boldsymbol{G}}^{\phi}_{2,l-n_p^{\phi}+1}(k) \\
\hat{\boldsymbol{G}}^{\phi}_{3,l+1}(k) &= \bar{\bar{\boldsymbol{\aleph}}}^{\phi}_1(k)\hat{\boldsymbol{G}}^{\phi}_{3,l}(k) + \bar{\bar{\boldsymbol{\aleph}}}^{\phi}_2(k)\hat{\boldsymbol{G}}^{\phi}_{3,l-1}(k) + \\
&\quad \ldots + \bar{\bar{\boldsymbol{\aleph}}}^{\phi}_{n_p^{\phi}}(k)\hat{\boldsymbol{G}}^{\phi}_{3,l-n_p^{\phi}+1}(k)
\end{aligned} \tag{5.20}$$

where $\boldsymbol{\aleph}^{\phi}_q(k)$, $\bar{\boldsymbol{\aleph}}^{\phi}_q(k)$, $\bar{\bar{\boldsymbol{\aleph}}}^{\phi}_q(k) \in \Re^{n\times n}$ $(q = 1, 2, \ldots, n_p^{\phi})$ are unknown parameter matrices. n_p^{ϕ} is an appropriate order.

Define $\hat{\boldsymbol{\aleph}}^{\phi}_{q,l}(k)$, $\hat{\bar{\boldsymbol{\aleph}}}^{\phi}_{q,l}(k)$ and $\hat{\bar{\bar{\boldsymbol{\aleph}}}}^{\phi}_{q,l}(k)$ as the estimation of the unknown parameter matrices $\boldsymbol{\aleph}^{\phi}_q(k)$, $\bar{\boldsymbol{\aleph}}^{\phi}_q(k)$ and $\bar{\bar{\boldsymbol{\aleph}}}^{\phi}_q(k)$, $(q = 1, 2, \ldots, n_p^{\phi})$ at lth iteration, and they are designed in the following (5.24). Then, we can get the following available $(l + i)$ $(i = 1, 2, \ldots, m)$ step iterative learning prediction algorithm

$$
\begin{aligned}
\hat{\boldsymbol{G}}_{1,l+i}^{\between}(k) &= \hat{\boldsymbol{\aleph}}_{1,l}^{\between}(k)\hat{\boldsymbol{G}}_{1,l+i-1}^{\between}(k) + \hat{\boldsymbol{\aleph}}_{2,l}^{\between}(k)\hat{\boldsymbol{G}}_{1,l+i-2}^{\between}(k) + \\
&\quad \ldots + \hat{\boldsymbol{\aleph}}_{n_p^{\between},l}^{\between}(k)\hat{\boldsymbol{G}}_{1,l+i-n_p^{\between}}^{\between}(k) \\
\hat{\boldsymbol{G}}_{2,l+i}^{\between}(k) &= \hat{\bar{\boldsymbol{\aleph}}}_{1,l}^{\between}(k)\hat{\boldsymbol{G}}_{2,l+i-1}^{\between}(k) + \hat{\bar{\boldsymbol{\aleph}}}_{2,l}^{\between}(k)\hat{\boldsymbol{G}}_{2,l+i-2}^{\between}(k) + \\
&\quad \ldots + \hat{\bar{\boldsymbol{\aleph}}}_{n_p^{\between},l}^{\between}(k)\hat{\boldsymbol{G}}_{2,l+i-n_p^{\between}}^{\between}(k) \\
\hat{\boldsymbol{G}}_{3,l+i}^{\between}(k) &= \hat{\bar{\bar{\boldsymbol{\aleph}}}}_{1,l}^{\between}(k)\hat{\boldsymbol{G}}_{3,l+i-1}^{\between}(k) + \hat{\bar{\boldsymbol{\aleph}}}_{2,l}^{\between}(k)\hat{\boldsymbol{G}}_{3,l+i-2}^{\between}(k) + \\
&\quad \ldots + \hat{\bar{\bar{\boldsymbol{\aleph}}}}_{n_p^{\between},l}^{\between}(k)\hat{\boldsymbol{G}}_{3,l+i-n_p^{\between}}^{\between}(k), \quad i = 1, 2, \ldots, m
\end{aligned} \tag{5.21}
$$

Now define that

$$
\begin{aligned}
\hat{\boldsymbol{\aleph}}_{l}^{\between n_p^{\between}}(k) &= \left[\hat{\boldsymbol{\aleph}}_{1,l}^{\between T}(k), \ldots, \hat{\boldsymbol{\aleph}}_{n_p^{\between},l}^{\between T}(k)\right]^T \in \Re^{nn_p^{\between} \times n} \\
\hat{\bar{\boldsymbol{\aleph}}}_{l}^{\between n_p^{\between}}(k) &= \left[\hat{\bar{\boldsymbol{\aleph}}}_{1,l}^{\between T}(k), \ldots, \hat{\bar{\boldsymbol{\aleph}}}_{n_p^{\between},l}^{\between T}(k)\right]^T \in \Re^{nn_p^{\between} \times n} \\
\hat{\bar{\bar{\boldsymbol{\aleph}}}}_{l}^{\between n_p^{\between}}(k) &= \left[\hat{\bar{\bar{\boldsymbol{\aleph}}}}_{1,l}^{\between T}(k), \ldots, \hat{\bar{\bar{\boldsymbol{\aleph}}}}_{n_p^{\between},l}^{\between T}(k)\right]^T \in \Re^{nn_p^{\between} \times n} \\
\hat{\boldsymbol{\Phi}}_{a,l-1}^{\between}(k) &= \left[\hat{\boldsymbol{G}}_{a,l-1}^{\between}(k), \ldots, \hat{\boldsymbol{G}}_{a,l-n_p^{\between}}^{\between}(k)\right]^T \in \Re^{nn_p^{\between} \times n} \quad, a = 1, 2, 3
\end{aligned} \tag{5.22}
$$

then (5.21) can be rewritten as:

$$
\begin{aligned}
\hat{\boldsymbol{G}}_{1,l+i}^{\between}(k) &= \hat{\boldsymbol{\Phi}}_{1,l+i-1}^{\between T}(k)\hat{\boldsymbol{\aleph}}_{l}^{\between n_p^{\between}}(k) \\
\hat{\boldsymbol{G}}_{2,l+i}^{\between}(k) &= \hat{\boldsymbol{\Phi}}_{2,l+i-1}^{\between T}(k)\hat{\bar{\boldsymbol{\aleph}}}_{l}^{\between n_p^{\between}}(k) \\
\hat{\boldsymbol{G}}_{3,l+i}^{\between}(k) &= \hat{\boldsymbol{\Phi}}_{3,l+i-1}^{\between T}(k)\hat{\bar{\bar{\boldsymbol{\aleph}}}}_{l}^{\between n_p^{\between}}(k), \quad i = 1, 2, \ldots, m
\end{aligned} \tag{5.23}
$$

Here, $\hat{\boldsymbol{\aleph}}_{l}^{\between n_p^{\between}}(k)$, $\hat{\bar{\boldsymbol{\aleph}}}_{l}^{\between n_p^{\between}}(k)$ and $\hat{\bar{\bar{\boldsymbol{\aleph}}}}_{l}^{\between n_p^{\between}}(k)$ are computed by the following iterative learning algorithm:

$$
\begin{aligned}
\hat{\boldsymbol{\aleph}}_{l}^{\between n_p^{\between}}(k) = \hat{\boldsymbol{\aleph}}_{l-1}^{\between n_p^{\between}}(k) + \ell_1^{\between}\hat{\boldsymbol{\Phi}}_{1,l-1}^{\between}(k)[\sigma_1^{\between}\boldsymbol{I} + \hat{\boldsymbol{\Phi}}_{1,l-1}^{\between T}(k)\hat{\boldsymbol{\Phi}}_{1,l-1}^{\between}(k)]^{-1} \\
[\hat{\boldsymbol{G}}_{1,l}^{\between}(k) - \hat{\boldsymbol{\Phi}}_{1,l-1}^{\between T}(k)\hat{\boldsymbol{\aleph}}_{l-1}^{\between n_p^{\between}}(k)]
\end{aligned}
$$

$$
\begin{aligned}
\hat{\bar{\boldsymbol{\aleph}}}_l^{\between n_p^{\between}}(k) &= \hat{\bar{\boldsymbol{\aleph}}}_{l-1}^{\between n_p^{\between}}(k) + \ell_2^{\between}\hat{\boldsymbol{\Phi}}_{2,l-1}^{\between}(k)[\sigma_2^{\between}\boldsymbol{I} + \hat{\boldsymbol{\Phi}}_{2,l-1}^{\between T}(k)\hat{\boldsymbol{\Phi}}_{2,l-1}^{\between}(k)]^{-1} \\
&\quad [\hat{\boldsymbol{G}}_{2,l}^{\between}(k) - \hat{\boldsymbol{\Phi}}_{2,l-1}^{\between T}(k)\hat{\bar{\boldsymbol{\aleph}}}_{l-1}^{\between n_p^{\between}}(k)] \\
\hat{\bar{\bar{\boldsymbol{\aleph}}}}_l^{\between n_p^{\between}}(k) &= \hat{\bar{\bar{\boldsymbol{\aleph}}}}_{l-1}^{\between n_p^{\between}}(k) + \ell_3^{\between}\hat{\boldsymbol{\Phi}}_{3,l-1}^{\between}(k)[\sigma_3^{\between}\boldsymbol{I} + \hat{\boldsymbol{\Phi}}_{3,l-1}^{\between T}(k)\hat{\boldsymbol{\Phi}}_{3,l-1}^{\between}(k)]^{-1} \\
&\quad [\hat{\boldsymbol{G}}_{3,l}^{\between}(k) - \hat{\boldsymbol{\Phi}}_{3,l-1}^{\between T}(k)\hat{\bar{\bar{\boldsymbol{\aleph}}}}_{l-1}^{\between n_p^{\between}}(k)]
\end{aligned} \tag{5.24}
$$

where $\ell_1^{\between}, \ell_2^{\between}, \ell_3^{\between}, \sigma_1^{\between}, \sigma_2^{\between}, \sigma_3^{\between}$ are adjustable learning parameters.

Now based on the designed iterative learning prediction algorithm (5.21)–(5.24), the predicted values $\hat{\boldsymbol{G}}_{1,l+i}^{\between}(k)$, $\hat{\boldsymbol{G}}_{2,l+i}^{\between}(k)$, $\hat{\boldsymbol{G}}_{3,l+i}^{\between}(k)$, $(i = 1, 2, \ldots, m)$ are obtained. Denote

$$
\hat{\boldsymbol{G}}_{a,l+1}^{\between m}(k) = \left[\hat{\boldsymbol{G}}_{a,l+1}^{\between}(k), \hat{\boldsymbol{G}}_{a,l+2}^{\between}(k), \ldots, \hat{\boldsymbol{G}}_{a,l+m}^{\between}(k)\right], a = 1, 2, 3 \tag{5.25}
$$

then according to (5.14), we can get the following available tracking error prediction model:

$$
\begin{aligned}
\boldsymbol{e}_{l+m|l}^{\between}(k+1) &= \boldsymbol{e}_l^{\between}(k+1) - \hat{\boldsymbol{G}}_{1,l+1}^{\between m}(k)\Delta\boldsymbol{u}_{l+1}^m(k) - \hat{\boldsymbol{G}}_{2,l+1}^{\between m}(k)\Delta\boldsymbol{u}_{l+1}^m(k - \tau_1^{\between}(k)) \\
&\quad - \hat{\boldsymbol{G}}_{3,l+1}^{\between m}(k)\Delta\boldsymbol{u}_{l+1}^m(k - \tau_2^{\between}(k))
\end{aligned} \tag{5.26}
$$

Based on (5.26), $\Delta\boldsymbol{u}_{l+1}^m(k)$ can be obtained by computing the following quadratic cost function:

$$
J_{l+1}^{\between}(k) = \min_{\Delta\boldsymbol{u}_{l+1}^m(k)} \left\{\boldsymbol{e}_{l+m|l}^{\between T}(k+1)\,\boldsymbol{Q}^{\between}\boldsymbol{e}_{l+m|l}^{\between}(k+1) + \Delta\boldsymbol{u}_{l+1}^{mT}(k)\,\boldsymbol{R}^{\between}\Delta\boldsymbol{u}_{l+1}^m(k)\right\} \tag{5.27}
$$

where $\boldsymbol{Q}^{\between} \in \Re^{n\times n}$, $\boldsymbol{R}^{\between} \in \Re^{nm\times nm}$ are two symmetric positive definite matrices, for simplicity, define $\boldsymbol{Q}^{\between} = q^{\between}\boldsymbol{I}_{n\times n}$, $\boldsymbol{R}^{\between} = r^{\between}\boldsymbol{I}_{nm\times nm}$, where $q^{\between} > 0$, $r^{\between} > 0$. According to the optimization condition $\frac{\partial J_{l+1}^{\between}(k)}{\partial \Delta\boldsymbol{u}_{l+1}^m(k)} = \boldsymbol{0}$, the optimal control input sequence is obtained:

$$
\begin{aligned}
\Delta\boldsymbol{u}_{l+1}^m(k) &= \left[\hat{\boldsymbol{G}}_{1,l+1}^{\between mT}(k)\boldsymbol{Q}^{\between}\hat{\boldsymbol{G}}_{1,l+1}^{\between m}(k) + \boldsymbol{R}^{\between}\right]^{-1}\left[\hat{\boldsymbol{G}}_{1,l+1}^{\between mT}(k)\boldsymbol{Q}^{\between}\boldsymbol{e}_l^{\between}(k+1)\right. \\
&\quad - \hat{\boldsymbol{G}}_{1,l+1}^{\between mT}(k)\boldsymbol{Q}^{\between}\hat{\boldsymbol{G}}_{2,l+1}^{\between m}(k)\Delta\boldsymbol{u}_{l+1}^m(k - \tau_1^{\between}(k)) \\
&\quad \left. - \hat{\boldsymbol{G}}_{1,l+1}^{\between mT}(k)\boldsymbol{Q}^{\between}\hat{\boldsymbol{G}}_{3,l+1}^{\between m}(k)\Delta\boldsymbol{u}_{l+1}^m(k - \tau_2^{\between}(k))\right]
\end{aligned} \tag{5.28}
$$

Note that the control algorithm (5.28) contains matrix inverse operation. The inverse operation will be very time-consuming if the input and output dimensions of the system are large. In order to solve this problem, the following more practical control algorithm instead of (5.28) will be applied:

$$\Delta \boldsymbol{u}_{l+1}^{m}(k) = \frac{\gamma^{\phi} q^{\phi} \hat{\boldsymbol{G}}_{1,l+1}^{\phi m T}(k)}{r^{\phi} + q^{\phi} \left\| \hat{\boldsymbol{G}}_{1,l+1}^{\phi m}(k) \right\|^{2}} \Big[\boldsymbol{e}_{l}^{\phi}(k+1) - \hat{\boldsymbol{G}}_{2,l+1}^{\phi m}(k)\, \Delta \boldsymbol{u}_{l+1}^{m}\left(k - \tau_{1}^{\phi}(k)\right) - \hat{\boldsymbol{G}}_{3,l+1}^{\phi m}(k)\, \Delta \boldsymbol{u}_{l+1}^{m}\left(k - \tau_{2}^{\phi}(k)\right) \Big] \tag{5.29}$$

where the step factor $\gamma^{\phi} \in (0, 1]$ is introduced to make the control algorithm more general.

Using the receding optimization strategy, one can obtain the optimal control input for the upcoming $(l+1)$th iteration:

$$\boldsymbol{u}_{l+1}(k) = \boldsymbol{u}_{l}(k) + \boldsymbol{U}^{T} \Delta \boldsymbol{u}_{l+1}^{m}(k) \tag{5.30}$$

where $\boldsymbol{U} = \left[\boldsymbol{I}_{n\times n} \boldsymbol{0}_{n\times n(m-1)} \right]^{T}$.

To facilitate the subsequent analysis of the convergence property, the following lemma is given.

Lemma 5.2 Gerschgorin (1931) *Let* $\mathbf{A}^{\phi} = (a_{ij}^{\phi}) \in \mathbb{C}^{\phi n\times n}$, *define the Gerschgorin disk:* $D_{i}^{\phi} = \{ z^{\phi} \big| |z^{\phi} - a_{ii}^{\phi}| \le \sum_{j=1, j\neq i}^{n} |a_{ij}^{\phi}| \}$, $z^{\phi} \in \mathbb{C}^{\phi}$, $1 \le i \le n$. *Then, all the eigenroots of matrix* $\mathbf{A}^{\phi}$ *are in the disk and all the eigenroots of the matrix satisfy* $z_{1}^{\phi}, z_{2}^{\phi}, \ldots, z_{n}^{\phi} \in D_{A}^{\phi} = \bigcup_{i=1}^{n} D_{i}^{\phi}$.

Theorem 5.1 *If system (5.1) satisfies Assumptions 5.1–5.4, the proposed time delay compensation-based PILC method (5.29)–(5.30), together with the estimation algorithm (5.17)–(5.19) and prediction algorithm (5.21)–(5.24) satisfies the following condition:*

$$0 < \eta_{1}^{\phi} \frac{\partial \tilde{e}_{i,l}^{\phi}(k+1)}{\partial \hat{\boldsymbol{G}}_{1,i,l}^{\phi}(k)} \left(\frac{\partial \tilde{e}_{i,l}^{\phi}(k+1)}{\partial \hat{\boldsymbol{G}}_{1,i,l}^{\phi}(k)} \right)^{T} + \eta_{2}^{\phi} \frac{\partial \tilde{e}_{i,l}^{\phi}(k+1)}{\partial \hat{\boldsymbol{G}}_{2,i,l}^{\phi}(k)} \left(\frac{\partial \tilde{e}_{i,l}^{\phi}(k+1)}{\partial \hat{\boldsymbol{G}}_{2,i,l}^{\phi}(k)} \right)^{T} + \eta_{3}^{\phi} \frac{\partial \tilde{e}_{i,l}^{\phi}(k+1)}{\partial \hat{\boldsymbol{G}}_{3,i,l}^{\phi}(k)} \left(\frac{\partial \tilde{e}_{i,l}^{\phi}(k+1)}{\partial \hat{\boldsymbol{G}}_{3,i,l}^{\phi}(k)} \right)^{T} < 1, i = 1, 2, \ldots, n \tag{5.31}$$

where $\hat{\boldsymbol{G}}_{a,i,l}^{\phi}(k)$, $(i = 1, 2, \ldots, n)$ *defined in (5.33) is the* i*th row of* $\hat{\boldsymbol{G}}_{a,l}^{\phi}(k)$, *(a=1,2,3).* $\tilde{e}_{i,l}^{\phi}(k+1)$, $(i = 1, 2, \ldots, n)$ *defined in (5.34) is the* i*th element of* $\tilde{\boldsymbol{e}}_{l+1}^{\phi}(k+1)$. *Then, we have:*

(5.1.1) The modeling error converges to zero when $l \to \infty$, *that is,* $\lim_{l\to\infty} \tilde{\boldsymbol{e}}_{l+1}^{\phi}(k+1) = \boldsymbol{0}$.

(5.1.2) The tracking control error can be converged to a residue when $l \to \infty$, *that is,* $\lim_{l\to\infty} \left\| \boldsymbol{e}_{l+1}^{\eth}(k+1) \right\|_{\upsilon} \le \frac{c^{\eth}}{1-d_1^{\eth}}$, *where* $0 \le d_1^{\eth} < 1$ *is defined in (5.60),* $c^{\eth} \ge 0$ *is defined in (5.61).*

Proof **Proof of (5.1.1):**

Note that

$$\boldsymbol{y}_l(k+1) = \left[y_{1,l}(k+1), \ldots, y_{n,l}(k+1)\right]^T \tag{5.32}$$

$$\hat{\boldsymbol{G}}_{a,l}^{\eth}(k) = \left[\hat{\boldsymbol{G}}_{a,1,l}^{\eth T}(k), \ldots, \hat{\boldsymbol{G}}_{a,n,l}^{\eth T}(k)\right]^T, (a = 1, 2, 3) \tag{5.33}$$

$$\tilde{\boldsymbol{e}}_{l+1}^{\eth}(k+1) = \left[\tilde{e}_{1,l+1}^{\eth}(k+1), \ldots, \tilde{e}_{n,l+1}^{\eth}(k+1)\right]^T \tag{5.34}$$

Then according to (5.32)–(5.34) and (5.15)–(5.16), we have:

$$\begin{aligned} \tilde{e}_{i,l+1}^{\eth}(k+1) &= y_{i,l+1}(k+1) - \hat{y}_{i,l+1}(k+1) \\ &= y_{i,l+1}(k+1) - y_{i,l}(k+1) - \hat{\boldsymbol{G}}_{1,i,l+1}^{\eth}(k)\Delta\boldsymbol{u}_{l+1}(k) \\ &\quad -\hat{\boldsymbol{G}}_{2,i,l+1}^{\eth}(k)\Delta\boldsymbol{u}_{l+1}(k-\tau_1^{\eth}(k)) - \hat{\boldsymbol{G}}_{3,i,l+1}^{\eth}(k)\Delta\boldsymbol{u}_{l+1}(k-\tau_2^{\eth}(k)) \end{aligned} \tag{5.35}$$

Define the following composite energy function:

$$E_{i,l+1}^{\eth}(k+1) = \tilde{e}_{i,l+1}^{\eth}(k+1)^2, i = 1, 2, \ldots, n \tag{5.36}$$

Then, its difference on the iterative learning axis can be expressed as :

$$\begin{aligned} \Delta E_{i,l+1}^{\eth}(k+1) &= E_{i,l+1}^{\eth}(k+1) - E_{i,l}^{\eth}(k+1) \\ &= \tilde{e}_{i,l+1}^{\eth}(k+1)^2 - \tilde{e}_{i,l}^{\eth}(k+1)^2 \\ &= \Delta\tilde{e}_{i,l+1}^{\eth}(k+1)^2 + 2\tilde{e}_{i,l}^{\eth}(k+1)\Delta\tilde{e}_{i,l+1}^{\eth}(k+1) \end{aligned} \tag{5.37}$$

where $\Delta\tilde{e}_{i,l+1}^{\eth}(k+1) = \tilde{e}_{i,l+1}^{\eth}(k+1) - \tilde{e}_{i,l}^{\eth}(k+1)$.

Applying Taylor expansion to $\Delta\tilde{e}^{\between}_{i,l+1}(k+1)$ yields:

$$\Delta\tilde{e}^{\between}_{i,l+1}(k+1) = \frac{\partial\tilde{e}^{\between}_{i,l}(k+1)}{\partial\hat{\boldsymbol{G}}^{\between}_{1,i,l}(k)}\Delta\hat{\boldsymbol{G}}^{\between T}_{1,i,l+1}(k) + \frac{\partial\tilde{e}^{\between}_{i,l}(k+1)}{\partial\hat{\boldsymbol{G}}^{\between}_{2,i,l}(k)}\Delta\hat{\boldsymbol{G}}^{\between T}_{2,i,l+1}(k) + \frac{\partial\tilde{e}^{\between}_{i,l}(k+1)}{\partial\hat{\boldsymbol{G}}^{\between}_{3,i,l}(k)}\Delta\hat{\boldsymbol{G}}^{\between T}_{3,i,l+1}(k) \tag{5.38}$$

where $\Delta\hat{\boldsymbol{G}}^{\between}_{a,i,l+1}(k) = \hat{\boldsymbol{G}}^{\between}_{a,i,l+1}(k) - \hat{\boldsymbol{G}}^{\between}_{a,i,l}(k)$, $(a = 1, 2, 3)$.

Taking (5.17) into (5.38) yields:

$$\begin{aligned}\Delta\tilde{e}^{\between}_{i,l+1}(k+1) = {} & \eta^{\between}_{1}\frac{\partial\tilde{e}^{\between}_{i,l}(k+1)}{\partial\hat{\boldsymbol{G}}^{\between}_{1,i,l}(k)}\tilde{e}^{\between}_{i,l}(k+1)\Delta\boldsymbol{u}_l(k) \\ & +\eta^{\between}_{2}\frac{\partial\tilde{e}^{\between}_{i,l}(k+1)}{\partial\hat{\boldsymbol{G}}^{\between}_{2,i,l}(k)}\tilde{e}^{\between}_{i,l}(k+1)\Delta\boldsymbol{u}_l(k-\tau^{\between}_{1}(k)) \\ & +\eta^{\between}_{3}\frac{\partial\tilde{e}^{\between}_{i,l}(k+1)}{\partial\hat{\boldsymbol{G}}^{\between}_{3,i,l}(k)}\tilde{e}^{\between}_{i,l}(k+1)\Delta\boldsymbol{u}_l(k-\tau^{\between}_{2}(k))\end{aligned} \tag{5.39}$$

According to (5.35) and (5.39), we have:

$$\begin{aligned}\Delta\tilde{e}^{\between}_{i,l+1}(k+1) = {} & -\eta^{\between}_{1}\tilde{e}^{\between}_{i,l}(k+1)\frac{\partial\tilde{e}^{\between}_{i,l}(k+1)}{\partial\hat{\boldsymbol{G}}^{\between}_{1,i,l}(k)}\left(\frac{\partial\tilde{e}^{\between}_{i,l}(k+1)}{\partial\hat{\boldsymbol{G}}^{\between}_{1,i,l}(k)}\right)^{T} \\ & -\eta^{\between}_{2}\tilde{e}^{\between}_{i,l}(k+1)\frac{\partial\tilde{e}^{\between}_{i,l}(k+1)}{\partial\hat{\boldsymbol{G}}^{\between}_{2,i,l}(k)}\left(\frac{\partial\tilde{e}^{\between}_{i,l}(k+1)}{\partial\hat{\boldsymbol{G}}^{\between}_{2,i,l}(k)}\right)^{T} \\ & -\eta^{\between}_{3}\tilde{e}^{\between}_{i,l}(k+1)\frac{\partial\tilde{e}^{\between}_{i,l}(k+1)}{\partial\hat{\boldsymbol{G}}^{\between}_{3,i,l}(k)}\left(\frac{\partial\tilde{e}^{\between}_{i,l}(k+1)}{\partial\hat{\boldsymbol{G}}^{\between}_{3,i,l}(k)}\right)^{T}\end{aligned} \tag{5.40}$$

Let

$$\begin{aligned}l^{\between}_{i,l}(k+1) = {} & \eta^{\between}_{1}\frac{\partial\tilde{e}^{\between}_{i,l}(k+1)}{\partial\hat{\boldsymbol{G}}^{\between}_{1,i,l}(k)}\left(\frac{\partial\tilde{e}^{\between}_{i,l}(k+1)}{\partial\hat{\boldsymbol{G}}^{\between}_{1,i,l}(k)}\right)^{T} \\ & +\eta^{\between}_{2}\frac{\partial\tilde{e}^{\between}_{i,l}(k+1)}{\partial\hat{\boldsymbol{G}}^{\between}_{2,i,l}(k)}\left(\frac{\partial\tilde{e}^{\between}_{i,l}(k+1)}{\partial\hat{\boldsymbol{G}}^{\between}_{2,i,l}(k)}\right)^{T} + \eta^{\between}_{3}\frac{\partial\tilde{e}^{\between}_{i,l}(k+1)}{\partial\hat{\boldsymbol{G}}^{\between}_{3,i,l}(k)}\left(\frac{\partial\tilde{e}^{\between}_{i,l}(k+1)}{\partial\hat{\boldsymbol{G}}^{\between}_{3,i,l}(k)}\right)^{T}\end{aligned} \tag{5.41}$$

Then, (5.40) leads to:

$$\Delta\tilde{e}_{i,l+1}^{\Diamond}(k+1) = -\tilde{e}_{i,l}^{\Diamond}(k+1)l_{i,l}^{\Diamond}(k+1) \tag{5.42}$$

Substituting (5.42) into (5.37) yields:

$$\begin{aligned}\Delta E_{i,l+1}^{\Diamond}(k+1) &= \Delta\tilde{e}_{i,l+1}^{\Diamond}(k+1)^2 + 2\tilde{e}_{i,l}^{\Diamond}(k+1)\Delta\tilde{e}_{i,l+1}^{\Diamond}(k+1) \\ &= \tilde{e}_{i,l}^{\Diamond}(k+1)^2\left[l_{i,l}^{\Diamond}(k+1)^2 - 2l_{i,l}^{\Diamond}(k+1)\right]\end{aligned} \tag{5.43}$$

According to the condition (5.31), we know that $0 < l_{i,l}^{\Diamond}(k+1) < 1$, then from (5.37) and (5.43), we have:

$$\Delta E_{i,l+1}^{\Diamond}(k+1) \le -l_{i,l}^{\Diamond}(k+1)\tilde{e}_{i,l}^{\Diamond}(k+1)^2 \tag{5.44}$$

Furthermore, according to (5.37), we have:

$$\tilde{e}_{i,k+1}^{\Diamond}(k+1)^2 \le \left[1 - l_{i,l}^{\Diamond}(k+1)\right]\tilde{e}_{i,l}^{\Diamond}(k+1)^2 \tag{5.45}$$

According to the condition (5.31), there exists a positive constant $\rho^{\Diamond}$ satisfying $0 < 1 - l_{i,l}^{\Diamond}(k+1) \le \rho^{\Diamond} < 1$, then (5.45) can be rewritten as:

$$\tilde{e}_{i,l+1}^{\Diamond}(k+1)^2 \le \rho^{\Diamond}\tilde{e}_{i,l}^{\Diamond}(k+1)^2 \tag{5.46}$$

thus we get $\lim\limits_{l\to\infty}\tilde{\boldsymbol{e}}_{l+1}^{\Diamond}(k+1) = \mathbf{0}$.

Proof of (5.1.2):

Substituting the proposed control algorithm (5.29)–(5.30) into (5.8) yields:

$$\begin{aligned}\boldsymbol{e}_{l+1}^{\Diamond}(k+1) &= \boldsymbol{e}_{l}^{\Diamond}(k+1) - \boldsymbol{G}_{1,l+1}^{\Diamond}(k)\Delta\boldsymbol{u}_{l+1}(k) - \boldsymbol{G}_{2,l+1}^{\Diamond}(k)\Delta\boldsymbol{u}_{l+1}(k-\tau_1^{\Diamond}(k)) \\ &\quad - \boldsymbol{G}_{3,l+1}^{\Diamond}(k)\Delta\boldsymbol{u}_{l+1}(k-\tau_2^{\Diamond}(k)) \\ &= \boldsymbol{e}_{l}^{\Diamond}(k+1) - \frac{\gamma^{\Diamond}q^{\Diamond}\boldsymbol{G}_{1,l+1}^{\Diamond}(k)\boldsymbol{U}^T\hat{\boldsymbol{G}}_{1,l+1}^{\Diamond mT}(k)}{r^{\Diamond} + q^{\Diamond}\left\|\hat{\boldsymbol{G}}_{1,l+1}^{\Diamond m}(k)\right\|^2}\boldsymbol{e}_{l}^{\Diamond}(k+1) \\ &\quad + \frac{\gamma^{\Diamond}q^{\Diamond}\boldsymbol{G}_{1,l+1}^{\Diamond}(k)\boldsymbol{U}^T\hat{\boldsymbol{G}}_{1,l+1}^{\Diamond mT}(k)}{r^{\Diamond} + q^{\Diamond}\left\|\hat{\boldsymbol{G}}_{1,l+1}^{\Diamond m}(k)\right\|^2} \times \\ &\quad \left[\hat{\boldsymbol{G}}_{2,l+1}^{\Diamond m}(k)\Delta\boldsymbol{u}_{l+1}^{m}(k-\tau_1^{\Diamond}(k)) + \hat{\boldsymbol{G}}_{3,l+1}^{\Diamond m}(t)\Delta\boldsymbol{u}_{l+1}^{m}(k-\tau_2^{\Diamond}(k))\right]\end{aligned}$$

$$
\begin{aligned}
&- \boldsymbol{G}^{\Diamond}_{2,l+1}(k)\Delta \boldsymbol{u}_{l+1}(k-\tau^{\Diamond}_1(k)) - \boldsymbol{G}^{\Diamond}_{3,l+1}(k)\Delta \boldsymbol{u}_{l+1}(k-\tau^{\Diamond}_2(k)) \\
&= \boldsymbol{e}^{\Diamond}_l(k+1) - \frac{\gamma^{\Diamond} q^{\Diamond} \boldsymbol{G}^{\Diamond}_{1,l+1}(k)\hat{\boldsymbol{G}}^{\Diamond T}_{1,l+1}(k)}{r^{\Diamond} + q^{\Diamond}\left\|\hat{\boldsymbol{G}}^{\Diamond m}_{1,l+1}(k)\right\|^2}\boldsymbol{e}^{\Diamond}_l(k+1) + \frac{\gamma^{\Diamond} q^{\Diamond} \boldsymbol{G}^{\Diamond}_{1,l+1}(k)\hat{\boldsymbol{G}}^{\Diamond T}_{1,l+1}(k)}{r^{\Diamond} + q^{\Diamond}\left\|\hat{\boldsymbol{G}}^{\Diamond m}_{1,l+1}(k)\right\|^2} \\
&\quad \left[\hat{\boldsymbol{G}}^{\Diamond m}_{2,l+1}(t)\Delta \boldsymbol{u}^m_{l+1}(k-\tau^{\Diamond}_1(k)) + \hat{\boldsymbol{G}}^{\Diamond m}_{3,l+1}(k)\Delta \boldsymbol{u}^m_{l+1}(k-\tau^{\Diamond}_2(k))\right] \\
&\qquad - \boldsymbol{G}^{\Diamond}_{2,l+1}(k)\Delta \boldsymbol{u}_{l+1}(k-\tau^{\Diamond}_1(k)) - \boldsymbol{G}^{\Diamond}_{3,l+1}(k)\Delta \boldsymbol{u}_{l+1}(k-\tau^{\Diamond}_2(k)) \\
&= \left\{\boldsymbol{I} - \frac{\gamma^{\Diamond} q^{\Diamond} \boldsymbol{G}^{\Diamond}_{1,l+1}(k)\hat{\boldsymbol{G}}^{\Diamond T}_{1,l+1}(k)}{r^{\Diamond} + q^{\Diamond}\left\|\hat{\boldsymbol{G}}^{\Diamond m}_{1,l+1}(k)\right\|^2}\right\}\boldsymbol{e}^{\Diamond}_l(k+1) + \frac{\gamma^{\Diamond} q^{\Diamond} \boldsymbol{G}^{\Diamond}_{1,l+1}(k)\hat{\boldsymbol{G}}^{\Diamond T}_{1,l+1}(k)}{r^{\Diamond} + q^{\Diamond}\left\|\hat{\boldsymbol{G}}^{\Diamond m}_{1,l+1}(k)\right\|^2} \\
&\quad \left[\hat{\boldsymbol{G}}^{\Diamond m}_{2,l+1}(k)\Delta \boldsymbol{u}^m_{l+1}(k-\tau^{\Diamond}_1(k)) + \hat{\boldsymbol{G}}^{\Diamond m}_{3,l+1}(k)\Delta \boldsymbol{u}^m_{l+1}(k-\tau^{\Diamond}_2(k))\right] \\
&\quad - \boldsymbol{G}^{\Diamond}_{2,l+1}(k)\Delta \boldsymbol{u}_{l+1}(k-\tau^{\Diamond}_1(k)) - \boldsymbol{G}^{\Diamond}_{3,l+1}(k)\Delta \boldsymbol{u}_{l+1}(k-\tau^{\Diamond}_2(k))
\end{aligned} \tag{5.47}
$$

Let

$$
\begin{aligned}
\boldsymbol{C}^{\Diamond} = &\frac{\gamma^{\Diamond} q^{\Diamond} \boldsymbol{G}^{\Diamond}_{1,l+1}(k)\hat{\boldsymbol{G}}^{\Diamond T}_{1,l+1}(k)}{r^{\Diamond} + q^{\Diamond}\left\|\hat{\boldsymbol{G}}^{\Diamond m}_{1,l+1}(k)\right\|^2}\left[\hat{\boldsymbol{G}}^{\Diamond m}_{2,l+1}(k)\Delta \boldsymbol{u}^m_{l+1}(k-\tau^{\Diamond}_1(k))\right. \\
&\left.+\hat{\boldsymbol{G}}^{\Diamond m}_{3,l+1}(k)\Delta \boldsymbol{u}^m_{l+1}(k-\tau^{\Diamond}_2(k))\right] - \boldsymbol{G}^{\Diamond}_{2,l+1}(k)\Delta \boldsymbol{u}_{l+1}(k-\tau^{\Diamond}_1(k)) \\
&-\boldsymbol{G}^{\Diamond}_{3,l+1}(k)\Delta \boldsymbol{u}_{l+1}(k-\tau^{\Diamond}_2(k))
\end{aligned} \tag{5.48}
$$

Then (5.47) can be rewritten as:

$$
\boldsymbol{e}^{\Diamond}_{l+1}(k+1) = \left\{\boldsymbol{I} - \frac{\gamma^{\Diamond} q^{\Diamond} \boldsymbol{G}^{\Diamond}_{1,l+1}(k)\hat{\boldsymbol{G}}^{T}_{1,l+1}(k)}{r^{\Diamond} + q^{\Diamond}\left\|\hat{\boldsymbol{G}}^{\Diamond m}_{1,l+1}(k)\right\|^2}\right\}\boldsymbol{e}^{\Diamond}_l(k+1) + \boldsymbol{C}^{\Diamond} \tag{5.49}
$$

According to Lemma 5.2, it follows that:

$$
\begin{aligned}
D^{\Diamond}_j = &\left\{z^{\Diamond}\middle|\left|z^{\Diamond} - (1 - \frac{\gamma^{\Diamond} q^{\Diamond}\sum_{i=1}^n g^{\Diamond}_{1,ji,l+1}(k)\hat{g}^{\Diamond}_{1,ji,l+1}(k)}{r^{\Diamond} + q^{\Diamond}\left\|\hat{\boldsymbol{G}}^{\Diamond m}_{1,l+1}(k)\right\|^2})\right|\right. \\
&\left.\le \sum_{h=1,h\neq j}^{n}\left|\frac{\gamma^{\Diamond} q^{\Diamond}\sum_{i=1}^n g^{\Diamond}_{1,ji,l+1}(k)\hat{g}^{\Diamond}_{1,hi,l+1}(k)}{r^{\Diamond} + q^{\Diamond}\left\|\hat{\boldsymbol{G}}^{\Diamond m}_{1,l+1}(k)\right\|^2}\right|\right\}
\end{aligned} \tag{5.50}
$$

where $z^{\Diamond}$ is the characteristic root of matrix $\boldsymbol{I} - \frac{\gamma^{\Diamond} q^{\Diamond} \boldsymbol{G}^{\Diamond}_{1,l+1}(k)\hat{\boldsymbol{G}}^{\Diamond T}_{1,l+1}(k)}{r^{\Diamond} + q^{\Diamond}\left\|\hat{\boldsymbol{G}}^{\Diamond m}_{1,l+1}(k)\right\|^2}$, $D^{\Diamond}_w, w = 1, \ldots, m$ is the Gerschgorin disk.

Using the trigonometric inequality, (5.50) leads to:

$$D_w^{\Diamond} = \left\{ z^{\Diamond} \Big| |z^{\Diamond}| \le \Big| 1 - \frac{\gamma^{\Diamond} q^{\Diamond} \sum_{i=1}^{n} g_{1,ji,l+1}^{\Diamond}(k) \hat{g}_{1,ji,l+1}^{\Diamond}(k)}{r^{\Diamond} + q^{\Diamond} \left\| \hat{\boldsymbol{G}}_{1,l+1}^{\Diamond m}(k) \right\|^2} \Big| + \sum_{h=1,h\neq j}^{n} \Big| \frac{\gamma^{\Diamond} q^{\Diamond} \sum_{i=1}^{n} g_{1,ji,l+1}^{\Diamond}(k) \hat{g}_{1,hi,l+1}^{\Diamond}(k)}{r^{\Diamond} + q^{\Diamond} \left\| \hat{\boldsymbol{G}}_{1,l+1}^{\Diamond m}(k) \right\|^2} \Big| \right\} \tag{5.51}$$

On the other hand, the following two inequalities can be obtained from the reset algorithm (5.18)–(5.19) and Assumption 5.4:

$$\begin{aligned} 1 - \frac{\gamma^{\Diamond} q^{\Diamond} \sum_{i=1}^{n} \left|g_{1,ji,l+1}^{\Diamond}(k)\right| \left|\hat{g}_{1,ji,l+1}^{\Diamond}(k)\right|}{r^{\Diamond} + q^{\Diamond} \left\| \hat{\boldsymbol{G}}_{1,l+1}^{\Diamond m}(k) \right\|^2} &\le 1 - \frac{\gamma^{\Diamond} q^{\Diamond} \left|g_{1,jj,l+1}^{\Diamond}(k)\right| \left|\hat{g}_{1,jj,l+1}^{\Diamond}(k)\right|}{r^{\Diamond} + q^{\Diamond} \left\| \hat{\boldsymbol{G}}_{1,l+1}^{\Diamond m}(k) \right\|^2} \\ &\le 1 - \frac{\gamma^{\Diamond} q^{\Diamond} b_2^{\Diamond 2}}{r^{\Diamond} + q^{\Diamond} \left\| \hat{\boldsymbol{G}}_{1,l+1}^{\Diamond m}(k) \right\|^2} \end{aligned} \tag{5.52}$$

and

$$\begin{aligned} &\sum_{h=1,h\neq j}^{n} \frac{\gamma^{\Diamond} q^{\Diamond} \sum_{i=1}^{n} g_{1,ji,l+1}^{\Diamond}(k)\, \hat{g}_{1,hi,l+1}^{\Diamond}(k)}{r^{\Diamond} + q^{\Diamond} \left\| \hat{\boldsymbol{G}}_{1,l+1}^{\Diamond m}(k) \right\|^2} \\ &\le \gamma^{\Diamond} q^{\Diamond} \sum_{h=1,h\neq j}^{n} \frac{\sum_{i=1}^{n} \left|g_{1,ji,l+1}^{\Diamond}(k)\right| \left|\hat{g}_{1,hi,l+1}^{\Diamond}(k)\right|}{r^{\Diamond} + q^{\Diamond} \left\| \hat{\boldsymbol{G}}_{1,l+1}^{\Diamond m}(k) \right\|^2} \\ &\le \gamma^{\Diamond} q^{\Diamond} \frac{\sum_{h=1,h\neq j}^{n} \left|g_{1,jj,l+1}^{\Diamond}(k)\right| \left|\hat{g}_{1,hj,l+1}^{\Diamond}(k)\right|}{r^{\Diamond} + q^{\Diamond} \left\| \hat{\boldsymbol{G}}_{1,l+1}^{\Diamond m}(k) \right\|^2} \\ &\quad + \gamma^{\Diamond} q^{\Diamond} \sum_{h=1,h\neq j}^{n} \frac{\sum_{i=1,i\neq j}^{n} \left|g_{1,ji,l+1}^{\Diamond}(k)\right| \left|\hat{g}_{1,hi,l+1}^{\Diamond}(k)\right|}{r^{\Diamond} + q^{\Diamond} \left\| \hat{\boldsymbol{G}}_{1,l+1}^{\Diamond m}(k) \right\|^2} \\ &\le \gamma^{\Diamond} q^{\Diamond} \frac{\sum_{h=1,h\neq j}^{n} \left|g_{1,jj,l+1}^{\Diamond}(k)\right| \left|\hat{g}_{1,hj,l+1}^{\Diamond}(k)\right|}{r^{\Diamond} + q^{\Diamond} \left\| \hat{\boldsymbol{G}}_{1,l+1}^{\Diamond m}(k) \right\|^2} \\ &\quad + \gamma^{\Diamond} q^{\Diamond} \frac{\sum_{h=1,h\neq j}^{n} \left|g_{1,jh,l+1}^{\Diamond}(k)\right| \left|\hat{g}_{1,hh,l+1}^{\Diamond}(k)\right|}{r^{\Diamond} + q^{\Diamond} \left\| \hat{\boldsymbol{G}}_{1,l+1}^{\Diamond m}(k) \right\|^2} \\ &\quad + \gamma^{\Diamond} q^{\Diamond} \sum_{h=1,h\neq j}^{n} \frac{\sum_{i=1,i\neq j,l}^{n} \left|g_{1,ji,l+1}^{\Diamond}(k)\right| \left|\hat{g}_{1,hi,l+1}^{\Diamond}(k)\right|}{r^{\Diamond} + q^{\Diamond} \left\| \hat{\boldsymbol{G}}_{1,l+1}^{\Diamond m}(k) \right\|^2} \\ &\le \gamma^{\Diamond} q^{\Diamond} \frac{2\alpha^{\Diamond} b_1^{\Diamond} b_2^{\Diamond} (n-1) + b_1^{\Diamond 2} (n-1)(n-2)}{r^{\Diamond} + q^{\Diamond} \left\| \hat{\boldsymbol{G}}_{1,l+1}^{\Diamond m}(k) \right\|^2} \end{aligned} \tag{5.53}$$

According to (5.52) and (5.53) , we have:

$$\begin{aligned}
&1-\frac{\gamma^{\Diamond}q^{\Diamond}\sum_{i=1}^{n}\left|g_{1,ji,l+1}^{\Diamond}(k)\right|\left|\hat{g}_{1,ji,l+1}^{\Diamond}(k)\right|}{r^{\Diamond}+q^{\Diamond}\left\|\hat{\boldsymbol{G}}_{1,l+1}^{\Diamond m}(k)\right\|^{2}} \\
&\quad+\sum_{h=1,h\neq j}^{n}\frac{\gamma^{\Diamond}q^{\Diamond}\sum_{i=1}^{n}g_{1,ji,l+1}^{\Diamond}(k)\,\hat{g}_{1,hi,l+1}^{\Diamond}(k)}{r^{\Diamond}+q^{\Diamond}\left\|\hat{\boldsymbol{G}}_{1,l+1}^{\Diamond m}(k)\right\|^{2}} \\
&\leq 1-\frac{\gamma^{\Diamond}q^{\Diamond}b_{2}^{\Diamond 2}}{r^{\Diamond}+q^{\Diamond}\left\|\hat{\boldsymbol{G}}_{1,l+1}^{\Diamond m}(k)\right\|^{2}}+\gamma^{\Diamond}q^{\Diamond}\frac{2\alpha^{\Diamond}b_{1}^{\Diamond}b_{2}^{\Diamond}(n-1)+b_{1}^{\Diamond 2}(n-1)(n-2)}{r^{\Diamond}+q^{\Diamond}\left\|\hat{\boldsymbol{G}}_{1,l+1}^{\Diamond m}(k)\right\|^{2}} \\
&\leq 1-\gamma^{\Diamond}q^{\Diamond}\frac{b_{2}^{\Diamond 2}-2\alpha^{\Diamond}b_{1}^{\Diamond}b_{2}^{\Diamond}(n-1)-b_{1}^{\Diamond 2}(n-1)(n-2)}{r^{\Diamond}+q^{\Diamond}\left\|\hat{\boldsymbol{G}}_{1,l+1}^{\Diamond m}(k)\right\|^{2}}
\end{aligned}\tag{5.54}$$

According to Assumption 5.4, (5.54) can be rewritten as:

$$\begin{aligned}
&1-\frac{\gamma^{\Diamond}q^{\Diamond}\sum_{i=1}^{n}\left|g_{1,ji,l+1}^{\Diamond}(k)\right|\left|\hat{g}_{1,ji,l+1}^{\Diamond}(k)\right|}{r^{\Diamond}+q^{\Diamond}\left\|\hat{\boldsymbol{G}}_{1,l+1}^{\Diamond m}(k)\right\|^{2}} \\
&\quad+\sum_{h=1,h\neq j}^{n}\frac{\gamma^{\Diamond}q^{\Diamond}\sum_{i=1}^{n}g_{1,ji,l+1}^{\Diamond}(k)\,\hat{g}_{1,hi,l+1}^{\Diamond}(k)}{r^{\Diamond}+q^{\Diamond}\left\|\hat{\boldsymbol{G}}_{1,l+1}^{\Diamond m}(k)\right\|^{2}} \\
&\leq 1-\gamma^{\Diamond}q^{\Diamond}\frac{b_{2}^{\Diamond}\left(b_{2}^{\Diamond}-2\alpha^{\Diamond}b_{1}^{\Diamond}(n-1)\right)-b_{1}^{\Diamond 2}(n-1)(n-2)}{r^{\Diamond}+q^{\Diamond}\left\|\hat{\boldsymbol{G}}_{1,l+1}^{\Diamond m}(k)\right\|^{2}} \\
&\leq 1-\gamma^{\Diamond}q^{\Diamond}\frac{b_{1}^{\Diamond}b_{2}^{\Diamond}(n-1)-b_{1}^{\Diamond 2}(n-1)(n-2)}{r^{\Diamond}+q^{\Diamond}\left\|\hat{\boldsymbol{G}}_{1,l+1}^{\Diamond m}(k)\right\|^{2}} \\
&\leq 1-\gamma^{\Diamond}q^{\Diamond}\frac{b_{1}^{\Diamond}b_{2}^{\Diamond}(n-1)-b_{1}^{\Diamond 2}(n-1)^{2}}{r^{\Diamond}+q^{\Diamond}\left\|\hat{\boldsymbol{G}}_{1,l+1}^{\Diamond m}(k)\right\|^{2}} \\
&\leq 1-\gamma^{\Diamond}q^{\Diamond}\frac{b_{1}^{\Diamond}(n-1)\left(b_{2}^{\Diamond}-b_{1}^{\Diamond}(n-1)\right)}{r^{\Diamond}+q^{\Diamond}\left\|\hat{\boldsymbol{G}}_{1,l+1}^{\Diamond m}(k)\right\|^{2}} \\
&\leq 1-\gamma^{\Diamond}q^{\Diamond}\frac{2\alpha^{\Diamond}b_{1}^{\Diamond 2}(n-1)^{2}}{r^{\Diamond}+q^{\Diamond}\left\|\hat{\boldsymbol{G}}_{1,l+1}^{\Diamond m}(k)\right\|^{2}}
\end{aligned}\tag{5.55}$$

By the reset algorithm (5.18)–(5.19) and Assumption 5.4, we can obtain that $g_{1,ji,l+1}^{\Diamond}(k)\,\hat{g}_{1,ji,l+1}^{\Diamond}(k)>0$. Therefore, there exists a $r_{min}^{\Diamond}>0$ such that when $r^{\Diamond}>r_{min}^{\Diamond}$, we have:

$$\begin{aligned}
\frac{\sum_{i=1}^{n} g_{1,ji,l+1}^{\between}(k)\,\hat{g}_{1,ji,l+1}^{\between}(k)}{r^{\between}+q^{\between}\left\|\hat{\boldsymbol{G}}_{1,l+1}^{\between m}(k)\right\|^{2}} &= \frac{\sum_{i=1}^{n}\left|g_{1,ji,l+1}^{\between}(k)\right|\left|\hat{g}_{1,ji,l+1}^{\between}(k)\right|}{r^{\between}+q^{\between}\left\|\hat{\boldsymbol{G}}_{1,l+1}^{\between m}(k)\right\|^{2}} \\
&\le \frac{\alpha^{\between 2}b_{2}^{\between 2}+b_{1}^{\between 2}(n-1)}{r^{\between}+q^{\between}\left\|\hat{\boldsymbol{G}}_{1,l+1}^{\between m}(k)\right\|^{2}} \le \frac{\alpha^{\between 2}b_{2}^{\between 2}+b_{1}^{\between 2}(n-1)}{r_{min}^{\between}+q^{\between}\left\|\hat{\boldsymbol{G}}_{1,l+1}^{\between m}(k)\right\|^{2}} < 1
\end{aligned} \tag{5.56}$$

when $r^{\between} > r_{min}^{\between} > 0$ and $0 < \gamma^{\between} q^{\between} \le 1$, one has:

$$\begin{aligned}
0 < M^{\between} &\le \frac{2\alpha^{\between}b_{1}^{\between 2}(n-1)^{2}}{r^{\between}+q^{\between}\left\|\hat{\boldsymbol{G}}_{1,l+1}^{\between m}(k)\right\|^{2}} < \frac{b_{2}^{\between 2}}{r^{\between}+q^{\between}\left\|\hat{\boldsymbol{G}}_{1,l+1}^{\between m}(k)\right\|^{2}} \\
&\le \frac{\alpha^{\between 2}b_{2}^{\between 2}+b_{1}^{\between 2}(n-1)}{r^{\between}+q^{\between}\left\|\hat{\boldsymbol{G}}_{1,l+1}^{\between m}(k)\right\|^{2}} < \frac{\alpha^{\between 2}b_{2}^{\between 2}+b_{1}^{\between 2}(n-1)}{r_{min}^{\between}+q^{\between}\left\|\hat{\boldsymbol{G}}_{1,l+1}^{\between m}(k)\right\|^{2}} < 1
\end{aligned} \tag{5.57}$$

where $M^{\between} > 0$ is a positive constant.

According to (5.55) and (5.57), one can get:

$$\begin{aligned}
&\left|1-\frac{\gamma^{\between}q^{\between}\sum_{i=1}^{n} g_{1,ji,l+1}^{\between}(k)\,\hat{g}_{1,ji,l+1}^{\between}(k)}{r^{\between}+q^{\between}\left\|\hat{\boldsymbol{G}}_{1,l+1}^{\between m}(k)\right\|^{2}}\right| \\
&\quad+\sum_{h=1,h\neq j}^{n}\frac{\gamma^{\between}q^{\between}\sum_{i=1}^{n} g_{1,ji,l+1}^{\between}(k)\,\hat{g}_{1,hi,l+1}^{\between}(k)}{r^{\between}+q^{\between}\left\|\hat{\boldsymbol{G}}_{1,l+1}^{\between m}(k)\right\|^{2}} \\
&=1-\frac{\gamma^{\between}q^{\between}\sum_{i=1}^{n}\left|g_{1,ji,l+1}^{\between}(k)\right|\left|\hat{g}_{1,ji,l+1}^{\between}(k)\right|}{r^{\between}+q^{\between}\left\|\hat{\boldsymbol{G}}_{1,l+1}^{\between m}(k)\right\|^{2}} \\
&\quad+\sum_{h=1,h\neq j}^{n}\frac{\gamma^{\between}q^{\between}\sum_{i=1}^{n} g_{1,ji,l+1}^{\between}(k)\,\hat{g}_{1,hi,l+1}^{\between}(k)}{r^{\between}+q^{\between}\left\|\hat{\boldsymbol{G}}_{1,l+1}^{\between m}(k)\right\|^{2}} \\
&<1-\frac{\gamma^{\between}q^{\between}2\alpha^{\between}b_{1}^{\between 2}(n-1)^{2}}{r^{\between}+q^{\between}\left\|\hat{\boldsymbol{G}}_{1,l+1}^{\between m}(k)\right\|^{2}} < 1-\gamma^{\between}q^{\between}M^{\between} < 1
\end{aligned} \tag{5.58}$$

Now according to (5.51) and (5.58), one can get:

$$s\left(\boldsymbol{I}-\frac{\gamma^{\between}q^{\between}\boldsymbol{G}_{1,l+1}^{\between}(k)\,\hat{\boldsymbol{G}}_{1,l+1}^{\between T}(k)}{r^{\between}+q^{\between}\left\|\hat{\boldsymbol{G}}_{1,l+1}^{\between m}(k)\right\|^{2}}\right) < 1-\gamma^{\between}q^{\between}M^{\between} < 1 \tag{5.59}$$

where $s\left(\mathbf{A}^{\between}\right)$ is the spectral radius of matrix $\mathbf{A}^{\between}$, that is $s\left(\mathbf{A}^{\between}\right) = \max_{s\in\{1,2,\ldots,m\}}\left|z_s^{\between}\right|.z_s^{\between}$ $(s = 1, 2, \ldots, m)$ is the eigenvalue of matrix $\mathbf{A}^{\between}$. From the knowledge of spectral radius of the matrix, it follows that there exists any small $\varepsilon^{\between} > 0$ and $0 < d_1^{\between} < 1$ such that:

$$0 \leq \left\| \boldsymbol{I} - \frac{\gamma^{\between} q^{\between} \boldsymbol{G}_{1,l+1}^{\between}(k)\, \hat{\boldsymbol{G}}_{1,l+1}^{\between T}(k)}{r^{\between} + q^{\between} \left\| \hat{\boldsymbol{G}}_{1,l+1}^{\between m}(k) \right\|^2} \right\|_{\upsilon} \leq s \left(\boldsymbol{I} - \frac{\gamma^{\between} q^{\between} \boldsymbol{G}_{1,l+1}^{\between}(k)\, \hat{\boldsymbol{G}}_{1,l+1}^{\between T}(k)}{r^{\between} + q^{\between} \left\| \hat{\boldsymbol{G}}_{1,l+1}^{\between m}(k) \right\|^2} \right) + \varepsilon^{\between}$$
$$\leq 1 - \gamma^{\between} q^{\between} M^{\between} + \varepsilon^{\between} \leq d_1^{\between} < 1 \tag{5.60}$$

where $\left\| \mathbf{A}^{\between} \right\|_{\upsilon}$ is the compatible norm of matrix $\mathbf{A}^{\between}$.

By taking compatible norm on both sides of (5.49), we can get:

$$\begin{aligned} \left\| \boldsymbol{e}_{l+1}^{\between}(k+1) \right\|_{\upsilon} &\leq \left\| \boldsymbol{I} - \frac{\gamma^{\between} q^{\between} \boldsymbol{G}_{1,l+1}^{\between}(k)\, \hat{\boldsymbol{G}}_{1,l+1}^{\between T}(k)}{r^{\between} + q^{\between} \left\| \hat{\boldsymbol{G}}_{1,l+1}^{\between m}(k) \right\|^2} \right\|_{\upsilon} \left\| \boldsymbol{e}_l^{\between}(k+1) \right\|_{\upsilon} + c^{\between} \\ &\leq d_1^{\between} \left\| \boldsymbol{e}_l^{\between}(k+1) \right\|_{\upsilon} + c^{\between} \leq d_1^{\between 2} \left\| \boldsymbol{e}_{l-1}^{\between}(k+1) \right\|_{\upsilon} + c^{\between} + c^{\between} d_1^{\between} \\ &\leq \ldots \leq d_1^{\between(l+1)} \left\| \boldsymbol{e}_0^{\between}(k+1) \right\|_{\upsilon} + \frac{c^{\between}\left(1 - d_1^{\between l}\right)}{1 - d_1^{\between}} \end{aligned} \tag{5.61}$$

where $c^{\between}$ is the result of (5.48) after taking the compatible norm. Then since $0 \leq d_1^{\between} < 1$, (5.61) can be rewritten as:

$$\lim_{l \to \infty} \left\| \boldsymbol{e}_{l+1}^{\between}(k+1) \right\|_{\upsilon} \leq \frac{c^{\between}}{1 - d_1^{\between}} \tag{5.62}$$

That is, the tracking error $\boldsymbol{e}_{l+1}^{\between}(k+1)$ can be converged to the residue $\frac{c^{\between}}{1-d_1^{\between}}$ when l tends to infinity. □

5.4 Simulation Validation

Consider the following linear motor model as in Chap. 4 Lee et al. (2000) with unknown time delay $h^{\between}(k)$

$$\begin{cases} \dot{x}(k) = \upsilon(k) \\ \dot{\upsilon}(k) = \frac{u(k-h^{\between}(k)) - f_{fric}(k) - f_{rip}(k)}{m^{\between}} \\ f_{fric}(k) = [f_c + (f_s - f_c)e^{-(\upsilon/\upsilon_s)^2} + f_\upsilon \upsilon] sgn(\upsilon) \\ f_{rip}(k) = A_r^{\between} sin(\omega_0^{\between} \int_0^k \upsilon(\tau) d\tau) \end{cases} \tag{5.63}$$

where $x(k)$, $v(k)$ and $u(k)$ are the position, speed and control input thrust of linear motor, respectively. $m^{\lozenge}$ is the mass of motor. $h^{\lozenge}(k)$ is the unknown time delay. f_{fric} is the frictional force, where f_s, f_c, v_s and f_v are the static friction, minimum value of coulomb friction, lubrication parameter, and coefficient of viscous friction, respectively. f_{rip} is the ripple force where $A_r^{\lozenge}$ and $\omega^{\lozenge}$ are two parameters of this force.

The model parameters of the considered linear motor in the simulation are set as follows: $m^{\lozenge}$=0.60kg, f_c=10N, f_s=20N, v_s= 0.01m/s, f_v=10$N{\cdot}\text{s}{\cdot}m^{-1}$, $A_r^{\lozenge} = 8.6$N, $\omega_0^{\lozenge} = 314$rad/s, and the sampling period is 1s. The time delay is set as:

$$h^{\lozenge}(k) = \begin{cases} 4 & 0 \le k < 300 \\ 5 & 300 \le k < 600 \\ 2 & 600 \le k < 900 \\ 3 & 900 \le k < 1200 \\ 4 & 1200 \le k < 1600 \\ 3 & 1600 \le k < 2000 \\ 1 & 2000 \le k \le 2500 \end{cases} \tag{5.64}$$

Note that the above linear motor model is only used to produce input and output data and does not participate in the design of the controller.

In the simulation, we take the speed $v(k)$ as the system output. The motor is controlled to track the following desired trajectory repeatedly and iteratively.

$$y_d^{\lozenge}(k) = 1.5\sin^2(\pi k/1000), (0 \le k \le 2500) \tag{5.65}$$

The control parameters of the proposed method (5.17)–(5.24), (5.29)–(5.30) are as follows: $r^{\lozenge} = 3.5 \times 10^{-1}$, $q^{\lozenge} = 0.9$, $\alpha^{\lozenge} = 10^3$, $b_1^{\lozenge} = 10^{-7}$, $b_2^{\lozenge} = 0.9 \times 10^{-3}$, $\eta_1^{\lozenge} = 5.5 \times 10^{-1}$, $\eta_2^{\lozenge}$=5.5×10^{-1}, $\eta_3^{\lozenge}$=5.5×10^{-1}, $\gamma^{\lozenge}$=5×10^{-2}, $l_1^{\lozenge} = 1 \times 10^{-2}$, $l_2^{\lozenge} = 2 \times 10^{-2}$, $l_3^{\lozenge} = 1 \times 10^{-2}$, $\sigma_1^{\lozenge} = 6$, $\sigma_2^{\lozenge}$=6, $\sigma_3^{\lozenge}$=6, $n_p^{\lozenge}$=2, $\hat{G}_{1,0}^{\lozenge}$=4.5×10^{-1}, $\hat{G}_{1,1}^{\lozenge} = 4.5 \times 10^{-1}$, $\hat{G}_{2,0}^{\lozenge} = 5.5 \times 10^{-1}$, $\hat{G}_{2,1}^{\lozenge} = 5.5 \times 10^{-1}$, $\hat{G}_{3,0}^{\lozenge} = 7.5 \times 10^{-1}$, $\hat{G}_{3,1}^{\lozenge} = 7.5 \times 10^{-1}$, $\aleph_{1,0}^{\lozenge} = 0.2$, $\aleph_{1,1}^{\lozenge} = 0.2$, $\aleph_{2,0}^{\lozenge} = 0.2$, $\aleph_{2,1}^{\lozenge} = 0.2$, $\bar{\aleph}_{1,0}^{\lozenge} = 0.25$, $\bar{\aleph}_{1,1}^{\lozenge} = 0.25$, $\bar{\aleph}_{2,0}^{\lozenge} = 0.25$, $\bar{\aleph}_{2,1}^{\lozenge} = 0.25$, $\bar{\bar{\aleph}}_{1,0}^{\lozenge} = 0.3$, $\bar{\bar{\aleph}}_{1,1}^{\lozenge} = 0.3$, $\bar{\bar{\aleph}}_{2,0}^{\lozenge} = 0.3$, $\bar{\bar{\aleph}}_{2,1}^{\lozenge} = 0.3$. The lower and upper bounds of the unknown time delay are set as:

$$\tau_1^{\lozenge}(k) = \begin{cases} 2 & 0 \le k < 1000 \\ 1 & 1000 \le k \le 2500 \end{cases} \tag{5.66}$$

$$\tau_2^{\lozenge}(k) = \begin{cases} 5 & 0 \le k < 1000 \\ 4 & 1000 \le k \le 2500 \end{cases} \tag{5.67}$$

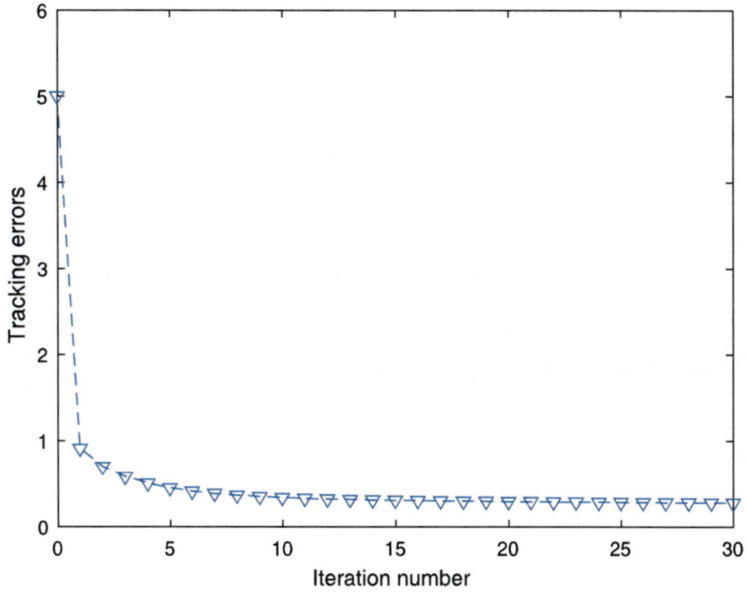

Fig. 5.1 Output tracking errors

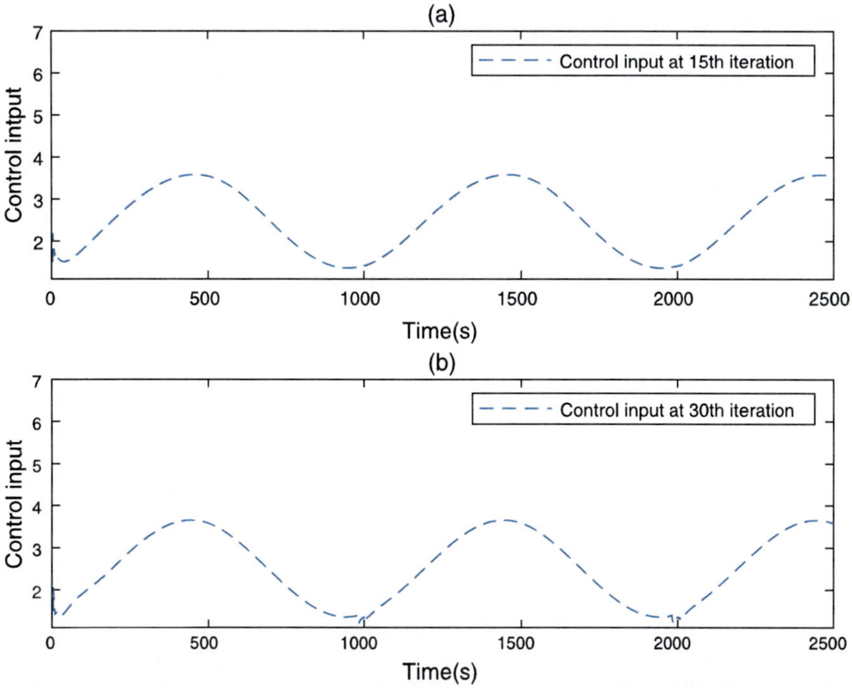

Fig. 5.2 Control input at different iterations

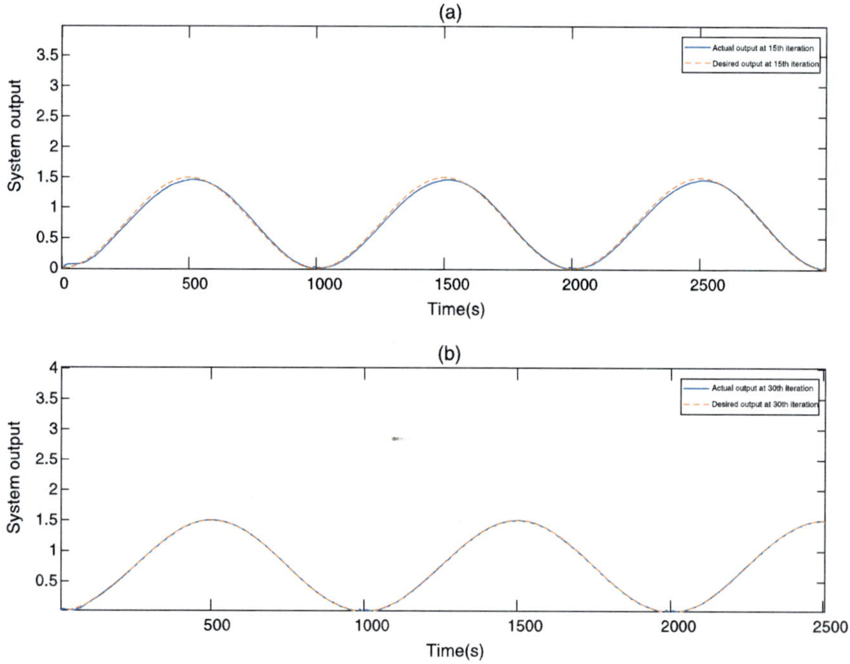

Fig. 5.3 System output at different iterations

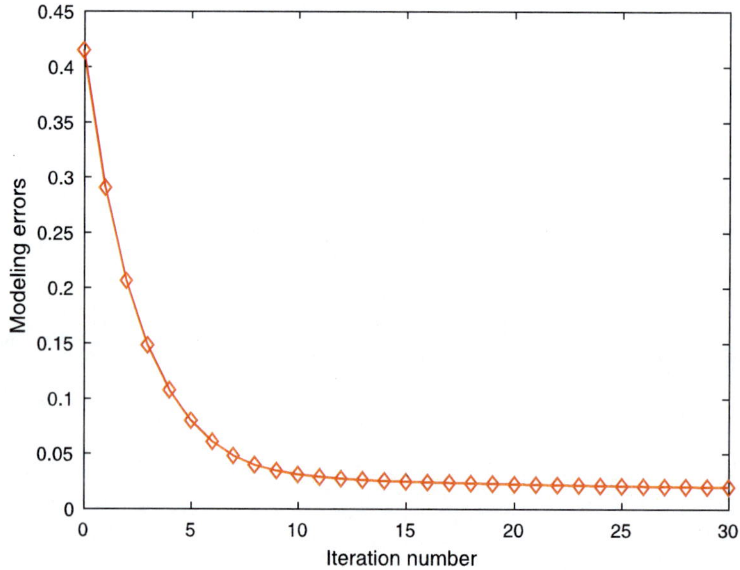

Fig. 5.4 Modeling errors

The simulation results are shown in Figs. 5.1, 5.2, 5.3 and 5.4. Figure 5.1 shows the output tracking errors defined in (5.7) along the iteration axis. Figure 5.2 shows the actual control input of the proposed time delay compensation-based predictive ILC at 15th and 30th iterations, respectively. Figure 5.3 displays the actual output speed of the linear motor at 15th and 30th iterations further. In addition, the modeling errors defined in (5.15) along the iteration axis is demonstrated in Fig. 5.4. As shown in Figs. 5.1, 5.2, 5.3 and 5.4, although the system suffers from unknown time delay, both reduced tracking control errors and modeling errors are still obtained by utilizing the proposed time delay compensation-based predictive ILC.

5.5 Conclusion

In this chapter, a time delay compensation-based PILC for tracking control of unknown nonaffine nonlinear systems with unknown time-varying time delay is proposed. Even though the considered unknown system is under unknown time-varying time delay, asymptotic and pointwise convergence property of the proposed PILC is guaranteed through theoretical analysis. Both the ability to handle unknown time-varying time delay and tracking control performance of the proposed method are verified by simulation.

References

Bentout S, Djilali S, Touaoula TM et al (2022) Bifurcation analysis for a double age dependence epidemic model with two delays. Nonlinear Dynam 108(2):1821–1835

Browne F, Rees B, Chiu GTC et al (2020) Iterative learning control with time-delay compensation: an application to twin-roll strip casting. IEEE Trans Cont Syst Technol 29(1):140–149

Feng L, Chai Y, Xu S et al (2017) Observer based fault estimators using iterative learning scheme for nonlinear time delay systems with intermittent faults. Int J Rob Nonlinear Cont 27(17):3412–3432

Gerschgorin S. Uber die abgrenzung der eigenwerte einer matrix [On the delimitation of the eigenvalues of a matrix]. Bulletin de l Academie des Sciences de l URSS. Classe des sciences mathematiques et na 7(6):749–754 (1931)

Hou Z, Jin S (2011) Data-driven model-free adaptive control for a class of MIMO nonlinear discrete-time systems. IEEE Trans Neural Netw 22(12):2173–2188

Hou ZS, Jin ST (2013) Model free adaptive control: theory and applications. CRC Press, Florida

Kali Y, Ayala M, Rodas J et al (2020) Time delay estimation based discrete-time super-twisting current control for a six-phase induction motor. IEEE Trans Power Electron 35(11):12570–12580

Lee TH, Tan KK, Lim SY, Dou HF (2000) Iterative learning control of permanent magnet linear motor with relay automatic tuning. Mechatronics 10(1–2):169–190

Liu S, Hou Z, Tian T et al (2020) Path tracking control of a self driving wheel excavator via an enhanced data driven model-free adaptive control approach. IET Cont Theo Appl 14(2):220–232

Qiang H, Lin Z, Zou X et al (2020) Synchronizing non-identical time-varying delayed neural network systems via iterative learning control. Neurocomputing 411:406–415

Ravelo B, Wan F, Yuan Z et al (2022) Pre-detection sensing with multi-stage low-pass type negative group delay circuit. IEEE Sens J 22(12):11835–11846

Tao H, Paszke W, Rogers E et al (2019) Finite frequency range iterative learning fault-tolerant control for discrete time-delay uncertain systems with actuator faults. ISA Trans 95:152–163
Wan Y, Cao J, Huang W et al (2018) Perimeter control of multiregion urban traffic networks with time-varying delays. IEEE Trans Syst Man Cybernet Syst 50(8):2795–2803
Wang Y, Zhou F, Yin L et al (2021) Iterative learning control for fractional order linear systems with time delay based on frequency analysis. Int J Cont Autom Syst 19(4):1588–1596
Wei J, Zhang Y, Sun M et al (2017) Adaptive iterative learning control of a class of nonlinear time-delay systems with unknown backlash-like hysteresis input and control direction. ISA Trans 70:79–92

Chapter 6
Predictive Iterative Learning Control for Systems with Full Available States

6.1 Introduction

By establishing the mapping relationship between output and input data, a series of PILC methods for unknown nonaffine nonlinear systems are proposed in Chaps. 2–5. In fact, in addition to the output data, system states can reflect the internal characteristics of a practical system more directly. In this chapter, we will exploit the relationship between state and input data, and full state information will be integrated into the controller design.

To exploit the characteristic of system states, the research effort on observers in time domain control methods is made (Zou et al. 2016; Yang et al. 2017; Zhang and Han 2018; Bobtsov et al. 2018), but most of these observers are model-based ones. Up to now, little work on observers in the iterative learning domain rather than the time domain is presented. Motivated by this consideration, in this chapter, a full-state observer-based robust constrained PILC method for unknown multiple-input multiple-output (MIMO) nonaffine nonlinear systems is designed, and in the following Chap. 7, immeasurable system states will be considered further.

The main contributions of this chapter are as follows.

- A new iterative learning observer is proposed to observe the unknown and nonlinear system states, and the convergence property is guaranteed without using any model information.
- By fully utilizing the observed system states, a full-state observer-based robust constrained PILC method is proposed.

This chapter is organized as follows. Section 6.2 presents the problem formulation. Section 6.3 shows the full-state observer-based predictive ILC method. Simulation results are presented in Sect. 6.4 to show the effectiveness of the proposed method. Some conclusions are given in Sect. 6.5.

© The Author(s), under exclusive license to Springer Nature Singapore Pte Ltd. 2023
Q. Yu et al., *Predictive Learning Control for Unknown Nonaffine Nonlinear Systems*,
Intelligent Control and Learning Systems 8,
https://doi.org/10.1007/978-981-19-8857-8_6

6.2 Problem Formulation

Consider the following general MIMO nonaffine nonlinear system (Chen et al. 2016; Kim et al. 2010; Kim and Rew 2013) that is widely used in various practical systems (Yang et al. 2014; Li et al. 2014; Yang et al. 2013):

$$\begin{cases} \mathbf{x}_l(k+1) = \mathbf{f}(\mathbf{x}_l(k), \mathbf{x}_l(k-1), \ldots, \mathbf{x}_l(k-n_x^{\pitchfork}), \\ \qquad\qquad \mathbf{u}_l(k), \mathbf{u}_l(k-1), \ldots, \mathbf{u}_l(k-n_u^{\pitchfork})) + \mathbf{d}_l(k) \\ \mathbf{y}_l(k+1) = \mathbf{C}\mathbf{x}_l(k+1) \end{cases} \tag{6.1}$$

where $l \in N^+$ is the iteration or repetitive operation index. $k \in \{0, 1, \ldots, K^{\pitchfork}\}$ is the operating time index. $n_x^{\pitchfork}, n_u^{\pitchfork}$ are unknown system orders. $\mathbf{f}(\ldots) \in \Re^m$ is an unknown nonlinear vector-valued function. $\mathbf{x}_l(k) \in \Re^m$, $\mathbf{u}_l(k) \in \Re^w$, $\mathbf{y}_l(k) \in \Re^m$ are the system state, input, and measurement output, respectively. $\mathbf{d}_l(k) \in \Re^m$ is unknown but bounded external disturbance. The output matrix $\mathbf{C} = \mathbf{I}_m$ which means full system states are available, and in the following Chap. 7, unavailable system states will be considered.

System (6.1) is under the following system constraints.

1. Input constraint:

$$\mathbf{u}_{\min} \leq \mathbf{u}_l(k) \leq \mathbf{u}_{\max} \tag{6.2}$$

2. State constraint:

$$\mathbf{y}_{\min} \leq \mathbf{y}_l(k+1) \leq \mathbf{y}_{\max} \tag{6.3}$$

where $\mathbf{u}_{\min}$, $\mathbf{u}_{\max}$ and $\mathbf{y}_{\min}$, $\mathbf{y}_{\max}$ are the constraints on the input and output vectors respectively.

System (6.1) satisfies the following assumptions.

Assumption 6.1 The partial derivatives of $\mathbf{f}(\cdots)$ with reference to $\mathbf{u}_l(k)$ are continuous.

Assumption 6.2 If $\mathbf{d}_l(k) = \mathbf{0}$ holds, the following generalized *Lipschitz* condition is satisfied

$$||\Delta\mathbf{x}_l(k+1)|| \leq \zeta^{\pitchfork}||\Delta\mathbf{u}_l(k)|| \tag{6.4}$$

where Δ is a difference operator between two consecutive iterations. $\Delta\mathbf{u}_l(k) = \mathbf{u}_l(k) - \mathbf{u}_{l-1}(k)$, $\Delta\mathbf{x}_l(k+1) = \mathbf{x}_l(k+1) - \mathbf{x}_{l-1}(k+1)$. $\zeta^{\pitchfork}$ is a positive constant.

Assumption 6.3 The initial values $\mathbf{x}_l(0)$ and $\mathbf{y}_l(0)$ can be randomly varying for $\forall l = 0, 1, 2 \ldots$.

Assumption 6.4 There exists a feasible input vector $\mathbf{u}_d(k)$ such that $\mathbf{e}_{y,l}^{\pitchfork}(\mathbf{u}_d(k)) = \mathbf{0}$, where $\mathbf{e}_{y,l}^{\pitchfork}(k)$ is the tracking error trajectory vector defined as $\mathbf{e}_l^{\pitchfork}(k) = \mathbf{y}_d^{\pitchfork}(k) - \mathbf{y}_l(k)$. $\mathbf{y}_d^{\pitchfork}(k)$ is the desired output trajectory.

Based on Assumptions 6.1–6.3, analogous to Lemma 4.1 in Chap. 4, an equivalent dynamic linearization model that serves for controller design is constructed in the following lemma.

Lemma 6.1 *The following dynamic linearization model (6.5) can equivalently describe the considered system (6.1) under Assumptions 6.1–6.3:*

$$\begin{cases} \Delta\mathbf{x}_l(k+1) = \boldsymbol{\Upsilon}_l^{\pitchfork}(k)\Delta\mathbf{u}_l(k) + \Delta\mathbf{d}_l(k) \\ \mathbf{y}_l(k+1) = \mathbf{C}\mathbf{x}_l(k+1) \end{cases} \tag{6.5}$$

for $\forall k \in \{0, 1, \ldots, K^{\pitchfork}\}, l=0, 1, 2\ldots$ *and* $||\Delta\mathbf{u}_l(k)|| \neq 0$. $\Delta\mathbf{d}_l(k) = \mathbf{d}_l(k) - \mathbf{d}_{l-1}(k)$. $\boldsymbol{\Upsilon}_l^{\pitchfork}(k) = \begin{bmatrix} \upsilon_{11,l}^{\pitchfork}(k) & \upsilon_{12,l}^{\pitchfork}(k) & \cdots & \upsilon_{1w,l}^{\pitchfork}(k) \\ \upsilon_{21,l}^{\pitchfork}(k) & \upsilon_{22,l}^{\pitchfork}(k) & \cdots & \upsilon_{2w,l}^{\pitchfork}(k) \\ \vdots & \vdots & \vdots & \vdots \\ \upsilon_{m1,l}^{\pitchfork}(k) & \upsilon_{m2,l}^{\pitchfork}(k) & \cdots & \upsilon_{mw,l}^{\pitchfork}(k) \end{bmatrix}_{m\times w}$ *is a newly introduced pseudo-Jacobian matrix. It is unknown but bounded.*

Proof Differencing the system (6.1) with respect to the iteration index l leads to

$$\begin{aligned} \Delta\mathbf{x}_l(k+1) &= \mathbf{f}(\mathbf{x}_l(k), \ldots, \mathbf{x}_l(k-n_x^{\pitchfork}), \mathbf{u}_l(k), \ldots, \mathbf{u}_l(k-n_u^{\pitchfork})) \\ &\quad - \mathbf{f}(\mathbf{x}_{l-1}(k), \ldots, \mathbf{x}_{l-1}(k-n_x^{\pitchfork}), \mathbf{u}_{l-1}(k), \ldots, \mathbf{u}_{l-1}(k-n_u^{\pitchfork})) + \mathbf{d}_l(k) - \mathbf{d}_{l-1}(k) \\ &= \mathbf{f}(\mathbf{x}_l(k), \ldots, \mathbf{x}_l(k-n_x^{\pitchfork}), \mathbf{u}_l(k), \ldots, \mathbf{u}_l(k-n_u^{\pitchfork})) \\ &\quad - \mathbf{f}(\mathbf{x}_l(k), \ldots, \mathbf{x}_l(k-n_x^{\pitchfork}), \mathbf{u}_{l-1}(k), \mathbf{u}_l(k-1), \ldots, \mathbf{u}_l(k-n_u^{\pitchfork})) \\ &\quad + \mathbf{f}(\mathbf{x}_l(k), \ldots, \mathbf{x}_l(k-n_x^{\pitchfork}), \mathbf{u}_{l-1}(k), \mathbf{u}_l(k-1), \ldots, \mathbf{u}_l(k-n_u^{\pitchfork})) \\ &\quad - \mathbf{f}(\mathbf{x}_{l-1}(k), \ldots, \mathbf{x}_{l-1}(k-n_x^{\pitchfork}), \mathbf{u}_{l-1}(k), \ldots, \mathbf{u}_{l-1}(k-n_u^{\pitchfork})) + \Delta\mathbf{d}_l(k) \end{aligned} \tag{6.6}$$

According to Assumption 6.1 and differential mean value theorem, (6.6) can be rewritten as

$$\Delta\mathbf{x}_l(k+1) = \frac{\partial\mathbf{f}^*}{\partial\mathbf{u}_l(k)}\Delta\mathbf{u}_l(k) + \varsigma_l^{\pitchfork}(k) + \Delta\mathbf{d}_l(k) \tag{6.7}$$

where $\frac{\partial\mathbf{f}^*}{\partial\mathbf{u}_l(k)} = \begin{bmatrix} \frac{\partial f_1^*}{\partial u_{1,l}(k)} & \frac{\partial f_1^*}{\partial u_{2,l}(k)} & \cdots & \frac{\partial f_1^*}{\partial u_{w,l}(k)} \\ \frac{\partial f_2^*}{\partial u_{1,l}(k)} & \frac{\partial f_2^*}{\partial u_{2,l}(k)} & \cdots & \frac{\partial f_2^*}{\partial u_{w,l}(k)} \\ \vdots & \vdots & \vdots & \vdots \\ \frac{\partial f_m^*}{\partial u_{1,l}(k)} & \frac{\partial f_m^*}{\partial u_{2,l}(k)} & \cdots & \frac{\partial f_m^*}{\partial u_{w,l}(k)} \end{bmatrix}$. $\partial f_i^*/\partial u_{j,l}(k)$ $(i = 1, \ldots m$ and $j = 1, \ldots w)$ is the partial derivative of f_i with respect to u_j at a certain point in $[u_{j,l-1}(k),\ u_{j,l}(k)]$. $\varsigma_l^{\pitchfork}(k) \triangleq \mathbf{f}(\mathbf{x}_l(k), \ldots, \mathbf{x}_l(k-n_x^{\pitchfork}), \mathbf{u}_{l-1}(k), \mathbf{u}_l(k-1), \ldots, \mathbf{u}_l(k-n_u^{\pitchfork})) - \mathbf{f}(\mathbf{x}_{l-1}(k), \ldots, \mathbf{x}_{l-1}(k-n_x^{\pitchfork}), \mathbf{u}_{l-1}(k), \ldots, \mathbf{u}_{l-1}(k-n_u^{\pitchfork}))$.

Consider the following equation in relation to $\varsigma_l^{\phi}(k)$

$$\varsigma_l^{\phi}(k) = \kappa_l^{\phi}(k)\Delta\mathbf{u}_l(k) \tag{6.8}$$

where $\kappa_l^{\phi}(k) \in \Re^{m\times w}$ is a numerical matrix. (6.8) must have at least one solution $\kappa_l^{\phi*}(k)$ for $\forall k \in \{0, 1, \ldots, K^{\phi}\}, l = 0, 1, 2\ldots$ and $||\Delta\mathbf{u}_l(k)|| \neq 0$. If $||\Delta\mathbf{u}_l(k)|| \neq 0$ cannot be satisfied, please refer to Remark 2.2 of Chap. 2.

Then (6.7) can be rewritten as

$$\Delta\mathbf{x}_l(k+1) = \boldsymbol{\Upsilon}_l^{\phi}(k)\Delta\mathbf{u}_l(k) + \Delta\mathbf{d}_l(k) \tag{6.9}$$

where $\boldsymbol{\Upsilon}_l^{\phi}(k) = \partial\mathbf{f}^*/\partial\mathbf{u}_l(k) + \kappa_l^{\phi*}(k)$. According to Assumption 6.2, it is derived that $\boldsymbol{\Upsilon}_l^{\phi}(k)$ is bounded.

Rewrite system (6.5) as:

$$\begin{cases} \mathbf{x}_l(k+1) = \mathbf{x}_{l-1}(k+1) + \boldsymbol{\Upsilon}_l^{\phi}(k)\Delta\mathbf{u}_l(k) + \Delta\mathbf{d}_l(k) \\ \mathbf{y}_l(k+1) = \mathbf{C}\mathbf{x}_l(k+1) \end{cases} \tag{6.10}$$

The control objectives for (6.10) are:

1. For $\mathbf{C} = \mathbf{I}_m$, designing a full-state observer to observe the unknown $\boldsymbol{\Upsilon}_l^{\phi}(k)$ and $\Delta\mathbf{d}_l(k)$ in (6.10) so as to fully capture the state information of the system.
2. Under system constraints (6.2)–(6.3), designing a full-state observer-based predictive ILC to make $\mathbf{y}_l(k+1)$ track the desired output $\mathbf{y}_d^{\phi}(k+1)$.

6.3 Full-State Observer-Based Predictive ILC Design

Since the parameters $\boldsymbol{\Upsilon}_l^{\phi}(k)$ and $\Delta\mathbf{d}_l(k)$ in the considered system (6.10) are unknown, in this chapter, both $\boldsymbol{\Upsilon}_l^{\phi}(k)$ and $\Delta\mathbf{d}_l(k)$ are estimated by a constructed full-state observer in Sect. 6.3.1 and be further predicted in Sect. 6.3.2. In Sect. 6.3.3, the corresponding predictive ILC will be designed.

6.3.1 *Full-State Observer Design*

Given two learning gain matrices $\boldsymbol{\Lambda}^{\phi} \in \Re^{w\times w}$, $\boldsymbol{\Psi}^{\phi} \in \Re^{m\times m}$ and a vector-valued function $\mathbf{H}_l^{\phi}(k) \in \Re^{1\times w}$, the following full-state observer is designed to estimate the unknown $\boldsymbol{\Upsilon}_l^{\phi}(k)$ and $\Delta\mathbf{d}_l(k)$ in system (6.10):

$$
\begin{cases}
\hat{\boldsymbol{\Upsilon}}_l^{\pitchfork}(k) = \mathbf{x}_{l-1}(k+1)\mathbf{H}_l^{\pitchfork}(k)\boldsymbol{\Lambda}^{\pitchfork} - \boldsymbol{\Pi}_l^{\pitchfork}(k) \\
\boldsymbol{\Pi}_{l+1}^{\pitchfork}(k) = \alpha^{\pitchfork}(k)\Big[\boldsymbol{\Pi}_l^{\pitchfork}(k) + \hat{\boldsymbol{\Upsilon}}_l^{\pitchfork}(k) - \hat{\boldsymbol{\Upsilon}}_l^{\pitchfork}(k)\Delta\mathbf{u}_l(k)\mathbf{H}_l^{\pitchfork}(k)\boldsymbol{\Lambda}^{\pitchfork} \\
\qquad\qquad + \Delta\hat{\mathbf{d}}_l(k)\mathbf{H}_l^{\pitchfork}(k)\boldsymbol{\Lambda}^{\pitchfork}\Big] \\
\Delta\hat{\mathbf{d}}_l(k) = \boldsymbol{\Psi}^{\pitchfork}\mathbf{x}_{l-1}(k+1) - \boldsymbol{\Xi}_l^{\pitchfork}(k) \\
\boldsymbol{\Xi}_{l+1}^{\pitchfork}(k) = \boldsymbol{\Xi}_l^{\pitchfork}(k) + \boldsymbol{\Psi}^{\pitchfork}\left[\hat{\boldsymbol{\Upsilon}}_l^{\pitchfork}(k)\Delta\mathbf{u}_l(k) + \Delta\hat{\mathbf{d}}_l(k)\right] + \Delta\hat{\mathbf{d}}_l(k)
\end{cases}
\tag{6.11}
$$

where $\hat{\boldsymbol{\Upsilon}}_l^{\pitchfork}(k)$ and $\Delta\hat{\mathbf{d}}_l(k)$ are the estimations of $\boldsymbol{\Upsilon}_l^{\pitchfork}(k)$ and $\Delta\mathbf{d}_l(k)$ in (6.10). $\boldsymbol{\Pi}_l^{\pitchfork}(k) \in \Re^{m\times w}$ and $\boldsymbol{\Xi}_l^{\pitchfork}(k) \in \Re^{m\times 1}$ are the state variables for observing $\boldsymbol{\Upsilon}_l^{\pitchfork}(k)$ and $\Delta\mathbf{d}_l(k)$ respectively. The given vector-valued function $\mathbf{H}_l^{\pitchfork}(k)$ satisfies $\mathbf{H}_l^{\pitchfork}(k) = \alpha^{\pitchfork}(k)\mathbf{H}_{l-1}^{\pitchfork}(k)$, where $\alpha^{\pitchfork}(k) \in \Re^1$ is an appropriate coefficient.

Denote the following parameter observer error:

$$
\mathbf{e}_{\tilde{\boldsymbol{\Upsilon}}_l^{\pitchfork}}^{\pitchfork}(k) \triangleq \boldsymbol{\Upsilon}_l^{\pitchfork}(k) - \hat{\boldsymbol{\Upsilon}}_l^{\pitchfork}(k)
\tag{6.12}
$$

and disturbance observer error:

$$
\mathbf{e}_{\Delta\tilde{\mathbf{d}},l}^{\pitchfork}(k) \triangleq \Delta\mathbf{d}_l(k) - \Delta\hat{\mathbf{d}}_l(k)
\tag{6.13}
$$

Then, the convergence property of the above designed full-state observer (6.11) is presented in the following theorem.

Theorem 6.1 *For the equivalent system (6.10) of the original considered system (6.1), if the learning gain matrices* $\boldsymbol{\Lambda}^{\pitchfork}$, $\boldsymbol{\Psi}^{\pitchfork}$ *and the vector-valued function* $\mathbf{H}_{l+1}^{\pitchfork}(k)$ *satisfy:*

$$
\varpi^{\pitchfork}\Big[\left\|\boldsymbol{\Psi}^{\pitchfork}\right\| + \left\|\mathbf{H}_{l+1}^{\pitchfork}(k)\right\|\left\|\boldsymbol{\Lambda}^{\pitchfork}\right\|\Big] \le \bar{\rho}^{\pitchfork} < 1
\tag{6.14}
$$

then the proposed full-state observer (6.11) can guarantee that

$$
\lim_{l\to\infty}\left\|\mathbf{e}_{\tilde{\boldsymbol{\Upsilon}}_{l+1}^{\pitchfork}}^{\pitchfork}(k)\right\| \le \frac{\bar{\zeta}^{\pitchfork}}{1-\bar{\rho}^{\pitchfork}}
\tag{6.15}
$$

and

$$
\lim_{l\to\infty}\left\|\mathbf{e}_{\Delta\tilde{\mathbf{d}},l+1}^{\pitchfork}(k)\right\| \le \frac{\bar{\zeta}^{\pitchfork}}{1-\bar{\rho}^{\pitchfork}}
\tag{6.16}
$$

where $0 \le \bar{\rho}^{\pitchfork} < 1$ *is a positive constant.* $\varpi^{\pitchfork}$ *is an upper bound related to the control input and is defined in the following (6.23).* $\bar{\zeta}^{\pitchfork}$ *is a positive constant defined in the subsequent (6.26).*

Proof Based on (6.11) and (6.13), one can obtain

$$
\begin{aligned}
\mathbf{e}^{\pitchfork}_{\Delta\tilde{\mathbf{d}},l+1}(k) &= \Delta\mathbf{d}_{l+1}(k) - \boldsymbol{\Psi}^{\pitchfork}\mathbf{x}_l(k+1) + \boldsymbol{\Xi}_l^{\pitchfork}(k) \\
&\quad + \boldsymbol{\Psi}^{\pitchfork}\left[\hat{\boldsymbol{\Upsilon}}_l^{\pitchfork}(k)\Delta\mathbf{u}_l(k) + \Delta\hat{\mathbf{d}}_l(k)\right] + \Delta\hat{\mathbf{d}}_l(k)
\end{aligned} \tag{6.17}
$$

According to (6.10), (6.17) can be rewritten as

$$
\begin{aligned}
\mathbf{e}^{\pitchfork}_{\Delta\tilde{\mathbf{d}},l+1}(k) &= \Delta\mathbf{d}_{l+1}(k) - \boldsymbol{\Psi}^{\pitchfork}\left[\mathbf{x}_{l-1}(k+1) + \boldsymbol{\Upsilon}_l^{\pitchfork}(k)\Delta\mathbf{u}_l(k) + \Delta\mathbf{d}_l(k)\right] \\
&\quad + \boldsymbol{\Xi}_l^{\pitchfork}(k) + \boldsymbol{\Psi}^{\pitchfork}\left[\hat{\boldsymbol{\Upsilon}}_l^{\pitchfork}(k)\Delta\mathbf{u}_l(k) + \Delta\hat{\mathbf{d}}_l(k)\right] + \Delta\hat{\mathbf{d}}_l(k)
\end{aligned} \tag{6.18}
$$

Based on the fact that $-\boldsymbol{\Psi}^{\pitchfork}\mathbf{x}_{l-1}(k+1) + \boldsymbol{\Xi}_l^{\pitchfork}(k) = -\Delta\hat{\mathbf{d}}_l(k)$ in the observer (6.11), (6.18) can be rewritten as

$$
\mathbf{e}^{\pitchfork}_{\Delta\tilde{\mathbf{d}},l+1}(k) = \Delta\mathbf{d}_{l+1}(k) - \boldsymbol{\Psi}^{\pitchfork}\mathbf{e}^{\pitchfork}_{\tilde{\boldsymbol{\Upsilon}}_l^{\pitchfork}}(k)\Delta\mathbf{u}_l(k) - \boldsymbol{\Psi}^{\pitchfork}\mathbf{e}^{\pitchfork}_{\Delta\tilde{\mathbf{d}},l}(k) \tag{6.19}
$$

On the other hand, based on (6.11) and (6.12), one can get

$$
\begin{aligned}
\mathbf{e}^{\pitchfork}_{\tilde{\boldsymbol{\Upsilon}}_{l+1}^{\pitchfork}}(k) &= \boldsymbol{\Upsilon}_{l+1}^{\pitchfork}(k) - [\mathbf{x}_l(k+1)\mathbf{H}_{l+1}^{\pitchfork}(k)\boldsymbol{\Lambda}^{\pitchfork} - \boldsymbol{\Pi}_{l+1}^{\pitchfork}(k)] \\
&= \boldsymbol{\Upsilon}_{l+1}^{\pitchfork}(k) - \mathbf{x}_l(k+1)\mathbf{H}_{l+1}^{\pitchfork}(k)\boldsymbol{\Lambda}^{\pitchfork} + \alpha^{\pitchfork}(k)\boldsymbol{\Pi}_l^{\pitchfork}(k) + \alpha^{\pitchfork}(k)\hat{\boldsymbol{\Upsilon}}_l^{\pitchfork}(k) \\
&\quad - \alpha^{\pitchfork}(k)\hat{\boldsymbol{\Upsilon}}_l^{\pitchfork}(k)\Delta\mathbf{u}_l(k)\mathbf{H}_l^{\pitchfork}(k)\boldsymbol{\Lambda}^{\pitchfork} + \alpha^{\pitchfork}(k)\Delta\hat{\mathbf{d}}_l(k)\mathbf{H}_l^{\pitchfork}(k)\boldsymbol{\Lambda}^{\pitchfork}
\end{aligned} \tag{6.20}
$$

By means of (6.10) and noting that $\mathbf{H}_l^{\pitchfork}(k) = \alpha^{\pitchfork}(k)\mathbf{H}_{l-1}^{\pitchfork}(k)$, (6.20) leads to

$$
\begin{aligned}
\mathbf{e}^{\pitchfork}_{\tilde{\boldsymbol{\Upsilon}}_{l+1}^{\pitchfork}}(k) &= \boldsymbol{\Upsilon}_{l+1}^{\pitchfork}(k) - \alpha^{\pitchfork}(k)\mathbf{x}_{l-1}(k+1)\mathbf{H}_l^{\pitchfork}(k)\boldsymbol{\Lambda}^{\pitchfork} + \alpha^{\pitchfork}(k)\boldsymbol{\Pi}_l^{\pitchfork}(k) \\
&\quad + \alpha^{\pitchfork}(k)\hat{\boldsymbol{\Upsilon}}_l^{\pitchfork}(k) - \alpha^{\pitchfork}(k)\boldsymbol{\Upsilon}_l^{\pitchfork}(k)\Delta\mathbf{u}_l(k)\mathbf{H}_l^{\pitchfork}(k)\boldsymbol{\Lambda}^{\pitchfork} - \alpha^{\pitchfork}(k)\Delta\mathbf{d}_l(k)\mathbf{H}_l^{\pitchfork}(k)\boldsymbol{\Lambda}^{\pitchfork} \\
&\quad + \alpha^{\pitchfork}(k)\Delta\hat{\mathbf{d}}_l(k)\mathbf{H}_l^{\pitchfork}(k)\boldsymbol{\Lambda}^{\pitchfork} - \alpha^{\pitchfork}(k)\hat{\boldsymbol{\Upsilon}}_l^{\pitchfork}(k)\Delta\mathbf{u}_l(k)\mathbf{H}_l^{\pitchfork}(k)\boldsymbol{\Lambda}^{\pitchfork}
\end{aligned} \tag{6.21}
$$

Based on the fact that $-\mathbf{x}_{l-1}(k+1)\mathbf{H}_l^{\pitchfork}(k)\boldsymbol{\Lambda}^{\pitchfork} + \boldsymbol{\Pi}_l^{\pitchfork}(k) = -\hat{\boldsymbol{\Upsilon}}_l^{\pitchfork}(k)$ in the observer (6.11), (6.21) can be rewritten as

$$
\begin{aligned}
\mathbf{e}^{\pitchfork}_{\tilde{\boldsymbol{\Upsilon}}_{l+1}^{\pitchfork}}(k) &= \boldsymbol{\Upsilon}_{l+1}^{\pitchfork}(k) - 2\alpha^{\pitchfork}(k)\boldsymbol{\Upsilon}_l^{\pitchfork}(k)\Delta\mathbf{u}_l(k)\mathbf{H}_l^{\pitchfork}(k)\boldsymbol{\Lambda}^{\pitchfork} \\
&\quad + 2\alpha^{\pitchfork}(k)\boldsymbol{\Upsilon}_l^{\pitchfork}(k)\Delta\mathbf{u}_l(k)\mathbf{H}_l^{\pitchfork}(k)\boldsymbol{\Lambda}^{\pitchfork} - \alpha^{\pitchfork}(k)\boldsymbol{\Upsilon}_l^{\pitchfork}(k)\Delta\mathbf{u}_l(k)\mathbf{H}_l^{\pitchfork}(k)\boldsymbol{\Lambda}^{\pitchfork} \\
&\quad - \alpha^{\pitchfork}(k)\hat{\boldsymbol{\Upsilon}}_l^{\pitchfork}(k)\Delta\mathbf{u}_l(k)\mathbf{H}_l^{\pitchfork}(k)\boldsymbol{\Lambda}^{\pitchfork} - \alpha^{\pitchfork}(k)\mathbf{e}^{\pitchfork}_{\Delta\tilde{\mathbf{d}},l}(k)\mathbf{H}_l^{\pitchfork}(k)\boldsymbol{\Lambda}^{\pitchfork} \\
&= \mathbf{e}^{\pitchfork}_{\tilde{\boldsymbol{\Upsilon}}_l^{\pitchfork}}(k)\Delta\mathbf{u}_l(k)\mathbf{H}_{l+1}^{\pitchfork}(k)\boldsymbol{\Lambda}^{\pitchfork} + \boldsymbol{\Upsilon}_{l+1}^{\pitchfork}(k) - 2\boldsymbol{\Upsilon}_l^{\pitchfork}(k)\Delta\mathbf{u}_l(k)\mathbf{H}_{l+1}^{\pitchfork}(k)\boldsymbol{\Lambda}^{\pitchfork} \\
&\quad - \mathbf{e}^{\pitchfork}_{\Delta\tilde{\mathbf{d}},l}(k)\mathbf{H}_{l+1}^{\pitchfork}(k)\boldsymbol{\Lambda}^{\pitchfork}
\end{aligned} \tag{6.22}
$$

According to (6.2), we can get

$$||\Delta \mathbf{u}_l(k)|| \leq 2||\mathbf{u}_{\max}|| \triangleq \varpi^{\pitchfork} \tag{6.23}$$

Then taking norm on both sides of (6.19) and (6.22), one has

$$\begin{aligned} &\left\|\mathbf{e}^{\pitchfork}_{\Delta\tilde{\mathbf{d}},l+1}(k)\right\| + \left\|\mathbf{e}^{\pitchfork}_{\tilde{\boldsymbol{\Upsilon}}^{\pitchfork}_{l+1}}(k)\right\| \leq \varpi^{\pitchfork}\Big[\|\boldsymbol{\Psi}^{\pitchfork}\| + \|\mathbf{H}^{\pitchfork}_{l+1}(k)\|\ \|\boldsymbol{\Lambda}^{\pitchfork}\|\Big]\left\|\mathbf{e}^{\pitchfork}_{\Delta\tilde{\mathbf{d}},l}(k)\right\| \\ &+\varpi^{\pitchfork}\Big[\|\boldsymbol{\Psi}^{\pitchfork}\| + \|\mathbf{H}^{\pitchfork}_{l+1}(k)\|\ \|\boldsymbol{\Lambda}^{\pitchfork}\|\Big]\left\|\mathbf{e}^{\pitchfork}_{\tilde{\boldsymbol{\Upsilon}}^{\pitchfork}_{l}}(k)\right\| + \|\Delta\mathbf{d}_{l+1}(k)\| \\ &+ \left\|\boldsymbol{\Upsilon}^{\pitchfork}_{l+1}(k) - 2\boldsymbol{\Upsilon}^{\pitchfork}_{l}(k)\Delta\mathbf{u}_l(k)\mathbf{H}^{\pitchfork}_{l+1}(k)\boldsymbol{\Lambda}^{\pitchfork}\right\| \end{aligned} \tag{6.24}$$

According to (6.14) in Theorem 6.1, (6.24) leads to

$$\begin{aligned} \left\|\mathbf{e}^{\pitchfork}_{\Delta\tilde{\mathbf{d}},l+1}(k)\right\| + \left\|\mathbf{e}^{\pitchfork}_{\tilde{\boldsymbol{\Upsilon}}^{\pitchfork}_{l+1}}(k)\right\| &\leq \bar{\rho}^{\pitchfork}\left(\left\|\mathbf{e}^{\pitchfork}_{\Delta\tilde{\mathbf{d}},l}(k)\right\| + \left\|\mathbf{e}^{\pitchfork}_{\tilde{\boldsymbol{\Upsilon}}^{\pitchfork}_{l}}(k)\right\|\right) \\ &+ \left\|\boldsymbol{\Upsilon}^{\pitchfork}_{l+1}(k) - 2\boldsymbol{\Upsilon}^{\pitchfork}_{l}(k)\Delta\mathbf{u}_l(k)\mathbf{H}^{\pitchfork}_{l+1}(k)\boldsymbol{\Lambda}^{\pitchfork}\right\| + \|\Delta\mathbf{d}_{l+1}(k)\| \end{aligned} \tag{6.25}$$

Based on Assumption 6.4, Lemma 6.1 and (6.23), $\Delta\mathbf{d}_l(k)$, $\boldsymbol{\Upsilon}^{\pitchfork}_l(k)$, $\Delta\mathbf{u}_l(k)$ are all bounded. Therefore, a positive constant $\bar{\zeta}^{\pitchfork}$ exists that satisfies

$$\left\|\boldsymbol{\Upsilon}^{\pitchfork}_{l+1}(k) - 2\boldsymbol{\Upsilon}^{\pitchfork}_{l}(k)\Delta\mathbf{u}_l(k)\mathbf{H}^{\pitchfork}_{l+1}(k)\boldsymbol{\Lambda}^{\pitchfork}\right\| + \|\Delta\mathbf{d}_{l+1}(k)\| \leq \bar{\zeta}^{\pitchfork} \tag{6.26}$$

Then, (6.25) leads to

$$\lim_{l\to\infty}\left(\left\|\mathbf{e}^{\pitchfork}_{\Delta\tilde{\mathbf{d}},l+1}(k)\right\| + \left\|\mathbf{e}^{\pitchfork}_{\tilde{\boldsymbol{\Upsilon}}^{\pitchfork}_{l+1}}(k)\right\|\right) \leq \frac{\bar{\zeta}^{\pitchfork}}{1-\bar{\rho}^{\pitchfork}} \tag{6.27}$$

□

Remark 6.1 The introduced vector-valued function $\mathbf{H}^{\pitchfork}_l(k)$ can increase the flexibility in parameter adjustment and facilitate the convergence of observer errors. For example, according to (6.26), if we select $\mathbf{H}^{\pitchfork}_{l+1}(k)\boldsymbol{\Lambda}^{\pitchfork} = \frac{1}{2}\Delta\mathbf{u}_l(k)^{+}$, the term on the left side of (6.26) will be zero for a class of relatively weak nonlinear systems with slowly iteration-varying system parameter $\boldsymbol{\Upsilon}^{\pitchfork}_l(k)$ and external disturbance $\mathbf{d}_l(k)$. And then, monotonic convergence can be achieved according to (6.25).

6.3.2 Predictive Model Construction

Define output tracking error $\mathbf{e}_{y,l}^{\pitchfork}(k+1)$ and state tracking error $\mathbf{e}_{x,l}^{\pitchfork}(k+1)$ as

$$\mathbf{e}_{y,l}^{\pitchfork}(k+1) = \mathbf{y}_d^{\pitchfork}(k+1) - \mathbf{y}_l(k+1) \tag{6.28}$$

$$\mathbf{e}_{x,l}^{\pitchfork}(k+1) = \mathbf{x}_d^{\pitchfork}(k+1) - \mathbf{x}_l(k+1) \tag{6.29}$$

where $\mathbf{y}_d^{\pitchfork}(k+1)$ and $\mathbf{x}_d^{\pitchfork}(k+1)$ are desired output and state, respectively.

Substituting (6.10) into (6.28), we have

$$\mathbf{e}_{y,l}^{\pitchfork}(k+1) = \mathbf{e}_{y,l-1}^{\pitchfork}(k+1) - \mathbf{C}\boldsymbol{\Upsilon}_l^{\pitchfork}(k)\Delta\mathbf{u}_l(k) - \mathbf{C}\Delta\mathbf{d}_l(k) \tag{6.30}$$

Since $\boldsymbol{\Upsilon}_l^{\pitchfork}(k)$ and $\Delta\mathbf{d}_l(k)$ in (6.30) are unknown, the following available error model for prediction is constructed

$$\mathbf{e}_{y,l}^{\pitchfork}(k+1) = \mathbf{e}_{y,l-1}^{\pitchfork}(k+1) - \mathbf{C}\hat{\boldsymbol{\Upsilon}}_l^{\pitchfork}(k)\Delta\mathbf{u}_l(k) - \mathbf{C}\Delta\hat{\mathbf{d}}_l(k) \tag{6.31}$$

where $\hat{\boldsymbol{\Upsilon}}_l^{\pitchfork}(k)$ and $\Delta\hat{\mathbf{d}}_l(k)$ are obtained by the designed observer (6.11).

Based on (6.31), the following prediction model for future v step iterative prediction is constructed

$$\begin{aligned}
\mathbf{e}_{y,l+1|l}^{\pitchfork}(k+1) &= \mathbf{e}_{y,l}^{\pitchfork}(k+1) - \mathbf{C}\hat{\boldsymbol{\Upsilon}}_{l+1|l}^{\pitchfork}(k)\Delta\mathbf{u}_{l+1}(k) - \mathbf{C}\Delta\hat{\mathbf{d}}_{l+1|l}(k)\\
\mathbf{e}_{y,l+2|l}^{\pitchfork}(k+1) &= \mathbf{e}_{y,l}^{\pitchfork}(k+1) - \mathbf{C}\hat{\boldsymbol{\Upsilon}}_{l+1|l}^{\pitchfork}(k)\Delta\mathbf{u}_{l+1}(k) - \mathbf{C}\Delta\hat{\mathbf{d}}_{l+1|l}(k)\\
&\quad -\mathbf{C}\hat{\boldsymbol{\Upsilon}}_{l+2|l}^{\pitchfork}(k)\Delta\mathbf{u}_{l+2}(k) - \mathbf{C}\Delta\hat{\mathbf{d}}_{l+2|l}(k)\\
&\ \ \vdots\\
\mathbf{e}_{y,l+v|l}^{\pitchfork}(k+1) &= \mathbf{e}_{y,l}^{\pitchfork}(k+1) - \mathbf{C}\hat{\boldsymbol{\Upsilon}}_{l+1|l}^{\pitchfork}(k)\Delta\mathbf{u}_{l+1}(k) - \mathbf{C}\Delta\hat{\mathbf{d}}_{l+1|l}(k)\\
&\quad -\mathbf{C}\hat{\boldsymbol{\Upsilon}}_{l+2|l}^{\pitchfork}(k)\Delta\mathbf{u}_{l+2}(k) - \mathbf{C}\Delta\hat{\mathbf{d}}_{l+2|l}(k) -\\
&\quad \cdots - \mathbf{C}\hat{\boldsymbol{\Upsilon}}_{l+v|l}^{\pitchfork}(k)\Delta\mathbf{u}_{l+v}(k) - \mathbf{C}\Delta\hat{\mathbf{d}}_{l+v|l}(k)
\end{aligned} \tag{6.32}$$

where the subscript $l+v|l$ implies the prediction for the future $(l+v)$th iteration at the current lth iteration.

Note that the terms $\hat{\boldsymbol{\Upsilon}}_{l+1|l}^{\pitchfork}(k), \ldots, \hat{\boldsymbol{\Upsilon}}_{l+v|l}^{\pitchfork}(k)$ and $\Delta\hat{\mathbf{d}}_{l+1|l}(k), \ldots, \Delta\hat{\mathbf{d}}_{l+v|l}(k)$ in the above prediction model (6.32) should be obtained before (6.32) is available. Next, iterative learning multi-level hierarchical predicting algorithms will be used to predict these unknown terms.

1. *Predict the terms* $\hat{\boldsymbol{\Upsilon}}^{\pitchfork}_{l+1|l}(k), \ldots, \hat{\boldsymbol{\Upsilon}}^{\pitchfork}_{l+v|l}(k)$

Based on the available values $\hat{\boldsymbol{\Upsilon}}^{\pitchfork}_{l}(k), \ldots, \hat{\boldsymbol{\Upsilon}}^{\pitchfork}_{l-h}(k)$ by means of (6.11), where h is a proper constant and can be selected as $h \geq v$, the terms $\hat{\boldsymbol{\Upsilon}}^{\pitchfork}_{l+1|l}(k), \ldots, \hat{\boldsymbol{\Upsilon}}^{\pitchfork}_{l+v|l}(k)$ is obtained by

$$\begin{aligned}
\hat{\boldsymbol{\Upsilon}}^{\pitchfork}_{l+1|l}(k) &= \Gamma^{\pitchfork}_{1}(k)\hat{\boldsymbol{\Upsilon}}^{\pitchfork}_{l}(k) + \Gamma^{\pitchfork}_{2}(k)\hat{\boldsymbol{\Upsilon}}^{\pitchfork}_{l-1}(k) + \cdots + \Gamma^{\pitchfork}_{h}(k)\hat{\boldsymbol{\Upsilon}}^{\pitchfork}_{l+1-h}(k)\\
\hat{\boldsymbol{\Upsilon}}^{\pitchfork}_{l+2|l}(k) &= \Gamma^{\pitchfork}_{1}(k)\hat{\boldsymbol{\Upsilon}}^{\pitchfork}_{l+1|l}(k) + \Gamma^{\pitchfork}_{2}(k)\hat{\boldsymbol{\Upsilon}}^{\pitchfork}_{l}(k) + \cdots + \Gamma^{\pitchfork}_{h}(k)\hat{\boldsymbol{\Upsilon}}^{\pitchfork}_{l+2-h}(k)\\
&\;\;\vdots\\
\hat{\boldsymbol{\Upsilon}}^{\pitchfork}_{l+v|l}(k) &= \Gamma^{\pitchfork}_{1}(k)\hat{\boldsymbol{\Upsilon}}^{\pitchfork}_{l+v-1|l}(k) + \Gamma^{\pitchfork}_{2}(k)\hat{\boldsymbol{\Upsilon}}^{\pitchfork}_{l+v-2|l}(k) + \cdots + \Gamma^{\pitchfork}_{h}(k)\hat{\boldsymbol{\Upsilon}}^{\pitchfork}_{l+v-h}(k)
\end{aligned} \tag{6.33}$$

where $\Gamma^{\pitchfork}_{j}(k) \in \Re^{m\times m}$ $(j = 1, \ldots, h)$ are unknown coefficient matrices.

Define

$$\hat{\Gamma}^{\pitchfork h}_{l}(k) = [\hat{\Gamma}^{\pitchfork}_{1,l}(k), \hat{\Gamma}^{\pitchfork}_{2,l}(k), \ldots, \hat{\Gamma}^{\pitchfork}_{h,l}(k)]^{T} \tag{6.34}$$

as the estimation of the unknown coefficient matrix, and it is obtained by the following iterative learning projection algorithm

$$\begin{aligned}
\hat{\Gamma}^{\pitchfork h}_{l}(k) = \hat{\Gamma}^{\pitchfork h}_{l-1}(k) + \kappa^{\pitchfork}\hat{\boldsymbol{\Upsilon}}^{\pitchfork h}_{l-1}(k)\Big[\nu^{\pitchfork}\mathbf{I} + \hat{\boldsymbol{\Upsilon}}^{\pitchfork h}_{l-1}(k)^{T}\hat{\boldsymbol{\Upsilon}}^{\pitchfork h}_{l-1}(k)\Big]^{-1}\\
\Big[\hat{\boldsymbol{\Upsilon}}^{\pitchfork}_{l}(k)^{T} - \hat{\boldsymbol{\Upsilon}}^{\pitchfork h}_{l-1}(k)^{T}\hat{\Gamma}^{\pitchfork h}_{l-1}(k)\Big]
\end{aligned} \tag{6.35}$$

where $\nu^{\pitchfork}$ and $\kappa^{\pitchfork}$ are known positive constants. $\mathbf{I} \in \Re^{\omega\times\omega}$ is the identity matrix, and

$$\hat{\boldsymbol{\Upsilon}}^{\pitchfork h}_{l-1}(k) = \Big[\hat{\boldsymbol{\Upsilon}}^{\pitchfork}_{l-1}(k)^{T}, \ldots, \hat{\boldsymbol{\Upsilon}}^{\pitchfork}_{l-h}(k)^{T}\Big]^{T} \tag{6.36}$$

2. *Predict the terms* $\Delta\hat{\mathbf{d}}_{l+1|l}(k), \ldots, \Delta\hat{\mathbf{d}}_{l+v|l}(k)$

Based on the available values $\Delta\hat{\mathbf{d}}_{l}(k), \ldots, \Delta\hat{\mathbf{d}}_{l-h}(k)$ by means of (6.11), the terms $\Delta\hat{\mathbf{d}}_{l+1|l}(k), \ldots, \Delta\hat{\mathbf{d}}_{l+v|l}(k)$ is obtained by

$$\begin{aligned}
\Delta\hat{\mathbf{d}}_{l+1|l}(k) &= \mathcal{H}^{\pitchfork}_{1}(k)\Delta\hat{\mathbf{d}}_{l}(k) + \mathcal{H}^{\pitchfork}_{2}(k)\Delta\hat{\mathbf{d}}_{l-1}(k) + \cdots + \mathcal{H}^{\pitchfork}_{h}(k)\Delta\hat{\mathbf{d}}_{l+1-h}(k)\\
\Delta\hat{\mathbf{d}}_{l+2|l}(k) &= \mathcal{H}^{\pitchfork}_{1}(k)\Delta\hat{\mathbf{d}}_{l+1|l}(k) + \mathcal{H}^{\pitchfork}_{2}(k)\Delta\hat{\mathbf{d}}_{l}(k) + \cdots + \mathcal{H}^{\pitchfork}_{h}(k)\Delta\hat{\mathbf{d}}_{l+2-h}(k)\\
&\;\;\vdots\\
\Delta\hat{\mathbf{d}}_{l+v|l}(k) &= \mathcal{H}^{\pitchfork}_{1}(k)\Delta\hat{\mathbf{d}}_{l+v-1|l}(k) + \mathcal{H}^{\pitchfork}_{2}(k)\Delta\hat{\mathbf{d}}_{l+v-2|l}(k) + \cdots + \mathcal{H}^{\pitchfork}_{h}(k)\Delta\hat{\mathbf{d}}_{l+v-h}(k)
\end{aligned} \tag{6.37}$$

where $\mathcal{H}^{\pitchfork}_{j}(k) \in \Re^{m\times m}$ $(j = 1, \ldots h)$ are unknown coefficient matrices.

Define

$$\hat{\mathcal{H}}_l^{\phi h}(k) = [\hat{\mathcal{H}}_{1,l}^{\phi}(k), \hat{\mathcal{H}}_{2,l}^{\phi}(k), \ldots, \hat{\mathcal{H}}_{h,l}^{\phi}(k)]^T \tag{6.38}$$

as the estimation of the unknown coefficient matrix, then it is obtained by the following iterative learning projection algorithm,

$$\begin{aligned} \hat{\mathcal{H}}_l^{\phi h}(k) = \hat{\mathcal{H}}_{l-1}^{\phi h}(k) + \bar{\kappa}^{\phi} \Delta\hat{\mathbf{d}}_{l-1}^{h}(k)\left[\bar{\nu}^{\phi} + \Delta\hat{\mathbf{d}}_{l-1}^{h}(k)^T \Delta\hat{\mathbf{d}}_{l-1}^{h}(k)\right]^{-1} \\ \left[\Delta\hat{\mathbf{d}}_l(k)^T - \Delta\hat{\mathbf{d}}_{l-1}^{h}(k)^T \hat{\mathcal{H}}_{l-1}^{\phi h}(k)\right] \end{aligned} \tag{6.39}$$

where $\bar{\nu}^{\phi}$ and $\bar{\kappa}^{\phi}$ are known positive constants, and

$$\Delta\hat{\mathbf{d}}_{l-1}^{h}(k) = \left[\Delta\hat{\mathbf{d}}_{l-1}(k)^T, \ldots, \Delta\hat{\mathbf{d}}_{l-h}(k)^T\right]^T \tag{6.40}$$

Remark 6.2 Based on the full-state observer (6.11) and predictor (6.33)–(6.40), the full system states can be predicted as follows.

$$\begin{aligned} \hat{\mathbf{x}}_l(k+1) &= \hat{\mathbf{x}}_{l-1}(k+1) - \hat{\boldsymbol{\Upsilon}}_l^{\phi}(k)\Delta\mathbf{u}_l(k) - \Delta\hat{\mathbf{d}}_l(k) \\ \hat{\mathbf{x}}_{l+1|l}(k+1) &= \hat{\mathbf{x}}_l(k+1) - \hat{\boldsymbol{\Upsilon}}_{l+1|l}^{\phi}(k)\Delta\mathbf{u}_{l+1}(k) - \Delta\hat{\mathbf{d}}_{l+1|l}(k) \\ \hat{\mathbf{x}}_{l+2|l}(k+1) &= \hat{\mathbf{x}}_l(k+1) - \hat{\boldsymbol{\Upsilon}}_{l+1|l}^{\phi}(k)\Delta\mathbf{u}_{l+1}(k) - \Delta\hat{\mathbf{d}}_{l+1|l}(k) \\ &\quad - \hat{\boldsymbol{\Upsilon}}_{l+2|l}^{\phi}(k)\Delta\mathbf{u}_{l+2}(k) - \Delta\hat{\mathbf{d}}_{l+2|l}(k) \\ &\vdots \\ \hat{\mathbf{x}}_{l+v|l}(k+1) &= \hat{\mathbf{x}}_l(k+1) - \hat{\boldsymbol{\Upsilon}}_{l+1|l}^{\phi}(k)\Delta\mathbf{u}_{l+1}(k) - \Delta\hat{\mathbf{d}}_{l+1|l}(k) \\ &\quad - \hat{\boldsymbol{\Upsilon}}_{l+2|l}^{\phi}(k)\Delta\mathbf{u}_{l+2}(k) - \Delta\hat{\mathbf{d}}_{l+2|l}(k) - \\ &\quad \cdots - \hat{\boldsymbol{\Upsilon}}_{l+v|l}^{\phi}(k)\Delta\mathbf{u}_{l+v}(k) - \Delta\hat{\mathbf{d}}_{l+v|l}(k) \end{aligned} \tag{6.41}$$

where $\hat{\mathbf{x}}_l(k+1)$ is the observation of $\mathbf{x}_l(k+1)$. The initial value of $\hat{\mathbf{x}}_l(k+1)$ at $l=0$, namely $\hat{\mathbf{x}}_0(k+1)$ can be estimated by using historical data information of the system. $\hat{\mathbf{x}}_{l+j|l}(k+1)$, $j = 1, \cdots, v$ is the predicted system state for the future $(l+j)$th iteration at the lth iteration and can be further used for system monitor, fault diagnosis, etc.

6.3.3 Predictive ILC Design

Rewrite (6.32) as the following compact form

$$\mathbf{e}^{\pitchfork}_{y,l+v|l}(k+1) = \mathbf{e}^{\pitchfork}_{y,l}(k+1) - \mathbf{C}\hat{\boldsymbol{\Upsilon}}^{\pitchfork v}_{l+1|l}(k)\Delta\mathbf{u}^{v}_{l+1}(k) - \mathbf{C}^{v}\Delta\hat{\mathbf{d}}^{v}_{l+1|l}(k) \quad (6.42)$$

where

$$\hat{\boldsymbol{\Upsilon}}^{\pitchfork v}_{l+1|l}(k) = \left[\hat{\boldsymbol{\Upsilon}}^{\pitchfork}_{l+1|l}(k), \ldots, \hat{\boldsymbol{\Upsilon}}^{\pitchfork}_{l+v|l}(k)\right] \in \Re^{m\times wv} \quad (6.43)$$

$$\Delta\mathbf{u}^{v}_{l+1}(k) = \left[\Delta\mathbf{u}_{l+1}(k)^{T}, \ldots, \Delta\mathbf{u}_{l+v}(k)^{T}\right]^{T} \in \Re^{wv\times 1} \quad (6.44)$$

$$\Delta\hat{\mathbf{d}}^{v}_{l+1|l}(k) = \left[\Delta\hat{\mathbf{d}}_{l+1|l}(k)^{T}, \ldots, \Delta\hat{\mathbf{d}}_{l+v|l}(k)\right]^{T} \in \Re^{mv\times 1} \quad (6.45)$$

$$\mathbf{C}^{v} = \left[\mathbf{C}, \ldots, \mathbf{C}\right] \in \Re^{m\times mv} \quad (6.46)$$

Define the following quadratic cost function:

$$\begin{aligned} J^{\pitchfork}_{l+1}(k) = \frac{1}{2}\Big\{&\mathbf{e}^{\pitchfork T}_{y,l+v|l}(k+1)\bar{\mathbf{Q}}^{\pitchfork}\mathbf{e}^{\pitchfork}_{y,l+v|l}(k+1) \\ &+\Delta\mathbf{u}^{v}_{l+1}(k)^{T}\bar{\mathbf{R}}^{\pitchfork}\Delta\mathbf{u}^{v}_{l+1}(k) + \varrho^{\pitchfork T}_{l}(k+1)\bar{\mathbf{S}}^{\pitchfork}\varrho^{\pitchfork}_{l}(k+1)\Big\} \end{aligned} \quad (6.47)$$

where $\bar{\mathbf{Q}}^{\pitchfork} \in \Re^{m\times m}$, $\bar{\mathbf{R}}^{\pitchfork} \in \Re^{wv\times wv}$, and $\bar{\mathbf{S}}^{\pitchfork} \in \Re^{m\times m}$ are positive definite and symmetric matrices. For simplicity, let $\bar{\mathbf{Q}}^{\pitchfork} = \bar{q}^{\pitchfork}\mathbf{I}_{m\times m}, \bar{\mathbf{R}}^{\pitchfork} = \bar{r}^{\pitchfork}\mathbf{I}_{wv\times wv}, \bar{q}^{\pitchfork} > 0, \bar{r}^{\pitchfork} > 0$. $\varrho^{\pitchfork}_{l}(k+1) \geq \mathbf{0}$ is a slack variable vector.

If there are no system constraints, minimizing (6.47) and using (6.42) will get the optimal $\Delta\mathbf{u}^{v}_{l+1}(k)$ as follows:

$$\begin{aligned} \Delta\mathbf{u}^{v}_{l+1}(k) = &\left[\hat{\boldsymbol{\Upsilon}}^{\pitchfork v}_{l+1|l}(k)^{T}\mathbf{C}^{T}\bar{\mathbf{Q}}^{\pitchfork}\mathbf{C}\hat{\boldsymbol{\Upsilon}}^{\pitchfork v}_{l+1|l}(k) + \bar{\mathbf{R}}^{\pitchfork}\right]^{-1}\hat{\boldsymbol{\Upsilon}}^{\pitchfork v}_{l+1|l}(k)^{T} \\ &\mathbf{C}^{T}\bar{\mathbf{Q}}^{\pitchfork}\left[\mathbf{e}^{\pitchfork}_{y,l}(k+1) - \mathbf{C}^{v}\Delta\hat{\mathbf{d}}^{v}_{l+1|l}(k)\right] \end{aligned} \quad (6.48)$$

The convergence of above predictive ILC (6.48) based on the proposed full-state observer (6.11) and prediction algorithm (6.33)–(6.40) is guaranteed by the following theorem.

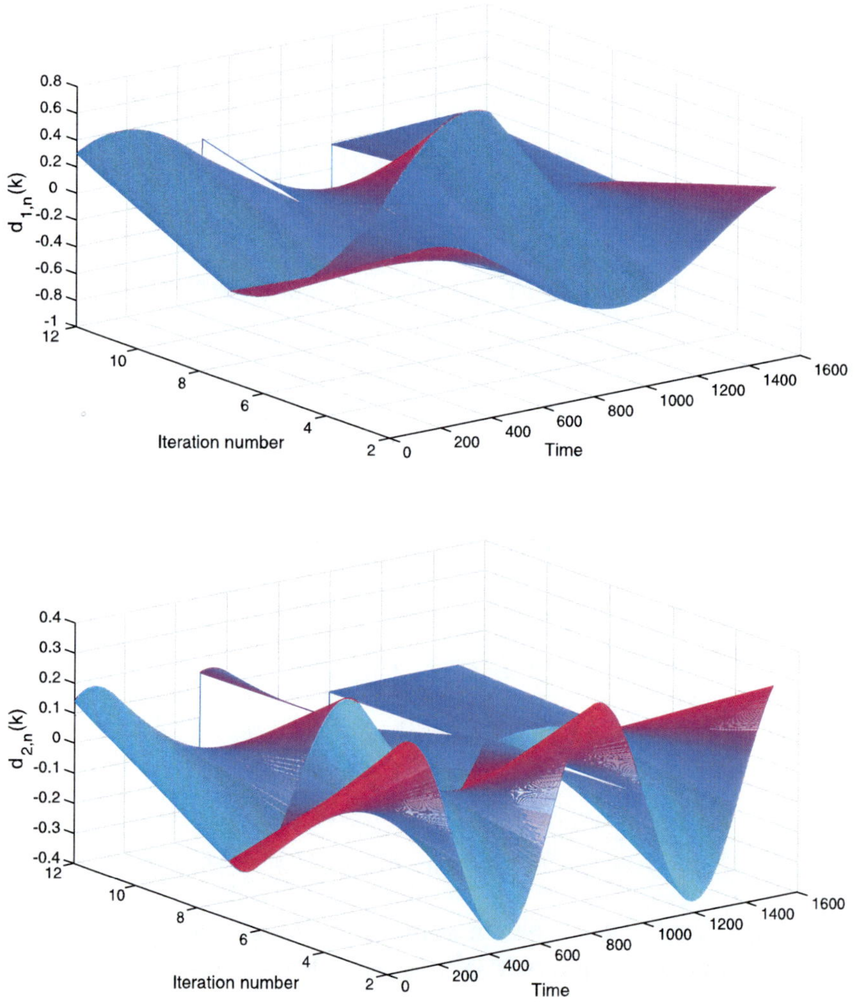

Fig. 6.1 External disturbance $\mathbf{d}_l(k)$ profiles

Theorem 6.2 *Considering system (6.2) under Assumptions 6.1–6.3, the proposed full-state observer-based predictive ILC method (6.11), (6.33)–(6.40), (6.48) guarantees that the tracking error* $\mathbf{e}_{y,l}^{\pitchfork}(k+1)$ *converges to a tunable residue.*

Proof The proof can be derived straightforwardly according the proof of Theorem 2.1 in Chap. 2 and hence is omitted here. □

Furthermore, if system constraints (6.2)–(6.3) are considered, the following constrained optimization problem is considered:

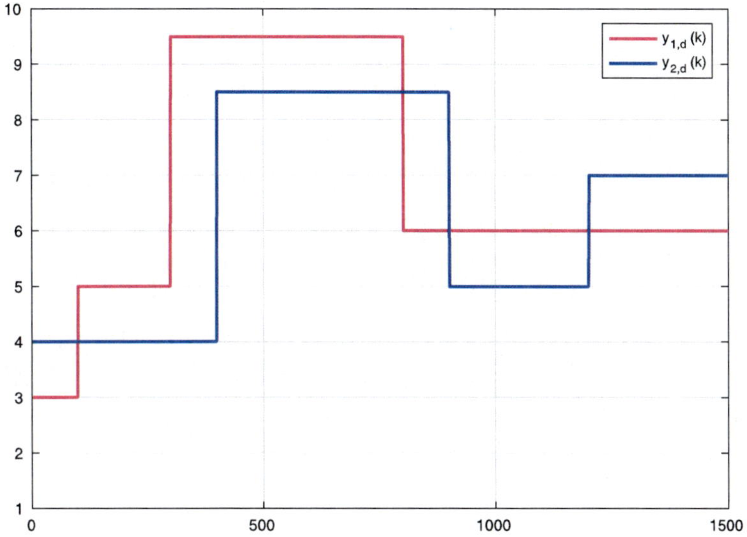

Fig. 6.2 Desired trajectory profiles

$$
\begin{aligned}
J_{l+1}^{\pitchfork}(k) = \min_{\Delta \mathbf{u}_{l+1}^{v}(k), \varrho_l^{\pitchfork}(k+1)} \frac{1}{2} \Big\{ & \mathbf{e}_{y,l+v|l}^{\pitchfork T}(k+1) \bar{\mathbf{Q}}^{\pitchfork} \mathbf{e}_{y,l+v|l}^{\pitchfork}(k+1) \\
& + \Delta \mathbf{u}_{l+1}^{v}(k)^T \bar{\mathbf{R}}^{\pitchfork} \Delta \mathbf{u}_{l+1}^{v}(k) + \varrho_l^{\pitchfork T}(k+1) \bar{\mathbf{S}}^{\pitchfork} \varrho_l^{\pitchfork}(k+1) \Big\} \\
& \text{subject to (6.50)}
\end{aligned}
\tag{6.49}
$$

where

$$
\begin{aligned}
\mathbf{u}_{\min} &\le \mathbf{u}_{l+1}(k) \le \mathbf{u}_{\max} \\
\Delta \mathbf{u}_{\min} &\le \Delta \mathbf{u}_{l+1}(k) \le \Delta \mathbf{u}_{\max} \\
\mathbf{y}_{\min} - \varrho_l^{\pitchfork}(k+1) &\le \hat{\mathbf{y}}_{l+v|l}(k+1) \le \mathbf{y}_{\max} + \varrho_l^{\pitchfork}(k+1)
\end{aligned}
\tag{6.50}
$$

Here (6.50) are the constraints on control input, change rate of control input, and predicted output. $\hat{\mathbf{y}}_{l+v|l}(k+1) = \mathbf{y}_d^{\pitchfork}(k+1) - \mathbf{e}_{y,l+v|l}^{\pitchfork}(k+1)$. $\mathbf{u}_{\min}$,$\mathbf{u}_{\max}$, $\mathbf{y}_{\min}$, $\mathbf{y}_{\max}$ are the restrictions defined in (6.2)–(6.3). $\Delta \mathbf{u}_{\min}$ and $\Delta \mathbf{u}_{\max}$ are the lower and upper restrictions on the change rate of control input.

Based on (6.11), (6.33)–(6.40), the above-constrained optimization problem (6.49) can be solved by QP algorithm in MATLAB and can be converged by the following theorem.

Theorem 6.3 *Considering system (6.1) under Assumptions 6.1–6.4, the proposed full-state observer-based constrained predictive ILC method (6.11), (6.33)–(6.40), (6.49)–(6.50) guarantees that the tracking error* $\mathbf{e}_{y,l}^{\pitchfork}(k+1)$ *converges to a tunable residue.*

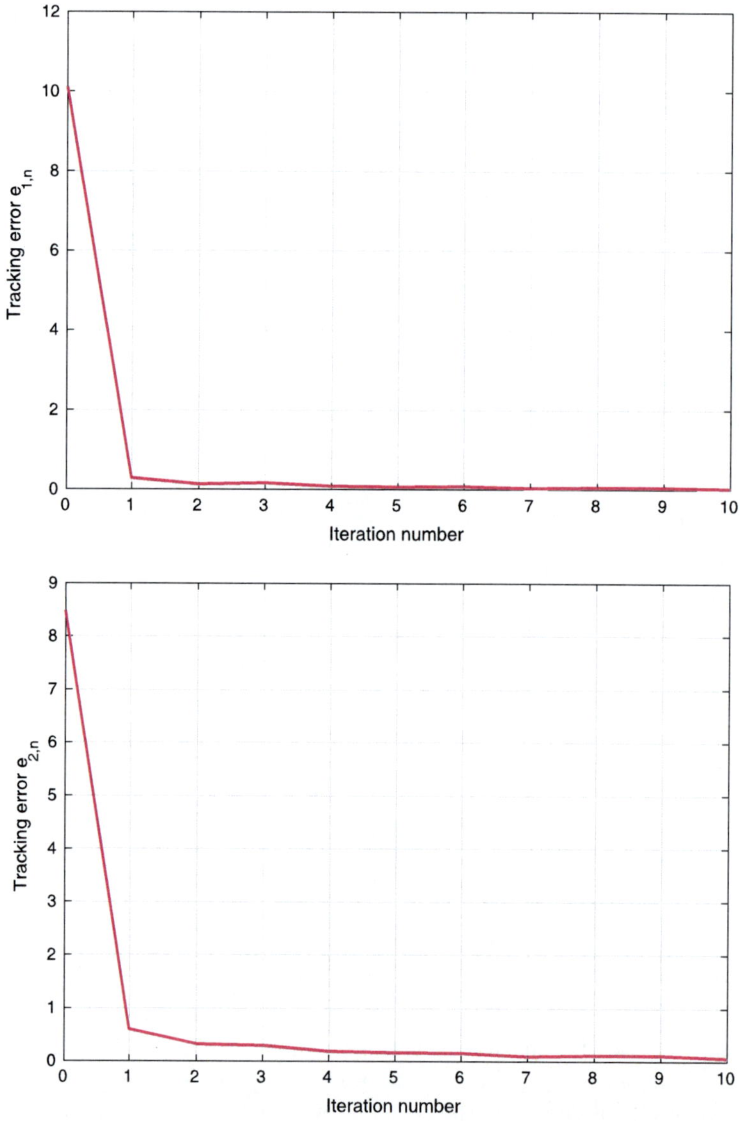

Fig. 6.3 Tracking error profiles

Proof The proof can be derived straightforwardly according the proof of Theorem 3.1 in Chap.3 and hence is omitted here. □

6.4 Simulation Validation

Consider the system (6.1) with $\mathbf{x}_l(k) = [x_{1,l}(k), x_{2,l}(k)]^T$, $\mathbf{u}_l(k) = [u_{1,l}(k), u_{2,l}(k)]^T$, $\mathbf{f}=[x_{1,l}(k)^2/(1+x_{1,l}(k)^2) + 0.5x_{2,l}(k) + a^{\pitchfork}(k)u_{1,l}(k)^3,\ \ x_{2,l}(k)^2/(1 + x_{2,l}(k)^2) + 0.8x_{2,l}(k) + b^{\pitchfork}(k)u_{2,l}(k)^3]^T$. $a^{\pitchfork}(k) = 2 + 0.2\sin(2\pi k/1500)$. $b^{\pitchfork}(k) = 3 + 0.3\cos(2\pi k/1500)$. $\mathbf{d}_l(k) = \left[\, d_{1,l}(k),\ \ d_{2,l}(k)\,\right]^T$ that is shown in Fig. 6.1. $\mathbf{y}_l(k) = [y_{1,l}(k), y_{2,l}(k)]^T$. $\mathbf{C} = \left[1\ 0;\ 0\ 1\right]$.

The desired trajectory $\mathbf{y}_d^{\pitchfork}(k) = [y_{1,d}^{\pitchfork}(k), y_{2,d}^{\pitchfork}(k)]^T$ is depicted in Fig. 6.2. The proposed full-state observer-based predictive ILC method (6.11), (6.33)-(6.40), (6.48) is simulated and the control parameters are selected as: $\mathbf{\Lambda}^{\pitchfork} = 3 \times 10^{-4}\mathbf{I}_{2\times 2}$, $\bar{\mathbf{\Psi}}^{\pitchfork} = 10^{-1} \times \mathbf{I}_{2\times 2}$. $\mathbf{H}_l^{\pitchfork}(k) = [\ 10\ \ 10\ ]$, $\bar{q}^{\pitchfork} = 0.9$, $\bar{r}^{\pitchfork} = 0.5$.

Figure 6.3 shows the maximum absolute value of output tracking errors $\mathbf{e}_{y,l}^{\pitchfork}(k)$ defined in (6.28). Note that even though the considered system is under randomly varying external disturbance, satisfactory control performance by applying the proposed full-state observer-based predictive ILC method (6.11), (6.33)–(6.40), (6.48) can be guaranteed.

6.5 Conclusion

This chapter designs a new full-state observer-based robust constrained PILC method for a class of unknown nonaffine nonlinear systems with unknown and iteration-varying external disturbances. The proposed method can fully capture the internal characteristics of unknown system states according to a newly constructed full-state observer, and then, the corresponding robust constrained PILC is designed without using any model information. The convergence of the proposed PILC method can also be guaranteed. Simulation results further demonstrate the effectiveness of the proposed method.

References

Bobtsov AA, Pyrkin AA, Ortega RS, Vedyakov AA (2018) A state observer for sensorless control of magnetic levitation systems. Automatica 97:263–270

Chen W-H, Yang J, Guo L, Li SH (2016) Disturbance-observer-based control and related methods—an overview. IEEE Trans Ind Electron 63(2):1083–1095

Kim K-S, Rew K-H (2013) Reduced order disturbance observer for discrete-time linear systems. Automatica 49(4):1968–975

Kim K-S, Rew K-H, Kim S (2010) Disturbance observer for estimating higher order disturbances in time series expansion. IEEE Trans Autom Control 55(8):1905–1911

Li K, Li DW, Xi YG, Yin DB (2014) Model predictive control with feedforward strategy for gas collectors of coke ovens. Chin J Chem Eng 7(22):769–773

Yang J, Li S, Sun C, Guo L (2013) Nonlinear-disturbance-observer-based robust flight control for airbreathing hypersonic vehicles. IEEE Trans Aerosp Electron Syst 1(49):160–169

Yang J, Su J, Li S, Yu X (2014) High-order mismatched disturbance compensation for motion control systems via a continuous dynamic sliding-mode approach. IEEE Trans Ind Inform 10(1):604–614

Yang JQ, Chen YT, Zhu FL, Wang FZ (2017) Simultaneous state and output disturbance estimations for a class of switched linear systems with unknown inputs. Int J Syst Sci 48(1):22–33

Zhang XM, Han QL (2018) State estimation for static neural networks with time-varying delays based on an improved reciprocally convex inequality. IEEE Trans Neural Netw Learn Syst 29(4):1376–1381

Zou L, Wang ZD, Gao HJ (2016) Observer-based $H\infty$ control of networked systems with stochastic communication protocol: the finite-horizon case. Automatica 63:366–373

Chapter 7
Predictive Iterative Learning Control for Systems with Unavailable States

7.1 Introduction

In Chap. 6, state information of the system is fully used for controller design since these states contain many inherent features of a practical system. However, if the system states cannot be completely measurable owing to the restrictions of sensor technology or other practical reasons, some state information will be unavailable.

In this chapter, we extend the nonlinear systems considered in Chap. 6 to those with unavailable system states, and a new reduced-order observer-based PILC method is designed. Firstly, an iterative learning reduced-order observer is designed to fully capture all the state information of the unknown system, and the convergence of observer errors is guaranteed with theoretical analysis. Based on the designed observer, corresponding predictive model and PILC algorithm are presented, and the convergence of output tracking errors can also also guaranteed.

The main contributions of this chapter are as follows.

- By using the proposed reduced-order observer and established predictive model, unavailable system states can not only be observed but also be predicted and can be further used for system prediction, monitor and fault diagnosis, etc.
- For the considered unknown nonaffine nonlinear system, the proposed reduced-order observer-based PILC method utilizes the measured input/output data only without using any model information.

This chapter is organized as follows. Section 7.2 presents the problem formulation. Section 7.3 shows the reduced-order observer-based predictive ILC method. In Sect. 7.4, simulation results are provided to show the effectiveness of the proposed method. Some conclusions are given in Sect. 7.5.

© The Author(s), under exclusive license to Springer Nature Singapore Pte Ltd. 2023
Q. Yu et al., *Predictive Learning Control for Unknown Nonaffine Nonlinear Systems*, Intelligent Control and Learning Systems 8,
https://doi.org/10.1007/978-981-19-8857-8_7

7.2 Problem Formulation

Consider the following widely used MIMO nonaffine nonlinear system (Chen et al. 2016; Kim et al. 2010; Kim and Rew 2013; Yang et al. 2014; Li et al. 2014; Yang et al. 2013):

$$\begin{cases} \mathbf{x}_l(k+1) = \mathbf{f}(\mathbf{x}_l(k), \mathbf{x}_l(k-1), \ldots, \mathbf{x}_l(k-n_x^{\ltimes}), \\ \qquad\qquad \mathbf{u}_l(k), \mathbf{u}_l(k-1), \ldots, \mathbf{u}_l(k-n_u^{\ltimes})) + \mathbf{d}_l(k) \\ \mathbf{y}_l(k+1) = \mathbf{C}\mathbf{x}_l(k+1) \end{cases} \tag{7.1}$$

where $l \in N^+$ is the iteration or repetitive operation index. $k \in \{0, 1, \ldots, K^{\ltimes}\}$ is the operating time index. $n_x^{\ltimes}$, $n_u^{\ltimes}$ are unknown system orders. $\mathbf{f}(\ldots) \in \Re^m$ is an unknown nonlinear vector-valued function. $\mathbf{x}_l(k) \in \Re^m$, $\mathbf{u}_l(k) \in \Re^w$, $\mathbf{y}_l(k) \in \Re^n$ are the system state, input, measurement output, respectively. $\mathbf{d}_l(k) \in \Re^m$ is unknown but bounded external disturbance. Different from Chap. 6 that $\mathbf{C} = \mathbf{I}_m$, here the output matrix $\mathbf{C} \in \Re^{n \times m}$ is subjected to $n < m$ which implies that some of the system states are unavailable.

Here, system (7.1) is under following input and output constraints.

1. Input constraint:

$$\mathbf{u}_{\min} \le \mathbf{u}_l(k) \le \mathbf{u}_{\max} \tag{7.2}$$

2. State constraint:

$$\mathbf{y}_{\min} \le \mathbf{y}_l(k+1) \le \mathbf{y}_{\max} \tag{7.3}$$

where $\mathbf{u}_{\min}$, $\mathbf{u}_{\max}$ and $\mathbf{y}_{\min}$, $\mathbf{y}_{\max}$ are the constraints on the input and output vectors respectively.

The considered system (7.1) satisfies the following assumptions.

Assumption 7.1 The partial derivatives of $\mathbf{f}(\cdots)$ with reference to $\mathbf{u}_l(k)$ are continuous.

Assumption 7.2 If $\mathbf{d}_l(k) = \mathbf{0}$ holds, the following generalized *Lipschitz* condition is satisfied

$$||\Delta\mathbf{x}_l(k+1)|| \le \zeta^{\ltimes}||\Delta\mathbf{u}_l(k)|| \tag{7.4}$$

where Δ is a difference operator between two consecutive iterations. $\Delta\mathbf{u}_l(k) = \mathbf{u}_l(k) - \mathbf{u}_{l-1}(k)$, $\Delta\mathbf{x}_l(k+1) = \mathbf{x}_l(k+1) - \mathbf{x}_{l-1}(k+1)$. $\zeta^{\ltimes}$ is a positive constant.

Assumption 7.3 The initial values $\mathbf{x}_l(0)$ and $\mathbf{y}_l(0)$ can be randomly varying for $\forall l = 0, 1, 2 \ldots$.

Assumption 7.4 There exists a feasible input vector $\mathbf{u}_d(k)$ such that $\mathbf{e}_{y,l}^{\ltimes}(\mathbf{u}_d(k)) = \mathbf{0}$, where $\mathbf{e}_{y,l}^{\ltimes}(k)$ is the tracking error trajectory vector defined as $\mathbf{e}_{y,l}^{\ltimes}(k) = \mathbf{y}_d^{\ltimes}(k) - \mathbf{y}_l(k)$. $\mathbf{y}_d^{\ltimes}(k)$is the desired output trajectory.

Based on Assumptions 7.1–7.3, analogous to Lemma 6.1 in Chap. 6, an equivalent dynamic linearization model that serves for controller design is constructed in the following lemma.

Lemma 7.1 *The following dynamic linearization model (7.5) can equivalently describe the considered system (7.1) under Assumptions 7.1–7.3:*

$$\begin{cases} \Delta\mathbf{x}_l(k+1) = \boldsymbol{\Upsilon}_l^{\ltimes}(k)\Delta\mathbf{u}_l(k) + \Delta\mathbf{d}_l(k) \\ \mathbf{y}_l(k+1) = \mathbf{C}\mathbf{x}_l(k+1) \end{cases} \tag{7.5}$$

for $\forall k \in \{0, 1, \ldots, K^{\ltimes}\}, l=0, 1, 2 \ldots$ *and* $||\Delta\mathbf{u}_l(k)|| \neq 0$. $\Delta\mathbf{d}_l(k) = \mathbf{d}_l(k) - \mathbf{d}_{l-1}(k)$. $\boldsymbol{\Upsilon}_l^{\ltimes}(k) = \begin{bmatrix} \upsilon_{11,l}^{\ltimes}(k) & \upsilon_{12,l}^{\ltimes}(k) & \cdots & \upsilon_{1w,l}^{\ltimes}(k) \\ \upsilon_{21,l}^{\ltimes}(k) & \upsilon_{22,l}^{\ltimes}(k) & \cdots & \upsilon_{2w,l}^{\ltimes}(k) \\ \vdots & \vdots & \vdots & \vdots \\ \upsilon_{m1,l}^{\ltimes}(k) & \upsilon_{m2,l}^{\ltimes}(k) & \cdots & \upsilon_{mw,l}^{\ltimes}(k) \end{bmatrix}_{m \times w}$ *is a newly introduced pseudo-Jacobian matrix. It is unknown but bounded.*

Proof Differencing the system (7.1) with respect to the iteration index l, one has

$$\begin{aligned} \Delta\mathbf{x}_l(k+1) &= \mathbf{f}(\mathbf{x}_l(k), \ldots, \mathbf{x}_l(k-n_x^{\ltimes}), \mathbf{u}_l(k), \ldots, \mathbf{u}_l(k-n_u^{\ltimes})) \\ &\quad -\mathbf{f}(\mathbf{x}_{l-1}(k), \ldots, \mathbf{x}_{l-1}(k-n_x^{\ltimes}), \mathbf{u}_{l-1}(k), \ldots, \mathbf{u}_{l-1}(k-n_u^{\ltimes})) \\ &\quad +\mathbf{d}_l(k) - \mathbf{d}_{l-1}(k) \\ &= \mathbf{f}(\mathbf{x}_l(k), \ldots, \mathbf{x}_l(k-n_x^{\ltimes}), \mathbf{u}_l(k), \ldots, \mathbf{u}_l(k-n_u^{\ltimes})) \\ &\quad -\mathbf{f}(\mathbf{x}_l(k), \ldots, \mathbf{x}_l(k-n_x^{\ltimes}), \mathbf{u}_{l-1}(k), \mathbf{u}_l(k-1), \ldots, \mathbf{u}_l(k-n_u^{\ltimes})) \\ &\quad +\mathbf{f}(\mathbf{x}_l(k), \ldots, \mathbf{x}_l(k-n_x^{\ltimes}), \mathbf{u}_{l-1}(k), \mathbf{u}_l(k-1), \ldots, \mathbf{u}_l(k-n_u^{\ltimes})) \\ &\quad -\mathbf{f}(\mathbf{x}_{l-1}(k), \ldots, \mathbf{x}_{l-1}(k-n_x^{\ltimes}), \mathbf{u}_{l-1}(k), \ldots, \mathbf{u}_{l-1}(k-n_u^{\ltimes})) + \Delta\mathbf{d}_l(k) \end{aligned} \tag{7.6}$$

According to Assumption 7.1 and differential mean value theorem, (7.6) leads to

$$\Delta\mathbf{x}_l(k+1) = \frac{\partial\mathbf{f}^*}{\partial\mathbf{u}_l(k)}\Delta\mathbf{u}_l(k) + \varsigma_l^{\ltimes}(k) + \Delta\mathbf{d}_l(k) \tag{7.7}$$

where $\frac{\partial\mathbf{f}^*}{\partial\mathbf{u}_l(k)} = \begin{bmatrix} \frac{\partial f_1^*}{\partial u_{1,l}(k)} & \frac{\partial f_1^*}{\partial u_{2,l}(k)} & \cdots & \frac{\partial f_1^*}{\partial u_{w,l}(k)} \\ \frac{\partial f_2^*}{\partial u_{1,l}(k)} & \frac{\partial f_2^*}{\partial u_{2,l}(k)} & \cdots & \frac{\partial f_2^*}{\partial u_{w,l}(k)} \\ \vdots & \vdots & \vdots & \vdots \\ \frac{\partial f_m^*}{\partial u_{1,l}(k)} & \frac{\partial f_m^*}{\partial u_{2,l}(k)} & \cdots & \frac{\partial f_m^*}{\partial u_{w,l}(k)} \end{bmatrix}$. $\partial f_i^*/\partial u_{j,l}(k)$ ($i = 1, \ldots m$ and $j = 1, \ldots w$) is the partial derivative of f_i with respect to u_j at a certain point in

$[u_{j,l-1}(k),\ u_{j,l}(k)]$. $\varsigma_l^{\ltimes}(k) \triangleq \mathbf{f}(\mathbf{x}_l(k), \ldots, \mathbf{x}_l(k-n_x^{\ltimes}), \mathbf{u}_{l-1}(k), \mathbf{u}_l(k-1), \ldots, \mathbf{u}_l(k-n_u^{\ltimes})) - \mathbf{f}(\mathbf{x}_{l-1}(k), \ldots, \mathbf{x}_{l-1}(k-n_x^{\ltimes}), \mathbf{u}_{l-1}(k), \ldots, \mathbf{u}_{l-1}(k-n_u^{\ltimes}))$.

Consider the following equation with respect to $\varsigma_l^{\ltimes}(k)$

$$\varsigma_l^{\ltimes}(k) = \kappa_l^{\ltimes}(k)\Delta\mathbf{u}_l(k) \tag{7.8}$$

where $\kappa_l^{\ltimes}(k) \in \Re^{m\times w}$ is a numerical matrix (7.8) must have at least one solution $\kappa_l^{\ltimes *}(k)$ for $\forall k \in \{0, 1, \ldots, K^{\ltimes}\}, l = 0, 1, 2\ldots$ and $||\Delta\mathbf{u}_l(k)|| \neq 0$. If $||\Delta\mathbf{u}_l(k)|| \neq 0$ cannot be satisfied, please refer to Remark 2.2 of Chap. 2.

Then (7.7) can be rewritten as

$$\Delta\mathbf{x}_l(k+1) = \boldsymbol{\Upsilon}_l^{\ltimes}(k)\Delta\mathbf{u}_l(k) + \Delta\mathbf{d}_l(k) \tag{7.9}$$

where $\boldsymbol{\Upsilon}_l^{\ltimes}(k) = \partial\mathbf{f}^*/\partial\mathbf{u}_l(k) + \kappa_l^{\ltimes *}(k)$. According to Assumption 7.2, it is derived that $\boldsymbol{\Upsilon}_l^{\ltimes}(k)$ is bounded. □

Rewrite system (7.5) as:

$$\begin{cases} \mathbf{x}_l(k+1) = \mathbf{x}_{l-1}(k+1) + \boldsymbol{\Upsilon}_l^{\ltimes}(k)\Delta\mathbf{u}_l(k) + \Delta\mathbf{d}_l(k) \\ \mathbf{y}_l(k+1) = \mathbf{C}\mathbf{x}_l(k+1) \end{cases} \tag{7.10}$$

The control objectives for (7.10) are:

1. For $\mathbf{C} \in \Re^{n\times m}$ with $n < m$, designing a reduced-order observer to observe the unknown $\boldsymbol{\Upsilon}_l^{\ltimes}(k)$ and $\Delta\mathbf{d}_l(k)$ in (7.10) through those measurable output data.
2. Under both input and output constraints (7.2)–(7.3), designing a reduced-order observer-based predictive ILC to make $\mathbf{y}_l(k+1)$ track the desired output $\mathbf{y}_d^{\ltimes}(k+1)$.

7.3 Reduced-Order Observer-Based Predictive ILC Design

Note that system (7.10) is subjected to not only unavailable system states due to $\mathbf{C} \in \Re^{n\times m}$ and $n < m$, but also unknown system parameters $\boldsymbol{\Upsilon}_l^{\ltimes}(k)$ and $\Delta\mathbf{d}_l(k)$. In this section, unknown $\boldsymbol{\Upsilon}_l^{\ltimes}(k)$ and $\Delta\mathbf{d}_l(k)$ will be estimated by a constructed reduced-order observer in Sect. 7.3.1 and be further predicted in Sect. 7.3.2. In Sect. 7.3.3, the corresponding predictive ILC is designed.

7.3.1 Reduced-Order Observer Design

By using minimal rank decomposition, define an unmeasurable state function of minimal order $\boldsymbol{\delta}_l^{\ltimes}(k) \in \Re^h$ as

$$\boldsymbol{\delta}_l^{\ltimes}(k) \triangleq \mathbf{B}^{\ltimes T}\mathbf{x}_{l-1}(k+1) \tag{7.11}$$

where $\mathbf{B}^{\ltimes T} \in \Re^{h\times m}$ satisfies

$$\hbar^{\ltimes} = \mathbf{I}_m - \mathbf{C}^{+}\mathbf{C} \tag{7.12}$$

$$\bar{\boldsymbol{\Lambda}}^{\ltimes}\hbar^{\ltimes} = \hbar_1^{\ltimes}\mathbf{B}^{\ltimes T} \tag{7.13}$$

$$\bar{\boldsymbol{\Psi}}^{\ltimes}\hbar^{\ltimes} = \hbar_2^{\ltimes}\mathbf{B}^{\ltimes T} \tag{7.14}$$

and $\bar{\boldsymbol{\Lambda}}^{\ltimes} \in \Re^{m\times m}$, $\bar{\boldsymbol{\Psi}}^{\ltimes} \in \Re^{m\times m}$ are gain matrices. $\hbar_1^{\ltimes} \in \Re^{m\times h}$, $\hbar_2^{\ltimes} \in \Re^{m\times h}$ are two appropriate matrices.

Then construct the following $\hat{\boldsymbol{\Upsilon}}_l^{\ltimes}(k)$ and $\Delta\hat{\mathbf{d}}_l(k)$ for observing $\boldsymbol{\Upsilon}_l^{\ltimes}(k)$ and $\Delta\mathbf{d}_l(k)$:

$$\begin{cases} \hat{\boldsymbol{\Upsilon}}_l^{\ltimes}(k) = \bar{\boldsymbol{\Lambda}}^{\ltimes}\mathbf{C}^{+}\mathbf{y}_{l-1}(k+1)\mathbf{H}_l^{\ltimes}(k) - \boldsymbol{\Phi}_l^{\ltimes}(k) + \hbar_1^{\ltimes}\hat{\boldsymbol{\delta}}_l^{\ltimes}(k)\mathbf{H}_l^{\ltimes}(k) \\ \boldsymbol{\Phi}_{l+1}^{\ltimes}(k) = \alpha^{\ltimes}(k)\Big[\boldsymbol{\Phi}_l^{\ltimes}(k) + \hat{\boldsymbol{\Upsilon}}_l^{\ltimes}(k) + \tau^{\ltimes}\bar{\boldsymbol{\Lambda}}^{\ltimes}\hat{\boldsymbol{\Upsilon}}_l^{\ltimes}(k)\Delta\mathbf{u}_l(k)\mathbf{H}_l^{\ltimes}(k) \\ \qquad\qquad + \bar{\boldsymbol{\Lambda}}^{\ltimes}\Delta\hat{\mathbf{d}}_l(k)\mathbf{H}_l^{\ltimes}(k)\Big] \\ \Delta\hat{\mathbf{d}}_l(k) = \bar{\boldsymbol{\Psi}}^{\ltimes}\mathbf{C}^{+}\mathbf{y}_{l-1}(k+1) - \boldsymbol{\Omega}_l^{\ltimes}(k) + \hbar_2^{\ltimes}\hat{\boldsymbol{\delta}}_l^{\ltimes}(k) \\ \boldsymbol{\Omega}_{l+1}^{\ltimes}(k) = \boldsymbol{\Omega}_l^{\ltimes}(k) + \bar{\boldsymbol{\Psi}}^{\ltimes}\left[\hat{\boldsymbol{\Upsilon}}_l^{\ltimes}(k)\Delta\mathbf{u}_l(k) + \Delta\hat{\mathbf{d}}_l(k)\right] + \Delta\hat{\mathbf{d}}_l(k) \\ \hat{\boldsymbol{\delta}}_l^{\ltimes}(k) = \boldsymbol{\eta}_l^{\ltimes}(k) + \mathbf{M}^{\ltimes}\mathbf{y}_{l-1}(k+1) \\ \boldsymbol{\eta}_{l+1}^{\ltimes}(k) = \boldsymbol{\eta}_l^{\ltimes}(k) + \left(\mathbf{B}^{\ltimes T} - \mathbf{M}^{\ltimes}\mathbf{C}\right)\hat{\boldsymbol{\Upsilon}}_l^{\ltimes}(k)\Delta\mathbf{u}_l(k) + \left(\mathbf{B}^{\ltimes T} - \mathbf{M}^{\ltimes}\mathbf{C}\right)\Delta\hat{\mathbf{d}}_l(k) \end{cases} \tag{7.15}$$

where $\hat{\boldsymbol{\delta}}_l^{\ltimes}(k) \in \Re^h$ is the observation of $\boldsymbol{\delta}_l^{\ltimes}(k)$ defined in (7.11). $\boldsymbol{\Phi}_l^{\ltimes}(k) \in \Re^{m\times w}$, $\boldsymbol{\Omega}_l^{\ltimes}(k) \in \Re^{m\times 1}$, $\boldsymbol{\eta}_l^{\ltimes}(k) \in \Re^h$ are the state variables of the observers $\hat{\boldsymbol{\Upsilon}}_l^{\ltimes}(k)$, $\Delta\hat{\mathbf{d}}_l(k)$ and $\hat{\boldsymbol{\delta}}_l^{\ltimes}(k)$ respectively. $\mathbf{H}_l^{\ltimes}(k) \in \Re^{1\times w}$ is a given vector-valued function satisfying $\mathbf{H}_l^{\ltimes}(k) = \alpha^{\ltimes}(k)\mathbf{H}_{l-1}^{\ltimes}(k)$. $\alpha^{\ltimes}(k) \in \Re^1$ is an appropriate coefficient. $\tau^{\ltimes}$ and $\mathbf{M}^{\ltimes} \in \Re^{h\times n}$ are tunable gain and gain matrix respectively.

Define the following two observer errors:

$$\mathbf{e}_{\tilde{\boldsymbol{\Upsilon}}_l^{\ltimes}}^{\ltimes}(k) \triangleq \boldsymbol{\Upsilon}_l^{\ltimes}(k) - \hat{\boldsymbol{\Upsilon}}_l^{\ltimes}(k) \tag{7.16}$$

$$\mathbf{e}_{\Delta\tilde{\mathbf{d}},l}^{\ltimes}(k) \triangleq \Delta\mathbf{d}_l(k) - \Delta\hat{\mathbf{d}}_l(k) \tag{7.17}$$

Then the following theorem shows the convergence property of the proposed reduced-order observer (7.15).

Theorem 7.1 *For system (7.1), if the tunable parameters* $\bar{\mathbf{\Lambda}}^{\ltimes}$, $\bar{\mathbf{\Psi}}^{\ltimes}$, $\mathbf{M}^{\ltimes}$ *and* $\tau^{\ltimes}$ *satisfy:*

$$\begin{aligned}
&\max\left\{\left\|\hbar_2^{\ltimes}\left(\mathbf{B}^{\ltimes T}-\mathbf{M}^{\ltimes}\mathbf{C}\right)-\bar{\mathbf{\Psi}}^{\ltimes}\right\|,\ \beta^{\ltimes}\left\|\hbar_2^{\ltimes}\left(\mathbf{B}^{\ltimes T}-\mathbf{M}^{\ltimes}\mathbf{C}\right)-\bar{\mathbf{\Psi}}^{\ltimes}\right\|\right\}\\
&\quad+\max\left\{\beta^{\ltimes}\left\|\hbar_1^{\ltimes}\left(\mathbf{B}^{\ltimes T}-\mathbf{M}^{\ltimes}\mathbf{C}\right)-\tau^{\ltimes}\bar{\mathbf{\Lambda}}^{\ltimes}\right\|\left\|\mathbf{H}_{l+1}^{\ltimes}(k)\right\|,\right.\\
&\quad\left.\left\|\hbar_1^{\ltimes}\left(\mathbf{B}^{\ltimes T}-\mathbf{M}^{\ltimes}\mathbf{C}\right)-\bar{\mathbf{\Lambda}}^{\ltimes}\right\|\left\|\mathbf{H}_{l+1}^{\ltimes}(k)\right\|\right\}\leq\bar{\chi}^{\ltimes}<1
\end{aligned}\tag{7.18}$$

then the proposed reduced-order observer (7.15) guarantees that

$$\lim_{l\to\infty}\left\|\mathbf{e}^{\ltimes}_{\tilde{\mathbf{\Upsilon}}^{\ltimes}_{l+1}}(k)\right\|\leq\frac{\bar{\phi}^{\ltimes}}{1-\bar{\chi}^{\ltimes}}\tag{7.19}$$

and

$$\lim_{l\to\infty}\left\|\mathbf{e}^{\ltimes}_{\Delta\tilde{\mathbf{d}},l+1}(k)\right\|\leq\frac{\bar{\phi}^{\ltimes}}{1-\bar{\chi}^{\ltimes}}\tag{7.20}$$

where $0\leq\bar{\chi}^{\ltimes}<1$ *is a positive constant.* $\beta^{\ltimes}$ *is the upper bound for the change rate of control input between two consecutive iterations, defined as*

$$||\Delta\mathbf{u}_l(k)||\leq\beta^{\ltimes}\tag{7.21}$$

$\bar{\phi}^{\ltimes}$ *is defined in the subsequent (7.38).*

Proof Based on (7.11), (7.12) and (7.14) , $\Delta\hat{\mathbf{d}}_l(k)$ in (7.15) leads to

$$\begin{aligned}
\Delta\hat{\mathbf{d}}_l(k)&=\bar{\mathbf{\Psi}}^{\ltimes}\mathbf{C}^{+}\mathbf{y}_{l-1}(k+1)-\mathbf{\Omega}_l^{\ltimes}(k)+\hbar_2^{\ltimes}\mathbf{B}^{\ltimes T}\mathbf{x}_{l-1}(k+1)-\hbar_2^{\ltimes}\mathbf{e}^{\ltimes}_{\tilde{\delta}_l^{\ltimes}}(k)\\
&=\bar{\mathbf{\Psi}}^{\ltimes}\mathbf{x}_{l-1}(k+1)-\mathbf{\Omega}_l^{\ltimes}(k)-\hbar_2^{\ltimes}\mathbf{e}^{\ltimes}_{\tilde{\delta}_l^{\ltimes}}(k)
\end{aligned}\tag{7.22}$$

where

$$\mathbf{e}^{\ltimes}_{\tilde{\delta}_l^{\ltimes}}(k)\triangleq\boldsymbol{\delta}_l^{\ltimes}(k)-\hat{\boldsymbol{\delta}}_l^{\ltimes}(k)\tag{7.23}$$

According to (7.10), (7.15) and (7.22), $\mathbf{e}^{\ltimes}_{\Delta\tilde{\mathbf{d}},n+1}(k)$ can be rewritten as

$$\begin{aligned}\mathbf{e}^{\ltimes}_{\Delta\tilde{\mathbf{d}},l+1}(k) &= \Delta\mathbf{d}_{l+1}(k) - \bar{\boldsymbol{\Psi}}^{\ltimes}\mathbf{x}_l(k+1) + \boldsymbol{\Omega}^{\ltimes}_{l+1}(k) + \hbar^{\ltimes}_2\mathbf{e}^{\ltimes}_{\tilde{\delta}^{\ltimes}_{l+1}}(k)\\ &= \Delta\mathbf{d}_{l+1}(k) - \bar{\boldsymbol{\Psi}}^{\ltimes}\mathbf{x}_{l-1}(k+1) - \bar{\boldsymbol{\Psi}}^{\ltimes}\boldsymbol{\Upsilon}^{\ltimes}_l(k)\Delta\mathbf{u}_l(k)\\ &\quad -\bar{\boldsymbol{\Psi}}^{\ltimes}\Delta\mathbf{d}_l(k) + \boldsymbol{\Omega}^{\ltimes}_l(k) + \bar{\boldsymbol{\Psi}}^{\ltimes}\hat{\boldsymbol{\Upsilon}}^{\ltimes}_l(k)\Delta\mathbf{u}_l(k)\\ &\quad +\bar{\boldsymbol{\Psi}}^{\ltimes}\Delta\hat{\mathbf{d}}_l(k) + \Delta\hat{\mathbf{d}}_l(k) + \hbar^{\ltimes}_2\mathbf{e}^{\ltimes}_{\tilde{\delta}^{\ltimes}_{l+1}}(k) \qquad (7.24)\end{aligned}$$

By means of (7.22), (7.24) leads to

$$\begin{aligned}\mathbf{e}^{\ltimes}_{\Delta\tilde{\mathbf{d}},l+1}(k) &= \Delta\mathbf{d}_{l+1}(k) - \hbar^{\ltimes}_2\mathbf{e}^{\ltimes}_{\tilde{\delta}_l}(k) - \bar{\boldsymbol{\Psi}}^{\ltimes}\mathbf{e}^{\ltimes}_{\tilde{\boldsymbol{\Upsilon}}^{\ltimes}_l}(k)\Delta\mathbf{u}_l(k)\\ &\quad -\bar{\boldsymbol{\Psi}}^{\ltimes}\mathbf{e}^{\ltimes}_{\Delta\tilde{\mathbf{d}},l}(k) + \hbar^{\ltimes}_2\mathbf{e}^{\ltimes}_{\tilde{\delta}^{\ltimes}_{l+1}}(k) \qquad (7.25)\end{aligned}$$

Similar to the analysis of (7.22), based on (7.11), (7.15) and (7.23), $\hat{\boldsymbol{\Upsilon}}^{\ltimes}_l(k)$ in (7.15) leads to

$$\begin{aligned}\hat{\boldsymbol{\Upsilon}}^{\ltimes}_l(k) &= \bar{\boldsymbol{\Lambda}}^{\ltimes}\mathbf{C}^{+}\mathbf{y}_{l-1}(k+1)\mathbf{H}^{\ltimes}_l(k) - \boldsymbol{\Phi}^{\ltimes}_l(k)\\ &\quad +\hbar^{\ltimes}_1\left(\mathbf{B}^{\ltimes T}\mathbf{x}_{l-1}(k+1) - \mathbf{e}^{\ltimes}_{\tilde{\delta}^{\ltimes}_l}(k)\right)\mathbf{H}^{\ltimes}_l(k) \qquad (7.26)\end{aligned}$$

By virtue of (7.12) and (7.13), (7.26) can be simplified as

$$\hat{\boldsymbol{\Upsilon}}^{\ltimes}_l(k) = \bar{\boldsymbol{\Lambda}}^{\ltimes}\mathbf{x}_{l-1}(k+1)\mathbf{H}^{\ltimes}_l(k) - \boldsymbol{\Phi}^{\ltimes}_l(k) - \hbar^{\ltimes}_1\mathbf{e}^{\ltimes}_{\tilde{\delta}^{\ltimes}_l}(k)\mathbf{H}^{\ltimes}_l(k) \qquad (7.27)$$

According to (7.16) and (7.27), the observer error $\mathbf{e}^{\ltimes}_{\tilde{\boldsymbol{\Upsilon}}_{l+1}}(k)$ can be rewritten as

$$\begin{aligned}\mathbf{e}^{\ltimes}_{\tilde{\boldsymbol{\Upsilon}}^{\ltimes}_{l+1}}(k) &= \boldsymbol{\Upsilon}^{\ltimes}_{l+1}(k) - \bar{\boldsymbol{\Lambda}}^{\ltimes}\mathbf{x}_l(k+1)\mathbf{H}^{\ltimes}_{l+1}(k)\\ &\quad +\boldsymbol{\Phi}^{\ltimes}_{l+1}(k) + \hbar^{\ltimes}_1\mathbf{e}^{\ltimes}_{\tilde{\delta}^{\ltimes}_{l+1}}(k)\mathbf{H}^{\ltimes}_{l+1}(k) \qquad (7.28)\end{aligned}$$

By means of (7.10) and (7.15), (7.28) leads to

$$\begin{aligned}\mathbf{e}^{\ltimes}_{\tilde{\boldsymbol{\Upsilon}}^{\ltimes}_{l+1}}(k) &= \boldsymbol{\Upsilon}^{\ltimes}_{l+1}(k) - \bar{\boldsymbol{\Lambda}}^{\ltimes}\left(\mathbf{x}_{l-1}(k+1) + \boldsymbol{\Upsilon}^{\ltimes}_l(k)\Delta\mathbf{u}_l(k) + \Delta\mathbf{d}_l(k)\right)\mathbf{H}^{\ltimes}_{l+1}(k)\\ &\quad +\alpha^{\ltimes}(k)[\boldsymbol{\Phi}^{\ltimes}_l(k) + \hat{\boldsymbol{\Upsilon}}^{\ltimes}_l(k) + \tau^{\ltimes}\bar{\boldsymbol{\Lambda}}^{\ltimes}\hat{\boldsymbol{\Upsilon}}^{\ltimes}_l(k)\Delta\mathbf{u}_l(k)\mathbf{H}^{\ltimes}_l(k)\\ &\quad +\bar{\boldsymbol{\Lambda}}^{\ltimes}\Delta\hat{\mathbf{d}}_l(k)\mathbf{H}^{\ltimes}_l(k)] + \hbar^{\ltimes}_1\mathbf{e}^{\ltimes}_{\tilde{\delta}^{\ltimes}_{l+1}}(k)\mathbf{H}^{\ltimes}_{l+1}(k)\\ &= \boldsymbol{\Upsilon}^{\ltimes}_{l+1}(k) - \alpha^{\ltimes}(k)\bar{\boldsymbol{\Lambda}}^{\ltimes}\mathbf{x}_{l-1}(k+1)\mathbf{H}^{\ltimes}_l(k) + \alpha^{\ltimes}(k)\boldsymbol{\Phi}^{\ltimes}_l(k) + \alpha^{\ltimes}(k)\hat{\boldsymbol{\Upsilon}}^{\ltimes}_l(k)\\ &\quad -\alpha^{\ltimes}(k)\bar{\boldsymbol{\Lambda}}^{\ltimes}\boldsymbol{\Upsilon}^{\ltimes}_l(k)\Delta\mathbf{u}_l(k)\mathbf{H}^{\ltimes}_l(k) + \alpha^{\ltimes}(k)\tau^{\ltimes}\bar{\boldsymbol{\Lambda}}^{\ltimes}\hat{\boldsymbol{\Upsilon}}^{\ltimes}_l(k)\Delta\mathbf{u}_l(k)\mathbf{H}^{\ltimes}_l(k)\\ &\quad -\alpha^{\ltimes}(k)\bar{\boldsymbol{\Lambda}}^{\ltimes}\mathbf{e}^{\ltimes}_{\Delta\tilde{\mathbf{d}},l}(k)\mathbf{H}^{\ltimes}_l(k) + \alpha^{\ltimes}(k)\hbar^{\ltimes}_1\mathbf{e}^{\ltimes}_{\tilde{\delta}^{\ltimes}_{l+1}}(k)\mathbf{H}^{\ltimes}_l(k) \qquad (7.29)\end{aligned}$$

Based on (7.16) and (7.27), (7.29) can be rewritten as

$$
\begin{aligned}
\mathbf{e}^{\ltimes}_{\tilde{\boldsymbol{\Upsilon}}^{\ltimes}_{l+1}}(k) = {} & \boldsymbol{\Upsilon}^{\ltimes}_{l+1}(k) - \alpha^{\ltimes}(k)\tau^{\ltimes}\bar{\boldsymbol{\Lambda}}^{\ltimes}\mathbf{e}^{\ltimes}_{\tilde{\boldsymbol{\Upsilon}}^{\ltimes}_{l}}(k)\Delta\mathbf{u}_l(k)\mathbf{H}^{\ltimes}_l(k) \\
& -\alpha^{\ltimes}(k)(1-\tau^{\ltimes})\bar{\boldsymbol{\Lambda}}^{\ltimes}\boldsymbol{\Upsilon}^{\ltimes}_l(k)\Delta\mathbf{u}_l(k)\mathbf{H}^{\ltimes}_l(k) \\
& -\alpha^{\ltimes}(k)\hbar^{\ltimes}_1\mathbf{e}^{\ltimes}_{\tilde{\delta}^{\ltimes}_l}(k)\mathbf{H}^{\ltimes}_l(k) - \alpha^{\ltimes}(k)\bar{\boldsymbol{\Lambda}}^{\ltimes}\mathbf{e}^{\ltimes}_{\Delta\tilde{\mathbf{d}},l}(k)\mathbf{H}^{\ltimes}_l(k) \\
& +\alpha^{\ltimes}(k)\hbar^{\ltimes}_1\mathbf{e}^{\ltimes}_{\tilde{\delta}^{\ltimes}_{l+1}}(k)\mathbf{H}^{\ltimes}_l(k)
\end{aligned}
\tag{7.30}
$$

Based on (7.10), (7.11) and (7.15), $\mathbf{e}^{\ltimes}_{\tilde{\delta}^{\ltimes}_l}(k)$ in (7.23) can be rewritten as

$$
\begin{aligned}
\mathbf{e}^{\ltimes}_{\tilde{\delta}^{\ltimes}_{l+1}}(k) &= \mathbf{B}^{\ltimes T}\mathbf{x}_l(k+1) - \boldsymbol{\eta}^{\ltimes}_{l+1}(k) - \mathbf{M}^{\ltimes}\mathbf{y}_l(k+1) \\
&= \big(\mathbf{B}^{\ltimes T} - \mathbf{M}^{\ltimes}\mathbf{C}\big)\big(\mathbf{x}_{l-1}(k+1) + \boldsymbol{\Upsilon}^{\ltimes}_l(k)\Delta\mathbf{u}_l(k) + \Delta\mathbf{d}_l(k)\big) \\
&\quad -\big(\boldsymbol{\eta}^{\ltimes}_l(k) + \big(\mathbf{B}^{\ltimes T} - \mathbf{M}^{\ltimes}\mathbf{C}\big)\hat{\boldsymbol{\Upsilon}}^{\ltimes}_l(k)\Delta\mathbf{u}_l(k) + \big(\mathbf{B}^{\ltimes T} - \mathbf{M}^{\ltimes}\mathbf{C}\big)\Delta\hat{\mathbf{d}}_l(k)\big)
\end{aligned}
\tag{7.31}
$$

According to (7.10), (7.11), (7.15) and (7.23), we have

$$
\begin{aligned}
\boldsymbol{\eta}^{\ltimes}_l(k) &= \hat{\boldsymbol{\delta}}^{\ltimes}_l(k) - \mathbf{M}^{\ltimes}\mathbf{C}\mathbf{x}_{l-1}(k+1) \\
&= \mathbf{B}^{\ltimes T}\mathbf{x}_{l-1}(k+1) - \mathbf{e}^{\ltimes}_{\tilde{\delta}^{\ltimes}_l}(k) - \mathbf{M}^{\ltimes}\mathbf{C}\mathbf{x}_{l-1}(k+1) \\
&= (\mathbf{B}^{\ltimes T} - \mathbf{M}^{\ltimes}\mathbf{C})\mathbf{x}_{l-1}(k+1) - \mathbf{e}^{\ltimes}_{\tilde{\delta}^{\ltimes}_l}(k)
\end{aligned}
\tag{7.32}
$$

Substituting (7.32) into (7.31) leads to

$$
\mathbf{e}^{\ltimes}_{\tilde{\delta}^{\ltimes}_{l+1}}(k) = \mathbf{e}^{\ltimes}_{\tilde{\delta}^{\ltimes}_l}(k) + (\mathbf{B}^{\ltimes T} - \mathbf{M}^{\ltimes}\mathbf{C})\mathbf{e}^{\ltimes}_{\tilde{\boldsymbol{\Upsilon}}^{\ltimes}_l}(k)\Delta\mathbf{u}_l(k) + \Big(\mathbf{B}^{\ltimes T} - \mathbf{M}^{\ltimes}\mathbf{C}\Big)\mathbf{e}^{\ltimes}_{\Delta\tilde{\mathbf{d}},l}(k)
\tag{7.33}
$$

Substituting (7.33) into (7.25) and (7.30), one has

$$
\begin{aligned}
\mathbf{e}^{\ltimes}_{\Delta\tilde{\mathbf{d}},l+1}(k) = {} & \Big[\hbar^{\ltimes}_2\left(\mathbf{B}^{\ltimes T} - \mathbf{M}^{\ltimes}\mathbf{C}\right) - \bar{\boldsymbol{\Psi}}^{\ltimes}\Big]\mathbf{e}^{\ltimes}_{\Delta\tilde{\mathbf{d}},l}(k) \\
& + \Big[\hbar^{\ltimes}_2\left(\mathbf{B}^{\ltimes T} - \mathbf{M}^{\ltimes}\mathbf{C}\right) - \bar{\boldsymbol{\Psi}}^{\ltimes}\Big]\mathbf{e}^{\ltimes}_{\tilde{\boldsymbol{\Upsilon}}^{\ltimes}_l}(k)\Delta\mathbf{u}_l(k) + \Delta\mathbf{d}_{l+1}(k)
\end{aligned}
\tag{7.34}
$$

and

$$
\begin{aligned}
\mathbf{e}^{\ltimes}_{\tilde{\boldsymbol{\Upsilon}}^{\ltimes}_{l+1}}(k) = {} & \Big[\hbar^{\ltimes}_1\left(\mathbf{B}^{\ltimes T} - \mathbf{M}^{\ltimes}\mathbf{C}\right) - \tau^{\ltimes}\bar{\boldsymbol{\Lambda}}^{\ltimes}\Big]\mathbf{e}^{\ltimes}_{\tilde{\boldsymbol{\Upsilon}}^{\ltimes}_l}(k)\Delta\mathbf{u}_l(k)\mathbf{H}^{\ltimes}_{l+1}(k) \\
& + \Big[\hbar^{\ltimes}_1\left(\mathbf{B}^{\ltimes T} - \mathbf{M}^{\ltimes}\mathbf{C}\right) - \bar{\boldsymbol{\Lambda}}^{\ltimes}\Big]\mathbf{e}^{\ltimes}_{\Delta\tilde{\mathbf{d}},l}(k)\mathbf{H}^{\ltimes}_{l+1}(k) \\
& + \boldsymbol{\Upsilon}^{\ltimes}_{l+1}(k) - (1-\tau^{\ltimes})\bar{\boldsymbol{\Lambda}}^{\ltimes}\boldsymbol{\Upsilon}^{\ltimes}_l(k)\Delta\mathbf{u}_l(k)\mathbf{H}^{\ltimes}_{l+1}(k)
\end{aligned}
\tag{7.35}
$$

Based on (7.34), (7.35) and (7.21), we have

$$\begin{aligned}&\left\|\mathbf{e}^{\ltimes}_{\Delta\tilde{\mathbf{d}},l+1}(k)\right\|+\left\|\mathbf{e}^{\ltimes}_{\tilde{\boldsymbol{\Upsilon}}^{\ltimes}_{l+1}}(k)\right\| \le \max\Big\{\left\|\hbar^{\ltimes}_2(\mathbf{B}^{\ltimes T}-\mathbf{M}^{\ltimes}\mathbf{C})-\bar{\boldsymbol{\Psi}}^{\ltimes}\right\|,\\ &\beta^{\ltimes}\left\|\hbar^{\ltimes}_2(\mathbf{B}^{\ltimes T}-\mathbf{M}^{\ltimes}\mathbf{C})-\bar{\boldsymbol{\Psi}}^{\ltimes}\right\|\Big\}\Big(\left\|\mathbf{e}^{\ltimes}_{\Delta\tilde{\mathbf{d}},l}(k)\right\|+\left\|\mathbf{e}^{\ltimes}_{\tilde{\boldsymbol{\Upsilon}}^{\ltimes}_{l}}(k)\right\|\Big)\\ &+\max\Big\{\beta^{\ltimes}\left\|\hbar^{\ltimes}_1(\mathbf{B}^{\ltimes T}-\mathbf{M}^{\ltimes}\mathbf{C})-\tau^{\ltimes}\bar{\boldsymbol{\Lambda}}^{\ltimes}\right\|\left\|\mathbf{H}^{\ltimes}_{l+1}(k)\right\|,\\ &\left\|\hbar^{\ltimes}_1(\mathbf{B}^{\ltimes T}-\mathbf{M}^{\ltimes}\mathbf{C})-\bar{\boldsymbol{\Lambda}}^{\ltimes}\right\|\left\|\mathbf{H}^{\ltimes}_{l+1}(k)\right\|\Big\}\Big(\left\|\mathbf{e}^{\ltimes}_{\Delta\tilde{\mathbf{d}},l}(k)\right\|+\left\|\mathbf{e}^{\ltimes}_{\tilde{\boldsymbol{\Upsilon}}^{\ltimes}_{l}}(k)\right\|\Big)\\ &+\left\|\Delta\mathbf{d}_{l+1}(k)\right\|+\left\|\boldsymbol{\Upsilon}^{\ltimes}_{l+1}(k)-(1-\tau^{\ltimes})\bar{\boldsymbol{\Lambda}}^{\ltimes}\boldsymbol{\Upsilon}^{\ltimes}_{l}(k)\Delta\mathbf{u}_l(k)\mathbf{H}^{\ltimes}_{l+1}(k)\right\| \end{aligned} \tag{7.36}$$

According to (7.18), one can obtain from (7.36)

$$\begin{aligned}\left\|\mathbf{e}^{\ltimes}_{\Delta\tilde{\mathbf{d}},l+1}(k)\right\|+\left\|\mathbf{e}^{\ltimes}_{\tilde{\boldsymbol{\Upsilon}}^{\ltimes}_{l+1}}(k)\right\| &\le \bar{\chi}^{\ltimes}\left(\left\|\mathbf{e}^{\ltimes}_{\Delta\tilde{\mathbf{d}},l}(k)\right\|+\left\|\mathbf{e}^{\ltimes}_{\tilde{\boldsymbol{\Upsilon}}^{\ltimes}_{l}}(k)\right\|\right)+\|\Delta\mathbf{d}_{l+1}(k)\|\\ &+\left\|\boldsymbol{\Upsilon}^{\ltimes}_{l+1}(k)-(1-\tau^{\ltimes})\bar{\boldsymbol{\Lambda}}^{\ltimes}\boldsymbol{\Upsilon}^{\ltimes}_{l}(k)\Delta\mathbf{u}_l(k)\mathbf{H}^{\ltimes}_{l+1}(k)\right\| \end{aligned} \tag{7.37}$$

Based on Lemma 7.1 and (7.21), there exists a positive constant $\bar{\phi}^{\ltimes}$ that satisfies

$$\|\Delta\mathbf{d}_{l+1}(k)\|+\left\|\boldsymbol{\Upsilon}^{\ltimes}_{l+1}(k)-(1-\tau^{\ltimes})\bar{\boldsymbol{\Lambda}}^{\ltimes}\boldsymbol{\Upsilon}^{\ltimes}_{l}(k)\Delta\mathbf{u}_l(k)\mathbf{H}^{\ltimes}_{l+1}(k)\right\|\le\bar{\phi}^{\ltimes} \tag{7.38}$$

Based on (7.38), (7.37) leads to

$$\lim_{l\to\infty}\left(\left\|\mathbf{e}^{\ltimes}_{\Delta\tilde{\mathbf{d}},l+1}(k)\right\|+\left\|\mathbf{e}^{\ltimes}_{\tilde{\boldsymbol{\Upsilon}}^{\ltimes}_{l+1}}(k)\right\|\right)\le\frac{\bar{\phi}^{\ltimes}}{1-\bar{\chi}^{\ltimes}} \tag{7.39}$$

□

7.3.2 Predictive Model Construction

Define output tracking error $\mathbf{e}^{\ltimes}_{y,l}(k+1)$ and state tracking error $\mathbf{e}^{\ltimes}_{x,l}(k+1)$ as

$$\mathbf{e}^{\ltimes}_{y,l}(k+1)=\mathbf{y}^{\ltimes}_d(k+1)-\mathbf{y}_l(k+1) \tag{7.40}$$

$$\mathbf{e}^{\ltimes}_{x,l}(k+1)=\mathbf{x}^{\ltimes}_d(k+1)-\mathbf{x}_l(k+1) \tag{7.41}$$

where the desired output and state trajectories $\mathbf{y}^{\ltimes}_d(k+1)$ and $\mathbf{x}^{\ltimes}_d(k+1)$ satisfy that $\mathbf{y}^{\ltimes}_d(k+1)=\mathbf{C}\mathbf{x}^{\ltimes}_d(k+1)$.

Substituting (7.10) into (7.40) leads to

$$\mathbf{e}_{y,l}^{\ltimes}(k+1) = \mathbf{e}_{y,l-1}^{\ltimes}(k+1) - \mathbf{C}\boldsymbol{\Upsilon}_{l}^{\ltimes}(k)\Delta\mathbf{u}_{l}(k) - \mathbf{C}\Delta\mathbf{d}_{l}(k) \tag{7.42}$$

Note that $\boldsymbol{\Upsilon}_{l}^{\ltimes}(k)$ and $\Delta\mathbf{d}_{l}(k)$ are unknown in (7.42). The following available error model is used for prediction

$$\mathbf{e}_{y,l}^{\ltimes}(k+1) = \mathbf{e}_{y,l-1}^{\ltimes}(k+1) - \mathbf{C}\hat{\boldsymbol{\Upsilon}}_{l}^{\ltimes}(k)\Delta\mathbf{u}_{l}(k) - \mathbf{C}\Delta\hat{\mathbf{d}}_{l}(k) \tag{7.43}$$

where $\hat{\boldsymbol{\Upsilon}}_{l}^{\ltimes}(k)$ and $\Delta\hat{\mathbf{d}}_{l}(k)$ are obtained by the observer (7.15).

Based on (7.43), the following prediction model for future v step iterative prediction is constructed

$$\begin{aligned}
\mathbf{e}_{y,l+1|l}^{\ltimes}(k+1) &= \mathbf{e}_{y,l}^{\ltimes}(k+1) - \mathbf{C}\hat{\boldsymbol{\Upsilon}}_{l+1|l}^{\ltimes}(k)\Delta\mathbf{u}_{l+1}(k) - \mathbf{C}\Delta\hat{\mathbf{d}}_{l+1|l}(k)\\
\mathbf{e}_{y,l+2|l}^{\ltimes}(k+1) &= \mathbf{e}_{y,l}^{\ltimes}(k+1) - \mathbf{C}\hat{\boldsymbol{\Upsilon}}_{l+1|l}^{\ltimes}(k)\Delta\mathbf{u}_{l+1}(k) - \mathbf{C}\Delta\hat{\mathbf{d}}_{l+1|l}(k)\\
&\quad -\mathbf{C}\hat{\boldsymbol{\Upsilon}}_{l+2|l}^{\ltimes}(k)\Delta\mathbf{u}_{l+2}(k) - \mathbf{C}\Delta\hat{\mathbf{d}}_{l+2|l}(k)\\
&\;\;\vdots\\
\mathbf{e}_{y,l+v|l}^{\ltimes}(k+1) &= \mathbf{e}_{y,l}^{\ltimes}(k+1) - \mathbf{C}\hat{\boldsymbol{\Upsilon}}_{l+1|l}^{\ltimes}(k)\Delta\mathbf{u}_{l+1}(k) - \mathbf{C}\Delta\hat{\mathbf{d}}_{l+1|l}(k)\\
&\quad -\mathbf{C}\hat{\boldsymbol{\Upsilon}}_{l+2|l}^{\ltimes}(k)\Delta\mathbf{u}_{l+2}(k) - \mathbf{C}\Delta\hat{\mathbf{d}}_{l+2|l}(k) -\\
&\quad \cdots - \mathbf{C}\hat{\boldsymbol{\Upsilon}}_{l+v|l}^{\ltimes}(k)\Delta\mathbf{u}_{l+v}(k) - \mathbf{C}\Delta\hat{\mathbf{d}}_{l+v|l}(k)
\end{aligned} \tag{7.44}$$

where the subscript $l+v|l$ means the prediction for the future $(l+v)$ th iteration at the l th iteration.

To make the above prediction model (7.44) available, the terms $\hat{\boldsymbol{\Upsilon}}_{l+1|l}^{\ltimes}(k), \ldots, \hat{\boldsymbol{\Upsilon}}_{l+v|l}^{\ltimes}(k)$ and $\Delta\hat{\mathbf{d}}_{l+1|l}(k), \ldots, \Delta\hat{\mathbf{d}}_{l+v|l}(k)$ will be predicted in the following by using iterative learning multi-level hierarchical predicting algorithms.

1. Predict the terms $\hat{\boldsymbol{\Upsilon}}_{l+1|l}^{\ltimes}(k), \ldots, \hat{\boldsymbol{\Upsilon}}_{l+v|l}^{\ltimes}(k)$

Based on the available values $\hat{\boldsymbol{\Upsilon}}_{l}^{\ltimes}(k), \ldots, \hat{\boldsymbol{\Upsilon}}_{l-h}^{\ltimes}(k)$ by virtue of (7.15),where h is a proper constant and can be selected as $h \geq v$, the terms $\hat{\boldsymbol{\Upsilon}}_{l+1|l}^{\ltimes}(k), \ldots, \hat{\boldsymbol{\Upsilon}}_{l+v|l}^{\ltimes}(k)$ is acquired by

$$\begin{aligned}
\hat{\boldsymbol{\Upsilon}}_{l+1|l}^{\ltimes}(k) &= \Gamma_{1}^{\ltimes}(k)\hat{\boldsymbol{\Upsilon}}_{l}^{\ltimes}(k) + \Gamma_{2}^{\ltimes}(k)\hat{\boldsymbol{\Upsilon}}_{l-1}^{\ltimes}(k) + \cdots + \Gamma_{h}^{\ltimes}(k)\hat{\boldsymbol{\Upsilon}}_{l+1-h}^{\ltimes}(k)\\
\hat{\boldsymbol{\Upsilon}}_{l+2|l}^{\ltimes}(k) &= \Gamma_{1}^{\ltimes}(k)\hat{\boldsymbol{\Upsilon}}_{l+1|l}^{\ltimes}(k) + \Gamma_{2}^{\ltimes}(k)\hat{\boldsymbol{\Upsilon}}_{l}^{\ltimes}(k) + \cdots + \Gamma_{h}^{\ltimes}(k)\hat{\boldsymbol{\Upsilon}}_{l+2-h}^{\ltimes}(k)\\
&\;\;\vdots\\
\hat{\boldsymbol{\Upsilon}}_{l+v|l}^{\ltimes}(k) &= \Gamma_{1}^{\ltimes}(k)\hat{\boldsymbol{\Upsilon}}_{l+v-1|l}^{\ltimes}(k) + \Gamma_{2}^{\ltimes}(k)\hat{\boldsymbol{\Upsilon}}_{l+v-2|l}^{\ltimes}(k) + \cdots + \Gamma_{h}^{\ltimes}(k)\hat{\boldsymbol{\Upsilon}}_{l+v-h}^{\ltimes}(k)
\end{aligned} \tag{7.45}$$

where $\Gamma_j^{\ltimes}(k) \in \Re^{m\times m}$ $(j = 1, \ldots, h)$ are unknown coefficient matrices.

Define

$$\hat{\Gamma}_l^{\ltimes h}(k) = [\hat{\Gamma}_{1,l}^{\ltimes}(k), \hat{\Gamma}_{2,l}^{\ltimes}(k), \ldots, \hat{\Gamma}_{h,l}^{\ltimes}(k)]^T \tag{7.46}$$

as the estimation of the unknown coefficient matrix, then it can be obtained by the following iterative learning projection algorithm

$$\begin{aligned}\hat{\Gamma}_l^{\ltimes h}(k) = \hat{\Gamma}_{l-1}^{\ltimes h}(k) + \kappa^{\ltimes}\hat{\mathbf{\Upsilon}}_{l-1}^{\ltimes h}(k)\Big[\nu^{\ltimes}\mathbf{I} + \hat{\mathbf{\Upsilon}}_{l-1}^{\ltimes h}(k)^T\hat{\mathbf{\Upsilon}}_{l-1}^{\ltimes h}(k)\Big]^{-1} \\ \Big[\hat{\mathbf{\Upsilon}}_l^{\ltimes}(k)^T - \hat{\mathbf{\Upsilon}}_{l-1}^{\ltimes h}(k)^T\hat{\Gamma}_{l-1}^{\ltimes h}(k)\Big]\end{aligned} \tag{7.47}$$

where $\nu^{\ltimes}$ and $\kappa^{\ltimes}$ are known positive constants. $\mathbf{I} \in \Re^{\omega\times\omega}$ is the identity matrix, and

$$\hat{\mathbf{\Upsilon}}_{l-1}^{\ltimes h}(k) = \Big[\hat{\mathbf{\Upsilon}}_{l-1}^{\ltimes}(k)^T, \ldots, \hat{\mathbf{\Upsilon}}_{l-h}^{\ltimes}(k)^T\Big]^T \tag{7.48}$$

2. Predict the terms $\Delta\hat{\mathbf{d}}_{l+1|l}(k), \ldots, \Delta\hat{\mathbf{d}}_{l+\upsilon|l}(k)$

Based on the available values $\Delta\hat{\mathbf{d}}_l(k), \ldots, \Delta\hat{\mathbf{d}}_{l-h}(k)$ by virtue of (7.15), the terms $\Delta\hat{\mathbf{d}}_{l+1|l}(k), \ldots, \Delta\hat{\mathbf{d}}_{l+\upsilon|l}(k)$ is acquired by

$$\begin{aligned}\Delta\hat{\mathbf{d}}_{l+1|l}(k) &= \mathcal{H}_1^{\ltimes}(k)\Delta\hat{\mathbf{d}}_l(k) + \mathcal{H}_2^{\ltimes}(k)\Delta\hat{\mathbf{d}}_{l-1}(k) + \cdots + \mathcal{H}_h^{\ltimes}(k)\Delta\hat{\mathbf{d}}_{l+1-h}(k) \\ \Delta\hat{\mathbf{d}}_{l+2|l}(k) &= \mathcal{H}_1^{\ltimes}(k)\Delta\hat{\mathbf{d}}_{l+1|l}(k) + \mathcal{H}_2^{\ltimes}(k)\Delta\hat{\mathbf{d}}_l(k) + \cdots + \mathcal{H}_h^{\ltimes}(k)\Delta\hat{\mathbf{d}}_{l+2-h}(k) \\ &\vdots \\ \Delta\hat{\mathbf{d}}_{l+\upsilon|l}(k) &= \mathcal{H}_1^{\ltimes}(k)\Delta\hat{\mathbf{d}}_{l+\upsilon-1|l}(k) + \mathcal{H}_2^{\ltimes}(k)\Delta\hat{\mathbf{d}}_{l+\upsilon-2|l}(k) + \cdots + \mathcal{H}_h^{\ltimes}(k)\Delta\hat{\mathbf{d}}_{l+\upsilon-h}(k)\end{aligned} \tag{7.49}$$

where $\mathcal{H}_j^{\ltimes}(k) \in \Re^{m\times m}$ $(j = 1, \ldots h)$ are unknown coefficient matrices.

Define

$$\hat{\mathcal{H}}_l^{\ltimes h}(k) = [\hat{\mathcal{H}}_{1,l}^{\ltimes}(k), \hat{\mathcal{H}}_{2,l}^{\ltimes}(k), \ldots, \hat{\mathcal{H}}_{h,l}(k)]^T \tag{7.50}$$

as the estimation of the unknown coefficient matrix, then it is obtained by the following iterative learning projection algorithm

$$\begin{aligned}\hat{\mathcal{H}}_l^{\ltimes h}(k) = \hat{\mathcal{H}}_{l-1}^{\ltimes h}(k) + \bar{\kappa}^{\ltimes}\Delta\hat{\mathbf{d}}_{l-1}^h(k)\Big[\bar{\nu}^{\ltimes} + \Delta\hat{\mathbf{d}}_{l-1}^h(k)^T\Delta\hat{\mathbf{d}}_{l-1}^h(k)\Big]^{-1} \\ \Big[\Delta\hat{\mathbf{d}}_l(k)^T - \Delta\hat{\mathbf{d}}_{l-1}^h(k)^T\hat{\mathcal{H}}_{l-1}^{\ltimes h}(k)\Big]\end{aligned} \tag{7.51}$$

where $\bar{\nu}^{\ltimes}$ and $\bar{\kappa}^{\ltimes}$ are known positive constants, and

$$\Delta\hat{\mathbf{d}}_{l-1}^{h}(k) = \left[\Delta\hat{\mathbf{d}}_{l-1}(k)^T, \ldots, \Delta\hat{\mathbf{d}}_{l-h}(k)^T\right]^T \tag{7.52}$$

Remark 7.1 Analogous to Remark 6.2 in Chap. 6, based on the reduced-order observer (7.15) and predictor (7.45)–(7.52), the full system states can also be predicted as follows.

$$\begin{aligned}
\hat{\mathbf{x}}_l(k+1) &= \hat{\mathbf{x}}_{l-1}(k+1) - \hat{\boldsymbol{\Upsilon}}_l^{\ltimes}(k)\Delta\mathbf{u}_l(k) - \Delta\hat{\mathbf{d}}_l(k) \\
\hat{\mathbf{x}}_{l+1|l}(k+1) &= \hat{\mathbf{x}}_l(k+1) - \hat{\boldsymbol{\Upsilon}}_{l+1|l}^{\ltimes}(k)\Delta\mathbf{u}_{l+1}(k) - \Delta\hat{\mathbf{d}}_{l+1|l}(k) \\
\hat{\mathbf{x}}_{l+2|l}(k+1) &= \hat{\mathbf{x}}_l(k+1) - \hat{\boldsymbol{\Upsilon}}_{l+1|l}^{\ltimes}(k)\Delta\mathbf{u}_{l+1}(k) - \Delta\hat{\mathbf{d}}_{n+1|n}(k) \\
&\quad - \hat{\boldsymbol{\Upsilon}}_{l+2|l}^{\ltimes}(k)\Delta\mathbf{u}_{l+2}(k) - \Delta\hat{\mathbf{d}}_{l+2|l}(k) \\
&\ \ \vdots \\
\hat{\mathbf{x}}_{l+v|l}(k+1) &= \hat{\mathbf{x}}_l(k+1) - \hat{\boldsymbol{\Upsilon}}_{l+1|l}^{\ltimes}(k)\Delta\mathbf{u}_{l+1}(k) - \Delta\hat{\mathbf{d}}_{l+1|l}(k) \\
&\quad - \hat{\boldsymbol{\Upsilon}}_{l+2|l}^{\ltimes}(k)\Delta\mathbf{u}_{l+2}(k) - \Delta\hat{\mathbf{d}}_{l+2|l}(k) - \\
&\quad \cdots - \hat{\boldsymbol{\Upsilon}}_{l+v|l}^{\ltimes}(k)\Delta\mathbf{u}_{l+v}(k) - \Delta\hat{\mathbf{d}}_{l+v|l}(k)
\end{aligned} \tag{7.53}$$

where $\hat{\mathbf{x}}_l(k+1)$ is the observation of $\mathbf{x}_l(k+1)$. The initial value of $\hat{\mathbf{x}}_l(k+1)$ at $l=0$, namely $\hat{\mathbf{x}}_0(k+1)$, can be estimated by using historical data information of the system. $\hat{\mathbf{x}}_{l+j|l}(k+1)$, $j = 1, \cdots, v$ is the predicted system state for the future $(l+j)$th iteration at the lth iteration, and it can be further used for system monitor, fault diagnosis, etc.

7.3.3 Predictive ILC Design

Rewrite (7.44) as the following compact form

$$\mathbf{e}_{y,l+v|l}^{\ltimes}(k+1) = \mathbf{e}_{y,l}^{\ltimes}(k+1) - \mathbf{C}\hat{\boldsymbol{\Upsilon}}_{l+1|l}^{\ltimes v}(k)\Delta\mathbf{u}_{l+1}^{v}(k) - \mathbf{C}^{v}\Delta\hat{\mathbf{d}}_{l+1|l}^{v}(k) \tag{7.54}$$

where

$$\hat{\boldsymbol{\Upsilon}}_{l+1|l}^{\ltimes v}(k) = \left[\hat{\boldsymbol{\Upsilon}}_{l+1|l}^{\ltimes}(k), \ldots, \hat{\boldsymbol{\Upsilon}}_{l+v|l}^{\ltimes}(k)\right] \in \Re^{m\times wv} \tag{7.55}$$

$$\Delta\mathbf{u}_{l+1}^{v}(k) = \left[\Delta\mathbf{u}_{l+1}(k)^T, \ldots, \Delta\mathbf{u}_{l+v}(k)^T\right]^T \in \Re^{wv\times 1} \tag{7.56}$$

$$\Delta\hat{\mathbf{d}}^{v}_{l+1|l}(k) = \left[\Delta\hat{\mathbf{d}}_{l+1|l}(k)^{T}, \ldots, \Delta\hat{\mathbf{d}}_{l+v|l}(k)\right]^{T} \in \Re^{mv\times 1} \tag{7.57}$$

$$\mathbf{C}^{v} = \left[\mathbf{C}, \ldots, \mathbf{C}\right] \in \Re^{n\times mv} \tag{7.58}$$

Define the following quadratic cost function:

$$\begin{aligned} J^{\ltimes}_{l+1}(k) = \frac{1}{2}\Big\{&\mathbf{e}^{\ltimes T}_{y,l+v|l}(k+1)\bar{\mathbf{Q}}^{\ltimes}\mathbf{e}^{\ltimes}_{y,l+v|l}(k+1) \\ &+\Delta\mathbf{u}^{v}_{l+1}(k)^{T}\bar{\mathbf{R}}^{\ltimes}\Delta\mathbf{u}^{v}_{l+1}(k) + \varrho^{\ltimes T}_{l}(k+1)\bar{\mathbf{S}}^{\ltimes}\varrho^{\ltimes}_{l}(k+1)\Big\} \end{aligned} \tag{7.59}$$

where $\bar{\mathbf{Q}}^{\ltimes} \in \Re^{n\times n}$, $\bar{\mathbf{R}}^{\ltimes} \in \Re^{wv\times wv}$, and $\bar{\mathbf{S}}^{\ltimes} \in \Re^{n\times n}$ are positive definite and symmetric matrices. For simplicity, let $\bar{\mathbf{Q}}^{\ltimes} = \bar{q}^{\ltimes}\mathbf{I}_{n\times n}$, $\bar{\mathbf{R}}^{\ltimes} = \bar{r}^{\ltimes}\mathbf{I}_{wv\times wv}$, $\bar{q}^{\ltimes} > 0$, $\bar{r}^{\ltimes} > 0$. $\varrho^{\ltimes}_{l}(k+1) \geq \mathbf{0}$ is a slack variable vector.

If there are no system constraints, the optimal $\Delta\mathbf{u}^{\ltimes v}_{l+1}(k)$ can be obtained via minimizing (7.59) and using (7.54) as follows:

$$\begin{aligned} \Delta\mathbf{u}^{v}_{l+1}(k) = &\left[\hat{\boldsymbol{\Upsilon}}^{\ltimes v}_{l+1|l}(k)^{T}\mathbf{C}^{T}\bar{\mathbf{Q}}^{\ltimes}\mathbf{C}\hat{\boldsymbol{\Upsilon}}^{\ltimes v}_{l+1|l}(k) + \bar{\mathbf{R}}^{\ltimes}\right]^{-1}\hat{\boldsymbol{\Upsilon}}^{\ltimes v}_{l+1|l}(k)^{T} \\ &\mathbf{C}^{T}\bar{\mathbf{Q}}^{\ltimes}\left[\mathbf{e}^{\ltimes}_{y,l}(k+1) - \mathbf{C}^{v}\Delta\hat{\mathbf{d}}^{v}_{l+1|l}(k)\right] \end{aligned} \tag{7.60}$$

The convergence of above predictive ILC (7.60) based on the the proposed reduced-order observer (7.15) and prediction algorithm (7.45)–(7.52) is guaranteed by the following theorem.

Theorem 7.2 *Considering system (7.1) under Assumptions 7.1–7.3, the proposed reduced-order observer-based predictive ILC method (7.15), (7.45)–(7.52), (7.60) guarantees that the tracking error* $\mathbf{e}^{\ltimes}_{y,l}(k+1)$ *converges to a tunable residue.*

Proof The proof can be derived straightforwardly according the proof of Theorem 2.1 in Chap.2 and hence is omitted here. □

Furthermore, if system constraints (7.2)–(7.3) are considered, the following constrained optimization problem is considered:

$$\begin{aligned} J^{\ltimes}_{l+1}(k) = \min_{\Delta\mathbf{u}^{v}_{l+1}(k),\varrho^{\ltimes}_{l}(k+1)} \frac{1}{2}\Big\{&\mathbf{e}^{\ltimes T}_{y,l+v|l}(k+1)\bar{\mathbf{Q}}^{\ltimes}\mathbf{e}^{\ltimes}_{y,l+v|l}(k+1) \\ &+\Delta\mathbf{u}^{v}_{l+1}(k)^{T}\bar{\mathbf{R}}^{\ltimes}\Delta\mathbf{u}^{v}_{l+1}(k) + \varrho^{\ltimes T}_{l}(k+1)\bar{\mathbf{S}}^{\ltimes}\varrho^{\ltimes}_{l}(k+1)\Big\} \\ &\text{subject to (7.62)} \end{aligned} \tag{7.61}$$

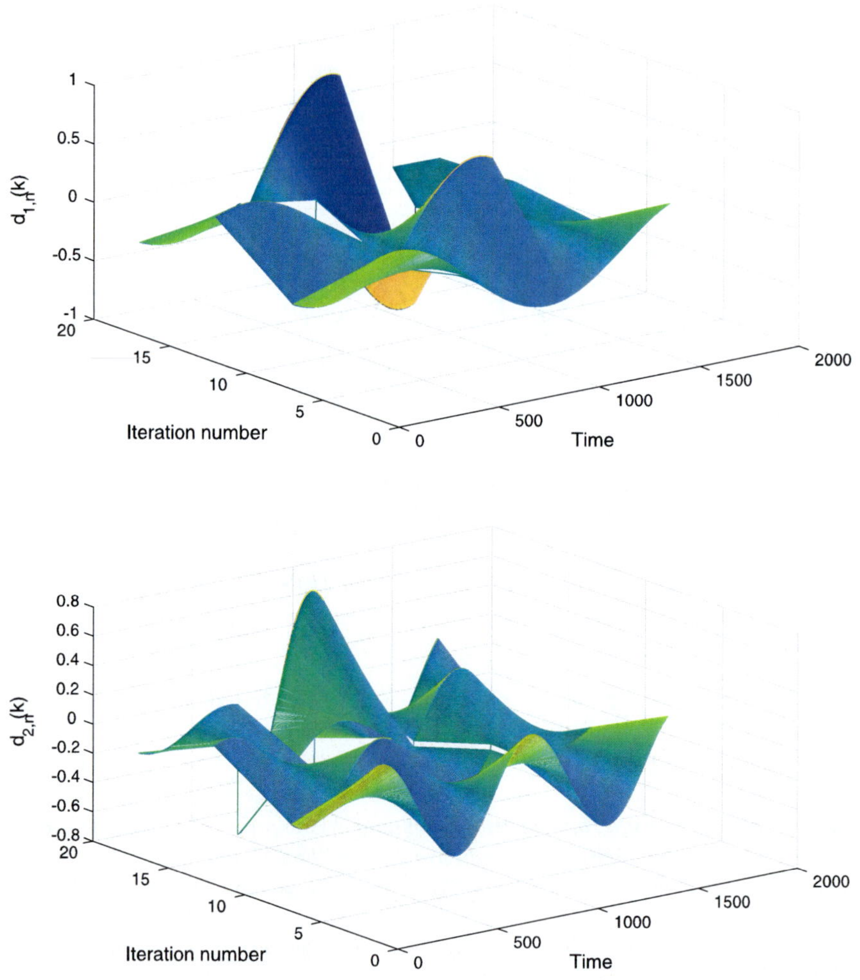

Fig. 7.1 External disturbance $\mathbf{d}_l(k)$ profiles

where

$$\begin{aligned} \mathbf{u}_{\min} &\le \mathbf{u}_{l+1}(k) \le \mathbf{u}_{\max} \\ \Delta\mathbf{u}_{\min} &\le \Delta\mathbf{u}_{l+1}(k) \le \Delta\mathbf{u}_{\max} \\ \mathbf{y}_{\min} - \varrho_l^{\ltimes}(k+1) &\le \hat{\mathbf{y}}_{l+v|l}(k+1) \le \mathbf{y}_{\max} + \varrho_l^{\ltimes}(k+1) \end{aligned} \tag{7.62}$$

Here (7.62) are the constraints on control input, change rate of control input and predicted output. $\hat{\mathbf{y}}_{l+v|l}(k+1) = \mathbf{y}_d^{\ltimes}(k+1) - \mathbf{e}_{y,l+v|l}^{\ltimes}(k+1)$. $\mathbf{u}_{\min}$,$\mathbf{u}_{\max}$, $\mathbf{y}_{\min}$, $\mathbf{y}_{\max}$ are the restrictions defined in (7.2)–(7.3). $\Delta\mathbf{u}_{\min}$ and $\Delta\mathbf{u}_{\max}$ are the lower and upper restrictions on the change rate of control input.

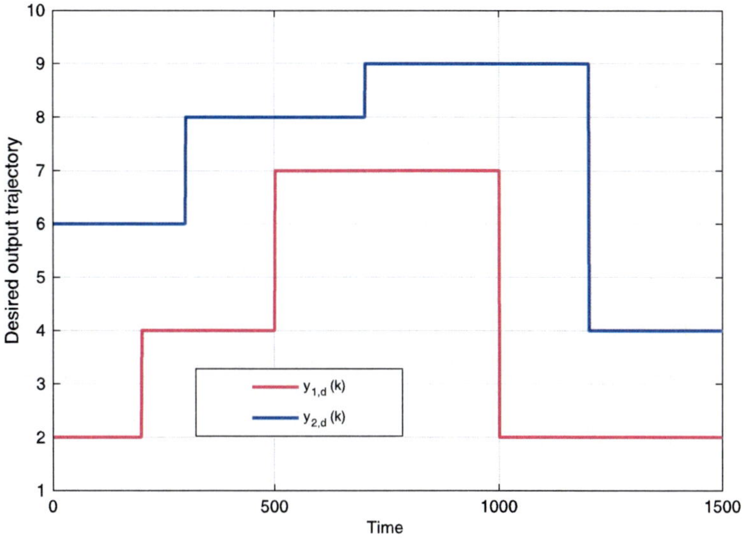

Fig. 7.2 Desired trajectory profiles

Based on (7.15), (7.45)–(7.54), the above constrained optimization problem (7.61) can be solved by QP algorithm in MATLAB and can be converged by the following theorem.

Theorem 7.3 *Considering system (7.1) under Assumptions 7.1–7.4, the proposed reduced-order observer-based constrained predictive ILC method (7.15), (7.45)–(7.52), (7.61)–(7.62) guarantees that the tracking error* $\mathbf{e}^{\ltimes}_{y,l}(k+1)$ *converges to a tunable residue.*

Proof The proof can be derived straightforwardly according the proof of Theorem 3.1 in Chap. 3 and hence is omitted here. □

7.4 Simulation Validation

Consider the system (7.1) with $\mathbf{x}_l(k) = [x_{11,l}(k), x_{12,l}(k), x_{21,l}(k), x_{22,l}(k)]^T$, $\mathbf{u}_l(k) = [0, u_{2,l}(k), 0, u_{4,l}(k)]^T$, $\mathbf{f} = [x_{11,l}(k)^2/\left(1 + x_{11,l}(k)^2\right) + 0.5x_{12,l}(k), \quad x_{11,l}(k)^2/\left(1 + x_{12,l}(k)^2 + x_{21,l}(k)^2 + x_{22,l}(k)^2\right) + a^{\ltimes}(k)u_{2,l}(k)^3, \; x_{21,l}(k)^2/\left(1 + x_{21,l}(k)^2\right) + 0.8x_{22,l}(k), \; x_{21,l}(k)^2/\left(1 + x_{11,l}(k)^2 + x_{12,l}(k)^2 + x_{22,l}(k)^2\right) + b^{\ltimes}(k) \quad u_{4,l}(k)^3]^T$. $a^{\ltimes}(k) = 2 + 0.2\sin(2\pi k/1500)$. $b^{\ltimes}(k) = 3 + 0.3\cos(2\pi k/1500)$. $\mathbf{d}_l(k) = \left[\, d_{1,l}(k),\; 0,\; d_{3,l}(k) \; 0 \,\right]^T$ that is shown in Fig. 7.1. $\mathbf{y}_l(k) = [y_{1,l}(k), y_{2,l}(k)]^T$. $\mathbf{C} = \left[1\;0\;0\;0\;;\; 0\;0\;1\;0\right]$ (Hou and Jin 2011).

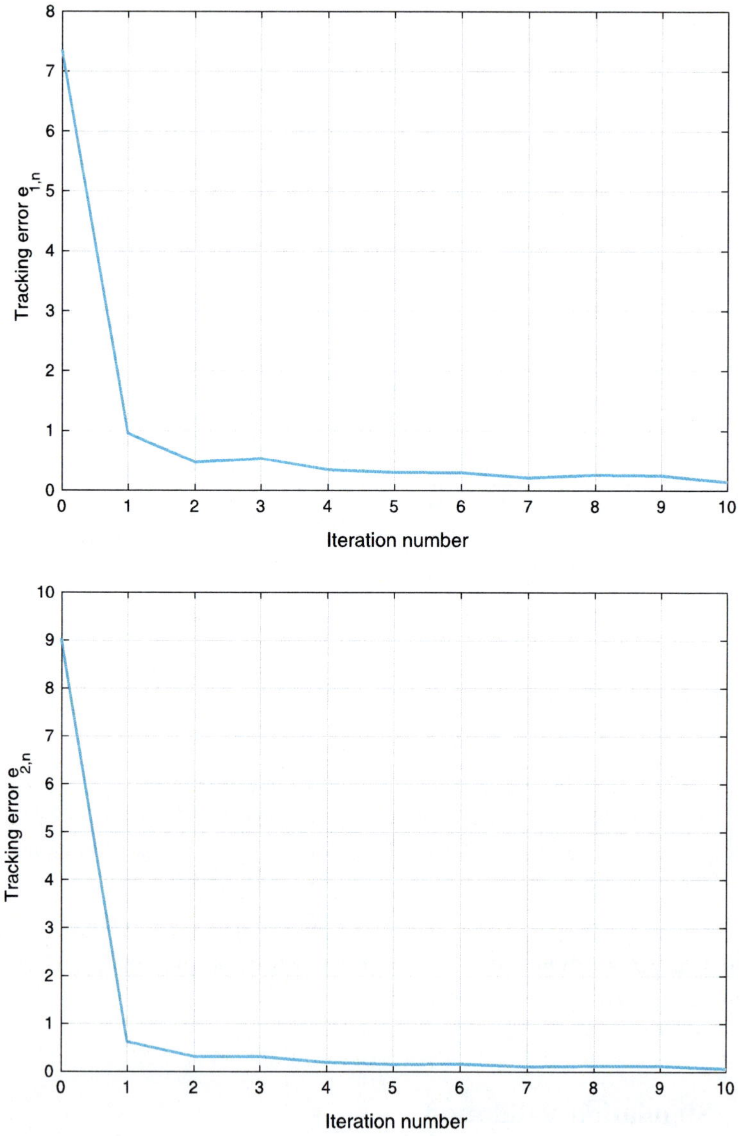

Fig. 7.3 Tracking error profiles

The desired trajectory $\mathbf{y}_d^{\ltimes}(k) = [y_{1,d}^{\ltimes}(k), y_{2,d}^{\ltimes}(k)]^T$ is depicted in Fig. 7.2. The proposed reduced-order observer-based predictive ILC method (7.15), (7.45)–(7.52), (7.60) is simulated and the control parameters are selected as: $\bar{\mathbf{\Lambda}}^{\ltimes} = 3 \times 10^{-4}\mathbf{I}_{4\times4}$, $\bar{\mathbf{\Psi}}^{\ltimes} = 10^{-1} \times \mathbf{I}_{4\times4}$. $\hbar_1^{\ltimes} = [\,0 \;\; 3\times10^{-4} \;\; 0 \;\; 0\,]^T$, $\hbar_2^{\ltimes} = \left[\,0 \;\; 1\times10^{-1} \;\; 0 \;\; 0\,\right]^T$, $\mathbf{B}^{\ltimes T} = \left[\,0 \;\; 1 \;\; 0 \;\; 0\,\right]$, $\mathbf{M}^{\ltimes} = [\,2 \;\; 2\,]$, $\mathbf{H}_l^{\ltimes}(k) = [\,10 \;\; 10\,]$, $\tau^{\ltimes} = 0.5$, $\bar{q}^{\ltimes} = 0.9$, $\bar{r}^{\ltimes} = 0.5$.

Figure 7.3 shows the maximum absolute value of output tracking errors $\mathbf{e}_{y,n}^{\ltimes}(k)$ defined in (7.40). Note that even though the considered system is under two unavailable system states and randomly varying external disturbance, satisfactory control performance by applying the proposed reduced-order observer-based predictive ILC method (7.15), (7.45)–(7.52), (7.60) can still be achieved.

7.5 Conclusion

This chapter designs a novel PILC method to deal with the control problem for a class of unknown MIMO nonaffine nonlinear systems with unavailable states. Not only that, the considered system is subjected to unknown external disturbances and system constraints. The proposed method can fully capture and predict the state information of the unknown system without using any prior model knowledge and can obtain an optimal control input sequence to guide the operation of the system in future iterations. The convergence of the proposed PILC method is guaranteed through theoretical analysis. Simulation results validate the effectiveness of the proposed PILC further.

References

Chen W-H, Yang J, Guo L, Li SH (2016) Disturbance-observer-based control and related methods-an overview. IEEE Trans Ind Electron 63(2):1083–1095

Hou ZS, Jin ST (2011) Data-driven model-free adaptive control for a class of MIMO nonlinear discrete-time systems. IEEE Trans Neural Netw Learn Syst 22(12):2173–2188

Kim K-S, Rew K-H (2013) Reduced order disturbance observer for discrete-time linear systems. Automatica 49(4):1968–1975

Kim K-S, Rew K-H, Kim S (2010) Disturbance observer for estimating higher order disturbances in time series expansion. IEEE Trans Autom Cont 55(8):1905–1911

Li K, Li DW, Xi YG, Yin DB (2014) Model predictive control with feedforward strategy for gas collectors of coke ovens. Chinese J Chem Eng 7(22):769–773

Yang J, Li S, Sun C, Guo L (2013) Nonlinear-disturbance-observer-based robust flight control for airbreathing hypersonic vehicles. IEEE Trans Aerospace Electron Syst 1(49):160–169

Yang J, Su J, Li S, Yu X (2014) High-order mismatched disturbance compensation for motion control systems via a continuous dynamic sliding-mode approach. IEEE Trans Ind Inf 10(1):604–614

Part II
Applications

Chapter 8
High-Speed Train Automatic Operation Systems

8.1 Introduction

High-speed train (HST) is a fast, comfortable, high-loading capacity and environmentally friendly transportation system and becomes more popular with modern people. In fact, HST operation systems have the significant characteristic of repeatability since the train always executes the same passenger transportation task in the same fixed time interval everyday. Therefore, ILC is an ideal control method to address the control problem of HST operation systems.

Up to date, a lot of time domain control methods have been proposed for HSTs (Dong et al. 2013; Song et al. 2014; Ye and Liu 2017; Xiao et al. 2020; Liu et al. 2022). These time domain control methods can guarantee convergence property only when the running time of the train goes to infinity, but in fact, the train always runs in the finite time interval. In addition, the transient behavior in the initial operating time interval may not be improved no matter how many times the train repeats. To overcome these problems, some ILC methods are developed for HSTs (Ji et al. 2015; Yu and Hou 2021; Li et al. 2021). These methods not only can achieve perfect tracking control in the finite operating time interval but also can eliminate the undesirable transient behavior in the initial operating time interval from learning; moreover, linearized parameter uncertainties in the train dynamics can also be addressed well.

However, the structural information of the HST model should be known in the above mentioned methods. In fact, actual HST operation system, as well as many other industry systems, is highly nonlinear and may experience many unknown external disturbances. For example, an 84th-order set of differential equations may be needed to describe a conventional seven-carriage train (Goodall and Kortüm 2002). In this chapter, we will consider the train model as an unknown nonaffine nonlinear system and design a new RBFNN-based PILC method.

This chapter is organized as follows. Section 8.2 presents the problem formulation. Section 8.3 shows the constructed RBFNN-based PILC. Simulation results on a HST operation system similar to CRH (China Railways High-speed)3 are shown in Sect. 8.4. Some conclusions are given in Sect. 8.5.

© The Author(s), under exclusive license to Springer Nature Singapore Pte Ltd. 2023

Q. Yu et al., *Predictive Learning Control for Unknown Nonaffine Nonlinear Systems*, Intelligent Control and Learning Systems 8,
https://doi.org/10.1007/978-981-19-8857-8_8

8.2 Train Dynamics and Problem Formation

8.2.1 Dynamics Description of HST

A widely used train model is described as follows (Ji et al. 2015; Yu and Hou 2021; Li et al. 2021)

$$M_l(t)\dot{v}_l(t) = u_l(t) - f_b(v_l(t)) - f_a(s_l(t)) \tag{8.1}$$

where l denotes the repeated operation number of the train from its starting station to the destination station. $t \in [0, T^{\rtimes}]$ (s) is the continuous operation time of the train. $M_l(t)$ (ton) denotes the total mass of the train, including different numbers of passengers and loads in each operation. $v_l(t)$ (m/s) and $s_l(t)$ (m) are the speed and position of the train respectively. $u_l(t)$ (N) represents the traction force or braking force when it becomes negative. $f_b(v_l(t))$ (N) represents basic resistance. $f_a(s_l(t))$ (N) represents additional resistance in special sections such as slops, curves, and tunnels.

To be convenient for control by a computer, the above continuous time train model (8.1) is discretized as follows by utilizing Euler formula:

$$v_l(k+1) = \frac{t_s}{Z_l(k)}u_l(k) + v_l(k) - \frac{t_s}{Z_l(k)}\left(f_b(v_l(k)) + f_a\left(\int_0^k v_l(\tau)d\tau\right)\right) \tag{8.2}$$

where $k \in \{0, 1, \ldots, K^{\rtimes}\}$ (s) is the discrete operation time of the train and t_s (s) is the sampling time. $T^{\rtimes} = t_s \cdot K^{\rtimes}$ holds.

Note that since the practical operation environment of HST grows increasingly complex, it is reasonable to assume that the basic resistance $f_b(v_l(k))$ and additional resistance $f_a\left(\int_0^k v_l(\tau)d\tau\right)$ in (8.2) are completely unknown. Then the widely used train model (8.2) can be rewritten as the following more general form:

$$v_l(k+1) = f(v_l(k), u_l(k)) + d_l(k) \tag{8.3}$$

where f is an unknown nonlinear function. $d_l(k)$ denotes unknown but bounded external disturbances.

In this chapter, the considered system (8.3) without external disturbances (namely $d_l(k) = 0$) is also under Assumptions 2.1–2.4 in Chap. 2. Then with external disturbances $d_l(k)$, the considered system (8.3) can be transformed into the following equivalent disturbance-based dynamic linearization data model.

Lemma 8.1 *For nonlinear system (8.3) satisfying Assumptions 2.1–2.4 in Chap. 2, with $|\Delta u_l(k)| \neq 0$ for each fixed k, l, there must exist a bounded pseudo-partial* *derivative $G_l^{\rtimes}(k)$ such that system (8.3) can be transformed into the following equivalent disturbance-based dynamic linearization data model:*

$$\Delta v_l(k+1) = G_l^{\rtimes}(k)\Delta u_l(k) + \Delta d_l(k) \tag{8.4}$$

where $\Delta u_l(k) = u_l(k) - u_{l-1}(k)$, $\Delta d_l(k) = d_l(k) - d_{l-1}(k)$.

Proof The proof can be derived straightforwardly according the proof of Lemma 2.1 in Chap. 2 and hence is omitted here. □

8.2.2 Control Objective

The objective of this chapter is to estimate the unknown $G_l^{\rtimes}(k)$ and unknown external disturbances $\Delta d_l(k)$ in (8.4) simultaneously by constructing a radial basis function neural network (RBFNN) along the iteration direction, and design a predictive iterative learning control input force $u_l(k)$ to make the actual speed of the train $v_l(k+1)$ track the desired speed trajectory $v_d^{\rtimes}(k+1)$ as the iteration number l tends to infinity.

8.3 RBFNN-Based PILC Design

As is well known, radial basis function neural network (RBFNN) is an effective technology for nonlinear approximation (Orr 1996). In this section, a new RBFNN along the iteration direction is constructed to estimate both $G_l^{\rtimes}(k)$ and $\Delta d_l(k)$ in (8.4).

The constructed RBFNN has one input layer, one hidden layer and one output layer in total. Denote $\psi_l^{\rtimes}(k) = [v_l(k),\ v_l(k-1),\ \ldots,\ v_l(k-L_y^{\rtimes}), v_{l-1}(k),\ \ldots,\ v_{l-l_y^{\rtimes}}(k), u_l(k),\ u_l(k-1),\ \ldots,\ u_l(k-L_u^{\rtimes}), u_{l-1}(k),\ \ldots,\ u_{l-l_u^{\rtimes}}(k)\ ]^T$ as the input of the RBFNN. Denote $\hat{G}_l^{\rtimes}(k)$ and $\Delta\hat{d}_l(k)$ as two outputs of the RBFNN, where they are the estimation of $G_l^{\rtimes}(k)$ and $\Delta d_l(k)$ in (8.4). Then the outputs $\hat{G}_l^{\rtimes}(k)$ and $\Delta\hat{d}_l(k)$ of the RBFNN can be described as

$$\hat{G}_l^{\rtimes}(k) = \mathbf{w}^{\rtimes}(k)^T\mathbf{h}_l^{\rtimes}(k) \tag{8.5}$$

and

$$\Delta\hat{d}_l(k) = \bar{\mathbf{w}}^{\rtimes}(k)^T\mathbf{h}_l^{\rtimes}(k) \tag{8.6}$$

where $\mathbf{w}^{\rtimes}(k) = [w_1^{\rtimes}(k), \ldots, w_{M^{\rtimes}}^{\rtimes}(k)]^T$ and $\bar{\mathbf{w}}^{\rtimes}(k) = [\bar{w}_1^{\rtimes}(k), \ldots, \bar{w}_{M^{\rtimes}}^{\rtimes}(k)]^T$ are the connecting weight vector between hidden layer and output layer. $\mathbf{h}_l^{\rtimes}(k) = \left[h_{1,l}^{\rtimes}(k), \ldots, h_{M^{\rtimes},l}^{\rtimes}(k)\right]^T$ is the output vector of the hidden layer. $M^{\rtimes}$ is the number of hidden nodes. Then the mth output value of the hidden layer can be obtained by

$$h_{m,l}^{\bowtie}(k) = \mathrm{e}^{(-\left\|\psi_l^{\bowtie}(k)-c_m^{\bowtie}(k)\right\| \quad -\left\|\psi_l^{\bowtie}(k)-c_m^{\bowtie}(k)\right\|\gamma_m^{\bowtie}(k)^2)}\gamma_m^{\bowtie}(k)^2),\ m = 1, \ldots, M^{\bowtie} \quad (8.7)$$

where $\mathbf{c}_m^{\bowtie}(k)$ denotes the center vector of the mth hidden node, $\left\|\psi_l^{\bowtie}(k) - \mathbf{c}_m^{\bowtie}(k)\right\|$ is the Euclidean distance between $\psi_l^{\bowtie}(k)$ and $\mathbf{c}_m^{\bowtie}(k)$, and $\gamma_m^{\bowtie}(k)$ is the radius of the mth hidden node.

In the above (8.5)–(8.7), the RBFNN parameters $\mathbf{w}^{\bowtie}(k)$, $\bar{\mathbf{w}}^{\bowtie}(k)$, $\mathbf{c}_m^{\bowtie}(k)$, and $\gamma_m^{\bowtie}(k)$ are updated by the following steepest descent iterative learning method:

$$\hat{\mathbf{w}}_l^{\bowtie}(k) = \hat{\mathbf{w}}_{l-1}^{\bowtie}(k) - \eta_1^{\bowtie} e_{p,l-1}^{\bowtie}(k+1)\Delta u_{l-1}(k)\mathbf{h}_{l-1}^{\bowtie}(k) \quad (8.8)$$

$$\hat{\bar{\mathbf{w}}}_l^{\bowtie}(k) = \hat{\bar{\mathbf{w}}}_{l-1}^{\bowtie}(k) - \eta_2^{\bowtie} e_{p,l-1}^{\bowtie}(k+1)\mathbf{h}_{l-1}^{\bowtie}(k) \quad (8.9)$$

$$\hat{\mathbf{c}}_{m,l}^{\bowtie}(k) = \hat{\mathbf{c}}_{m,l-1}^{\bowtie}(k) - \eta_3^{\bowtie} e_{p,l-1}^{\bowtie}(k+1)\left[\Delta u_{l-1}(k)\hat{w}_{m,l-1}^{\bowtie}(k)\left(\frac{\partial h_{m,l-1}^{\bowtie}(k)}{\partial \hat{\mathbf{c}}_{m,l-1}^{\bowtie}(k)}\right) + \hat{\bar{w}}_{m,l-1}^{\bowtie}(k)\left(\frac{\partial h_{m,l-1}^{\bowtie}(k)}{\partial \hat{\mathbf{c}}_{m,l-1}^{\bowtie}(k)}\right)\right] \quad (8.10)$$

$$\hat{\gamma}_{m,l}^{\bowtie}(k) = \hat{\gamma}_{m,l-1}^{\bowtie}(k) - \eta_4^{\bowtie} e_{p,l-1}^{\bowtie}(k+1)\left[\Delta u_{l-1}(k)\hat{w}_{m,l-1}^{\bowtie}(k)\left(\frac{\partial h_{m,l-1}^{\bowtie}(k)}{\partial \hat{\gamma}_{m,l-1}^{\bowtie}(k)}\right) + \hat{\bar{w}}_{m,l-1}^{\bowtie}(k)\left(\frac{\partial h_{m,l-1}^{\bowtie}(k)}{\partial \hat{\gamma}_{m,l-1}^{\bowtie}(k)}\right)\right] \quad (8.11)$$

where

$$e_{p,l}^{\bowtie}(k+1) = \hat{v}_l(k+1) - v_l(k+1) \quad (8.12)$$

and

$$\hat{v}_l(k+1) = v_{l-1}(k+1) + \hat{G}_l^{\bowtie}(k)\Delta u_l(k) + \Delta\hat{d}_l(k) \quad (8.13)$$

according to (8.8)–(8.13), $v_l(k+1)$ is the actual speed and $\hat{v}_l(k+1)$ is the estimated speed by the constructed RBFNN. $\hat{\mathbf{w}}_l^{\bowtie}(k)$, $\hat{\bar{\mathbf{w}}}_l^{\bowtie}(k)$, $\hat{\mathbf{c}}_{m,l}^{\bowtie}(k)$, $\hat{\gamma}_{m,l}^{\bowtie}(k)$ are the estimation of $\mathbf{w}^{\bowtie}(k)$, $\bar{\mathbf{w}}^{\bowtie}(k)$, $\mathbf{c}_m^{\bowtie}(k)$, $\gamma_m^{\bowtie}(k)$ at the lth operation of the HST. $\eta_1^{\bowtie}$, $\eta_2^{\bowtie}$, $\eta_3^{\bowtie}$ and $\eta_4^{\bowtie}$ are four tunable step sizes. $m \in \{1, \ldots, M^{\bowtie}\}$. $\hat{w}_{m,l-1}^{\bowtie}(k)$ and $\hat{\bar{w}}_{m,l-1}^{\bowtie}(k)$ are the estimation of the mth element of $\mathbf{w}^{\bowtie}(k)$ and $\bar{\mathbf{w}}^{\bowtie}(k)$ at the $(l-1)$th operation of the HST.

Then, the unknown $G_l^{\bowtie}(k)$ and $\Delta d_l(k)$ can be estimated as follows:

$$\hat{G}_l^{\bowtie}(k) = \hat{\mathbf{w}}_l^{\bowtie}(k)^T\mathbf{h}_l^{\bowtie}(k) \quad (8.14)$$

$$\Delta\hat{d}_l(k) = \hat{\bar{\mathbf{w}}}_l^{\bowtie}(k)^T\mathbf{h}_l^{\bowtie}(k) \quad (8.15)$$

Based on (8.14)–(8.15), the data model (8.4) is available. In the following, the RBFNN-based PILC scheme is designed.

Define the following tracking control error

$$e_l^{\rtimes}(k+1) = v_d^{\rtimes}(k+1) - v_l(k+1) \tag{8.16}$$

where $v_d^{\rtimes}(k+1)$ is the desired speed trajectory.

Subtracting $v_d^{\rtimes}(k+1)$ from both sides of (8.4), we have:

$$e_l^{\rtimes}(k+1) = e_{l-1}^{\rtimes}(k+1) - G_l^{\rtimes}(k)\Delta u_l(k) - \Delta d_l(k) \tag{8.17}$$

Based on the constructed RBFNN (8.14)–(8.15), we can get the following available error model for prediction:

$$e_l^{\rtimes}(k+1) = e_{l-1}^{\rtimes}(k+1) - \hat{G}_l^{\rtimes}(k)\Delta u_l(k) - \Delta \hat{d}_l(k) \tag{8.18}$$

According to (8.18), we can get the following error model for j step iterative prediction:

$$\begin{aligned}
e_{l+1}^{\rtimes}(k+1) &= e_l^{\rtimes}(k+1) - \hat{G}_{l+1}^{\rtimes}(k)\Delta u_{l+1}(k) - \Delta \hat{d}_{l+1}(k)\\
e_{l+2}^{\rtimes}(k+1) &= e_l^{\rtimes}(k+1) - \hat{G}_{l+1}^{\rtimes}(k)\Delta u_{l+1}(k) - \Delta \hat{d}_{l+1}(k)\\
&\quad - \hat{G}_{l+2}^{\rtimes}(k)\Delta u_{l+2}(k) - \Delta \hat{d}_{l+2}(k)\\
&\vdots\\
e_{l+j}^{\rtimes}(k+1) &= e_l^{\rtimes}(k+1) - \hat{G}_{l+1}^{\rtimes}(k)\Delta u_{l+1}(k)\\
&\quad - \hat{G}_{l+2}^{\rtimes}(k)\Delta u_{l+2}(k) - \cdots\\
&\quad - \hat{G}_{l+j}^{\rtimes}(k)\Delta u_{l+j}(k) - \Delta \hat{d}_{l+1}(k)\\
&\quad - \Delta \hat{d}_{l+2}(k) - \cdots - \Delta \hat{d}_{l+j}(k)
\end{aligned} \tag{8.19}$$

where $e_{l+j}^{\rtimes}(k+1)$, $\hat{G}_{l+j}^{\rtimes}(k)$, and $\Delta \hat{d}_{l+j}(k)$ are the future $(l+j)$step prediction of speed error, pseudo-partial derivative and external disturbances at the current iteration l.

In (8.19), $\hat{G}_{l+1}^{\rtimes}(k), \ldots, \hat{G}_{l+j}^{\rtimes}(k)$ and $\Delta \hat{d}_{l+1}(k), \ldots, \Delta \hat{d}_{l+j}(k)$ are unknown at the current iteration l, in the following, they will be predicted based on the obtained $\hat{G}_l^{\rtimes}(k)$ and $\Delta \hat{d}_l(k)$ by means of (8.14)–(8.15).

By utilizing $\hat{G}_1^{\rtimes}(k), \hat{G}_2^{\rtimes}(k), \ldots, \hat{G}_l^{\rtimes}(k)$ that have been available, the following autoregressive (AR) model along iteration direction is constructed:

$$\begin{aligned}
\hat{G}_{l+1}^{\rtimes}(k) &= \Gamma_1^{\rtimes}(k)\hat{G}_l^{\rtimes}(k) + \Gamma_2^{\rtimes}(k)\hat{G}_{l-1}^{\rtimes}(k)\\
&\quad + \ldots + \Gamma_{\rho^{\rtimes}}^{\rtimes}(k)\hat{G}_{l-\rho^{\rtimes}}^{\rtimes} + 1(k)
\end{aligned} \tag{8.20}$$

where $\Gamma_h^{\rtimes}(k)$, $h = 1, 2, \ldots, \rho^{\rtimes}$ are unknown coefficient matrices. $\rho^{\rtimes}$ is a proper constant.

Define

$$\hat{\boldsymbol{\Gamma}}_l^{\rtimes}(k) = [\hat{\Gamma}_{1,l}^{\rtimes}(k), \hat{\Gamma}_{2,n}^{\rtimes}(k), \ldots, \hat{\Gamma}_{\rho^{\rtimes},l}^{\rtimes}(k)]^T \tag{8.21}$$

as the estimation of the unknown coefficient matrix in (8.20). It can be updated by the following iterative learning least squares algorithm

$$\begin{aligned}\hat{\boldsymbol{\Gamma}}_l^{\rtimes}(k) = \hat{\boldsymbol{\Gamma}}_{l-1}^{\rtimes}(k) + \boldsymbol{\Phi}_{l-2}^{\rtimes}(k)\hat{\mathbf{G}}_{l-1}^{\rtimes}(k)\big[\varsigma^{\rtimes} + \hat{\mathbf{G}}_{l-1}^{\rtimes T}(k)\boldsymbol{\Phi}_{l-2}^{\rtimes}(k) \\ \hat{\mathbf{G}}_{l-1}^{\rtimes}(k)\big]^{-1}\left[\hat{G}_l^{\rtimes}(k) - \hat{\mathbf{G}}_{l-1}^{\rtimes T}(k)\hat{\boldsymbol{\Gamma}}_{l-1}^{\rtimes}(k)\right]\end{aligned} \tag{8.22}$$

$$\begin{aligned}\boldsymbol{\Phi}_{l-1}^{\rtimes}(k) = \boldsymbol{\Phi}_{l-2}^{\rtimes}(k) - \boldsymbol{\Phi}_{l-2}^{\rtimes}(k)\hat{\mathbf{G}}_{l-1}^{\rtimes}(k) \\ \left[\varpi^{\rtimes} + \hat{\mathbf{G}}_{l-1}^{\rtimes T}(k)\boldsymbol{\Phi}_{l-2}^{\rtimes}(k)\hat{\mathbf{G}}_{l-1}^{\rtimes}(k)\right]^{-1}\hat{\mathbf{G}}_{l-1}^{\rtimes T}(k)\boldsymbol{\Phi}_{l-2}^{\rtimes}(k)\end{aligned} \tag{8.23}$$

where $\varsigma^{\rtimes} > 0$ and $\varpi^{\rtimes} > 0$ are introduced to guarantee the feasibility of the inversion operation, and

$$\hat{\mathbf{G}}_{l-1}^{\rtimes}(k) = \left[\hat{G}_{l-1}^{\rtimes}(k), \hat{G}_{l-2}^{\rtimes}(k), \ldots, \hat{G}_{l-\rho^{\rtimes}}^{\rtimes}(k)\right]^T \tag{8.24}$$

Then based on (8.22)–(8.23), the predicted values $\hat{G}_{l+1}^{\rtimes}(k), \cdots, \hat{G}_{l+j}^{\rtimes}(k)$ in (8.19) can be obtained by

$$\hat{G}_{l+s}^{\rtimes}(k) = \hat{\mathbf{G}}_{l-1+s}^{\rtimes T}(k)\,\hat{\boldsymbol{\Gamma}}_l^{\rtimes}(k), \quad s = 1, 2, \ldots j \tag{8.25}$$

Now, based on (8.22)–(8.25), the predicted values $\hat{G}_{l+1}^{\rtimes}(k), \cdots, \hat{G}_{l+j}^{\rtimes}(k)$ are available. Similarly, the predicted values $\Delta\hat{d}_{l+1}(k), \cdots, \Delta\hat{d}_{l+j}(k)$ will be obtained as follows.

By utilizing $\Delta\hat{d}_1(k), \Delta\hat{d}_2(k), \ldots, \Delta\hat{d}_l(k)$ that have been available, the following AR model along iteration direction is constructed:

$$\Delta\hat{d}_{l+s}(k) = \Delta\hat{\mathbf{d}}_{l-1+s}^T(k)\,\boldsymbol{\Upsilon}_l^{\rtimes}(k) \quad s = 1, 2, \ldots j \tag{8.26}$$

$$\Delta\hat{\mathbf{d}}_{l-1}(k) = \left[\Delta\hat{d}_{l-1}(k), \Delta\hat{d}_{l-2}(k), \cdots, \Delta\hat{d}_{l-\rho_1^{\rtimes}}(k)\right]^T \tag{8.27}$$

$$\boldsymbol{\Upsilon}_l^{\rtimes}(k) = [\Upsilon_{1,l}^{\rtimes}(k), \Upsilon_{2,l}^{\rtimes}(k), \cdots, \Upsilon_{\rho_1^{\rtimes},l}^{\rtimes}(k)]^T \tag{8.28}$$

where $\rho_1^{\rtimes}$ is a proper constant.

The unknown coefficient vector $\boldsymbol{\Upsilon}_l^{\rtimes}(k)$ in above (8.28) can be updated by the following iterative learning least squares algorithm,

$$\begin{aligned}\boldsymbol{\Upsilon}_l^{\rtimes}(k) = \boldsymbol{\Upsilon}_{l-1}^{\rtimes}(k) + \boldsymbol{\Theta}_{l-2}^{\rtimes}(k)\Delta\hat{\mathbf{d}}_{l-1}(k)\big[\varsigma_1^{\rtimes} + \Delta\hat{\mathbf{d}}_{l-1}^T(k)\\ \boldsymbol{\Theta}_{l-2}^{\rtimes}(k)\Delta\hat{\mathbf{d}}_{l-1}(k)\big]^{-1}\big[\Delta\hat{d}_l(k) - \Delta\hat{\mathbf{d}}_{l-1}^T(k)\boldsymbol{\Upsilon}_{l-1}^{\rtimes}(k)\big]\end{aligned} \tag{8.29}$$

with

$$\begin{aligned}\boldsymbol{\Theta}_{l-1}^{\rtimes}(k) = \boldsymbol{\Theta}_{l-2}^{\rtimes}(k) - \boldsymbol{\Theta}_{l-2}^{\rtimes}(k)\Delta\hat{\mathbf{d}}_{l-1}(k)\\ \left[\varpi_1^{\rtimes} + \Delta\hat{\mathbf{d}}_{l-1}^T(k)\boldsymbol{\Theta}_{l-2}^{\rtimes}(k)\Delta\hat{\mathbf{d}}_{l-1}(k)\right]^{-1}\Delta\hat{\mathbf{d}}_{l-1}^T(k)\boldsymbol{\Theta}_{l-2}^{\rtimes}(k)\end{aligned} \tag{8.30}$$

where $\varsigma_1^{\rtimes} > 0$ and $\varpi_1^{\rtimes} > 0$ are introduced to guarantee the feasibility of the inversion operation.

Based on (8.26)–(8.30), the predicted values $\Delta\hat{d}_{l+1}(k), \cdots, \Delta\hat{d}_{l+j}(k)$ are also available.

Now on the basis of (8.21)–(8.30), the constructed error model for j step iterative prediction (8.19) is available. Then the predicted control input sequence $\Delta\mathbf{u}_{l+1}(k)$ is computed.

Denote

$$\hat{\mathbf{G}}_{l+1}^{\rtimes}(k) = \left[\hat{G}_{l+1}^{\rtimes}(k), \hat{G}_{l+2}^{\rtimes}(k), \cdots, \hat{G}_{l+j}^{\rtimes}(k)\right] \tag{8.31}$$

$$\Delta\hat{\mathbf{d}}_{l+1}(k) = \left[\Delta\hat{d}_{l+1}(k), \Delta\hat{d}_{l+2}(k), \cdots, \Delta\hat{d}_{l+j}(k)\right]^T \tag{8.32}$$

$$\Delta\mathbf{u}_{l+1}(k) = \left[\Delta u_{l+1}(k), \Delta u_{l+2}(k), \cdots, \Delta u_{l+j}(k)\right]^T \tag{8.33}$$

$$\mathbf{P}^{\rtimes} = \underbrace{[1, 1, \cdots, 1]}_{j} \tag{8.34}$$

Then, $e_{l+j}^{\rtimes}(k+1)$ can be rewritten as

$$e_{l+j}^{\rtimes}(k+1) = e_l^{\rtimes}(k+1) - \hat{\mathbf{G}}_{l+1}^{\rtimes}(k)\Delta\mathbf{u}_{l+1}(k) - \mathbf{P}^{\rtimes}\Delta\hat{\mathbf{d}}_{l+1}(k) \tag{8.35}$$

Based on (8.35), minimizing the following quadratic cost function

$$J_{l+1}^{\rtimes}(k) = q^{\rtimes} e_{l+j}^{\rtimes}(k+1)^2 + r^{\rtimes} \Delta \mathbf{u}_{l+1}^T(k) \Delta \mathbf{u}_{l+1}(k) \tag{8.36}$$

where $q^{\rtimes} > 0$ and $r^{\rtimes} > 0$ are two positive constants, and using the optimality condition $\frac{\partial J_{l+1}^{\rtimes}(k)}{\partial \Delta \mathbf{u}_{l+1}(k)} = \mathbf{0}$, then the predicted control input sequence $\Delta \mathbf{u}_{l+1}(k)$ and the optimal control input for next operation $u_{l+1}(k)$ are obtained as follows:

$$\Delta \mathbf{u}_{l+1}(k) = \frac{q^{\rtimes} \hat{\mathbf{G}}_{l+1}^{\rtimes T}(k)}{r^{\rtimes} + q^{\rtimes} ||\hat{\mathbf{G}}_{l+1}^{\rtimes}(k)||^2} [e_l^{\rtimes}(k+1) - \mathbf{P}^{\rtimes} \Delta \hat{\mathbf{d}}_{l+1}(k)] \tag{8.37}$$

and

$$u_{l+1}(k) = u_l(k) + \mathbf{U}^T \Delta \mathbf{u}_{l+1}(k) \tag{8.38}$$

where $\mathbf{U} = \left[1 \ \mathbf{0}_{1\times(j-1)}\right]^T$.

Theorem 8.1 *For the considered system (8.3) satisfying Assumptions 2.1–2.4 in Chap. 2, the proposed RBFNN-based PILC method (8.37)–(8.38), (8.25)–(8.30) and (8.14)–(8.15) guarantee that*

(8.1.1) The speed estimation error $e_{p,l}^{\rtimes}(k+1)$ defined in (8.12) by the proposed RBFNN (8.14)–(8.15) converges to zero when the operation number $l \to \infty$, namely

$$\lim_{l \to \infty} e_{p,l}^{\rtimes}(k+1) = 0 \tag{8.39}$$

if the following condition is met

$$\left(\eta_1^{\rtimes} \sigma_l^{\rtimes}(k) + \eta_2^{\rtimes} \bar{\sigma}_l^{\rtimes}(k) + \eta_3^{\rtimes} \frown{\sigma}_l^{\rtimes}(k) + \eta_4^{\rtimes} \breve{\sigma}_l^{\rtimes}(k)\right)^2 \\ -2\left(\eta_1^{\rtimes} \sigma_l^{\rtimes}(k) + \eta_2^{\rtimes} \bar{\sigma}_l^{\rtimes}(k) + \eta_3^{\rtimes} \frown{\sigma}_l^{\rtimes}(k) + \eta_4^{\rtimes} \breve{\sigma}_l^{\rtimes}(k)\right) < 0 \tag{8.40}$$

where

$$\sigma_l^{\rtimes}(k) = \left(\frac{\partial e_{p,l-1}^{o\rtimes}(k+1)}{\partial \hat{\mathbf{w}}_{l-1}^{\rtimes}(k)}\right)^T \frac{\partial \hat{v}_{l-1}(k+1)}{\partial \hat{\mathbf{w}}_{l-1}^{\rtimes}(k)} \tag{8.41}$$

$$\bar{\sigma}_l^{\rtimes}(k) = \left(\frac{\partial e_{p,l-1}^{o\rtimes}(k+1)}{\partial \hat{\hat{\mathbf{w}}}_{l-1}^{\rtimes}(k)}\right)^T \frac{\partial \hat{v}_{l-1}(k+1)}{\partial \hat{\hat{\mathbf{w}}}_{l-1}^{\rtimes}(k)} \tag{8.42}$$

$$\frown{\sigma}_l^{\rtimes}(k) = \sum_{m=1}^{M^{\rtimes}} \left(\frac{\partial e_{p,l-1}^{o\rtimes}(k+1)}{\partial \hat{\mathbf{c}}_{m,l-1}^{\rtimes}(k)}\right)^T \frac{\partial \hat{v}_{l-1}(k+1)}{\partial \hat{\mathbf{c}}_{m,l-1}^{\rtimes}(k)} \tag{8.43}$$

$$\breve{\sigma}_l^{\rtimes}(k) = \sum_{m=1}^{M^{\rtimes}} \frac{\partial e_{p,l-1}^{o\rtimes}(k+1)}{\partial \hat{\gamma}_{m,l-1}^{\rtimes}(k)} \frac{\partial \hat{v}_{l-1}(k+1)}{\partial \hat{\gamma}_{m,l-1}^{\rtimes}(k)} \tag{8.44}$$

(8.1.2) The speed tracking control error $e_l^{\rtimes}(k+1)$ defined in (8.16) can be converged to a residual set with respect to the randomly varying but bounded external disturbances.

Proof **proof of (8.1.1)**:

The proof is analogous to the proof of (4.1.1) in Theorem 4.1 in Chap. 4 and hence is omitted here.

proof of (8.1.2):

The proof is analogous to the proof of (2.1.2) in Theorem 2.1 in Chap. 2 and hence is omitted here. □

8.4 Simulation Validation

A HST operation system similar to CRH3 is considered in the simulation. The total operating time interval is $k \in \{0, 1, 2, \ldots, 3240\}$ (s). The nominal mass of the train is set as 350,000 (kg) while the actual total mass of the train is randomly varying in each operation, as described in Fig. 8.1.

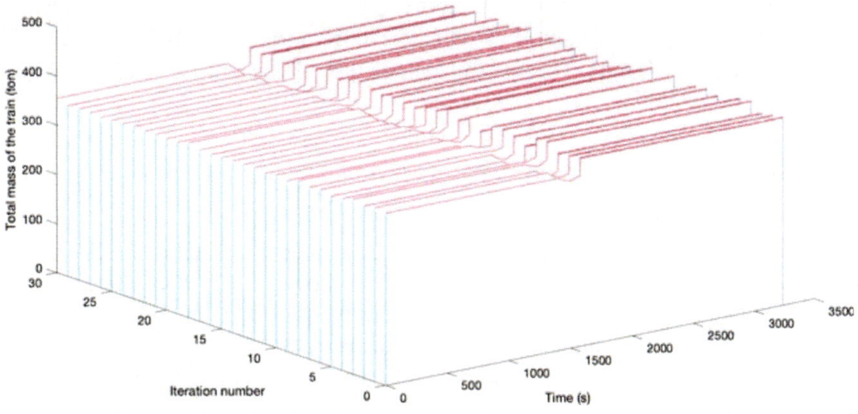

Fig. 8.1 Randomly varying total mass of the train

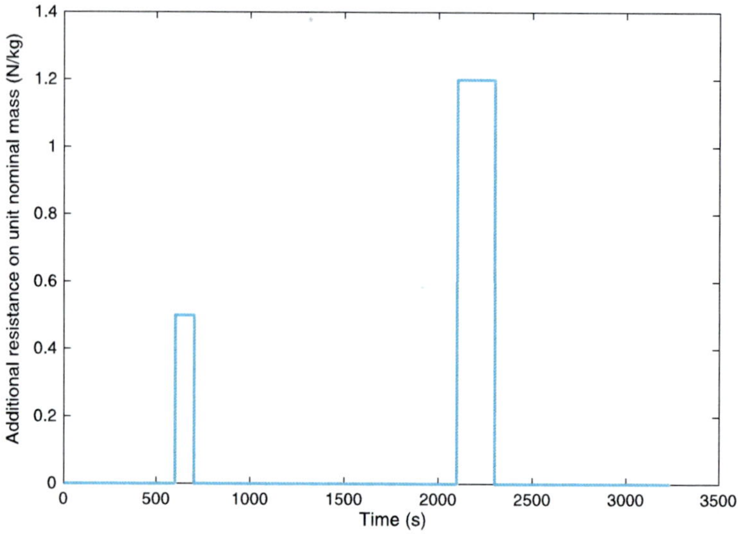

Fig. 8.2 The additional resistance profile

The basic resistance $f_b(v_l(k))$ in (8.2) is described as $f_b(v_l(k)) = a_1^{\rtimes}(k) + a_2^{\rtimes}(k)v_l(k) + a_3^{\rtimes}(k){v_l}^2(k)$ where the basic resistance coefficients $a_1^{\rtimes}(k)$, $a_2^{\rtimes}(k)$ and $a_3^{\rtimes}(k)$ are set as $a_1^{\rtimes}(k) = 375\ \sin(0.0039k) + 2980$, $a_2^{\rtimes}(k) = 2.5\ \sin(0.0039k) + 25.17$, and $a_3^{\rtimes}(k) = 0.05\ \sin(0.0039k) + 0.3864$. The additional resistance $f_a\left(\int_0^k v_l(\tau)d\tau\right)$ is described as shown in Fig. 8.2. The desired speed trajectory under 300/3.6 (m/s) is depicted in Fig. 8.3.

It is worth pointing out that the above given model parameters are only used to generate input and output data and no model information is used in the proposed controller design. The control parameters of the proposed RBFNN-based PILC are set as follows. $M^{\rtimes} = 7$, $\eta_1^{\rtimes} = 0.1$, $\eta_2^{\rtimes} = 0.1$, $\eta_3^{\rtimes} = 2$, $\eta_4^{\rtimes} = 2$, $L_y^{\rtimes} = 2$, $I_y^{\rtimes} = 2$, $L_u^{\rtimes} = 2$, $I_u^{\rtimes} = 2$. The initial estimations in (8.8)–(8.11) for $l = 0$ are $\hat{\mathbf{w}}_0^{\rtimes} = [0, 0, 0, 0, 0, 0, 0]^T$, $\hat{\hat{\mathbf{w}}}_0 = [0, 0, 0, 0, 0, 0, 0]^T$, $\hat{\mathbf{c}}_{m,0}^{\rtimes} = [10, 10, 10, 10, 10, 0.25, 0.25, 0.25, 0.25, 0.25]^T$, $m \in \{1, \ldots, 7\}$, $\hat{\gamma}_{m,0}^{\rtimes} = 0.01$, $m \in \{1, \ldots, 7\}$. $\rho^{\rtimes} = 3$, $\varsigma^{\rtimes} = 5$, $\varpi^{\rtimes} = 2$, $\rho_1^{\rtimes} = 3$, $\varsigma_1^{\rtimes} = 2$, $\varpi_1^{\rtimes} = 2$, $j = 2$, $r^{\rtimes} = 0.2$, $q^{\rtimes} = 0.7$.

The speed tracking control performance is shown in Fig. 8.4, where the root mean square of the tracking errors over the whole operating time interval $k \in \{0, 1, \ldots, 3240\}$ is evaluated. In Fig. 8.4, although the train undergoes randomly varying mass disturbances, the proposed RBFNN-based PILC can still achieve a gradually improved control performance as the iteration number increases.

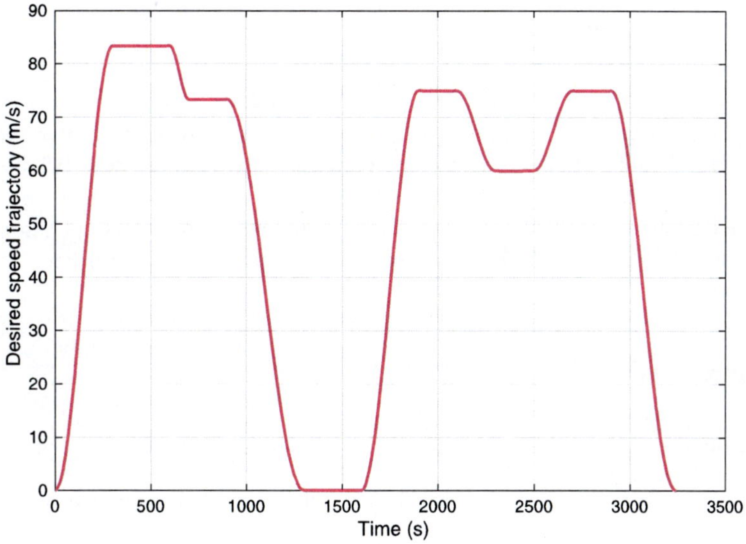

Fig. 8.3 Desired speed

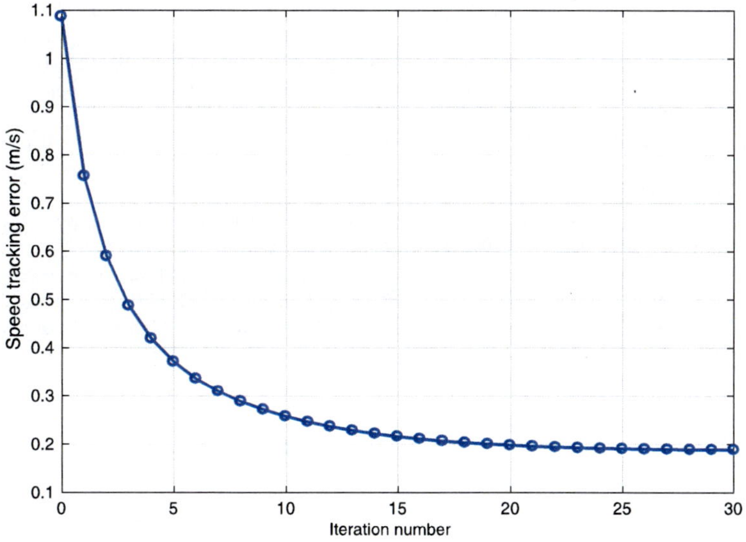

Fig. 8.4 Speed tracking errors

8.5 Conclusion

This chapter designs a RBFNN-based PILC method for complex and unknown HST operation systems with randomly varying external disturbances. An iterative learning RBFNN algorithm is constructed to approximate the unknown HST dynamics, and then, a corresponding PILC method is designed. The convergence of the proposed RBFNN-based PILC is guaranteed and simulations on a HST operation system similar to CRH3 show the effectiveness of the proposed method further.

References

Dong HR, Gao SG, Ning B, Li L (2013) Extended fuzzy logic controller for high speed train. Neural Comput 22(2): 321–328

Goodall RM, Kortüm W (2002) Mechatronic developments for railway vehicles of the future. Syst Sci Control Eng 10(8):887–898

Ji HH, Hou ZS, Zhang RK (2015) Adaptive iterative learning control for high-speed trains with unknown speed delays and input saturations. IEEE Trans Autom Sci Eng 13(1):260–273

Li ZX, Yin CK, Ji HH, Hou ZS (2021) Constrained spatial adaptive iterative learning control for trajectory tracking of high speed train. IEEE Trans Intell Transp Syst 23(8):11720–11728

Liu HG, Yang LJ, Yang H (2022) Cooperative optimal control of the following operation of high-speed trains. IEEE Trans Intell Transp Syst. https://doi.org/10.1109/TITS.2022.3163971

Orr MJL (1996) Introduction to radial basis function networks. Center for Cognitive Science, Scotland, UK

Song YD, Song Q, Cai WC (2014) Fault-tolerant adaptive control of high speed trains under traction/braking failures: a virtual parameter based approach. IEEE Trans Intell Transp Syst 15(2):737–748

Xiao Z, Wang Q, Sun P, You B, Feng X (2020) Modeling and energy-optimal control for high-speed trains. IEEE Trans Transp Electrification 6(2):797–807

Ye HB, Liu RH (2017) Nonlinear programming methods based on closed-form expressions for optimal train control. Transp Res Part C Emerging Technol 82:102–123

Yu QX, Hou ZS (2021) Adaptive fuzzy iterative learning control for high-speed trains with both randomly varying operation lengths and system constraints. IEEE Trans Fuzzy Syst 29(8):2408–2418

Chapter 9
Medium-Scale Two-Region Urban Road Networks

9.1 Introduction

Most of the existing urban traffic system control methods are based on the urban traffic model. However, the mathematical model of the urban transportation system is often difficult to establish, and the accuracy is difficult to meet the requirements. Therefore, it is very important to control the urban traffic system based on traffic data rather than urban traffic model. In this chapter, a series of PRC methods based on model free adaptive predictive learning control are proposed for perimeter control of the medium-scale urban traffic networks consist of two regions.

The main contributions of this chapter are that two new data-driven methods called one-step model free adaptive predictive learning control with input and output constraints (IOC-MFAPLC) and multi-step model free adaptive predictive learning control with input and output constraints (cMFAPLC) respectively are proposed to deal with the problem of perimeter control for medium-scale two-region urban traffic systems. Besides, the two perimeter control strategies are no longer just "perimeter restriction" but further optimizes the perimeter control according to the traffic conditions of each region. A remarkable advantage of the perimeter control methods proposed in this chapter is that in the design process of the boundary controller, it is no longer necessary to know the details of the urban transportation system and establish the accurate mathematical model of the urban traffic system. On the contrary, the proposed perimeter control strategies can be designed only by using the control input and system output data of the urban traffic system.

This chapter is organized as follows. Section 9.2 introduces the state of the art for control of urban road networks. Section 9.3 shows the one-step model free adaptive predictive learning perimeter control strategy. In Sect. 9.4, the multi-step model free adaptive predictive learning perimeter control method is proposed. Some conclusions are given in Sect. 9.5.

© The Author(s), under exclusive license to Springer Nature Singapore Pte Ltd. 2023

Q. Yu et al., *Predictive Learning Control for Unknown Nonaffine Nonlinear Systems*, Intelligent Control and Learning Systems 8,
https://doi.org/10.1007/978-981-19-8857-8_9

9.2 The State of the Art for Control of Urban Road Networks

Urban traffic control is the result of the desirability of the rapid development of urban traffic networks and the application of control theory and technology in traffic systems. It is an effective way recognized by traffic experts all over the world to improve the traffic capacity and ensure the smoothness and safety of urban traffic systems. In this section, the history and development process of urban road traffic control and its research purpose and content, the classification of control modes, and the technology involved in urban traffic control and its theoretical basis will be all summarized.

9.2.1 The Purpose of Urban Road Traffic Control

Urban road traffic control is closely related to the emergence and development of vehicles and urban road traffic systems. With the development of the society and urban traffic systems, the purpose of traffic control is also changing constantly. The main purpose of early urban traffic control is to ensure traffic safety. With the increase of the number of vehicles, traffic congestion began to appear in the urban road networks more and more frequently. Therefore, on the basis of ensuring traffic safety, it is also required to achieve the purpose of dredging and ensuring smooth traffic.

In addition, traffic congestion has greatly reduced the vehicle speed. Meanwhile, the actual situation of vehicles stopping and starting at the intersections increases fuel consumption, exhaust emission and noise, resulting in environmental pollution. Therefore, reducing fuel consumption, exhaust emissions and noise, and maintaining economic operation and a good living environment are also one of the purposes of traffic control. The purpose of modern traffic control is to take various traffic management, control, guidance and other measures to dredge traffic congestion and make the urban road network run safely, orderly and efficiently.

9.2.2 The History and Development of Urban Road Traffic Control

Since the nineteenth century, people have begun to study traffic signals and use signals to direct the passage of vehicles and control the order of vehicles entry and exit intersections. According to the literature, in 1868, red and green traffic lights were installed on Westminster Street in London, England. This kind of signal lamp is actually a kind of gas lamp, which is lit at night for pedestrians and vehicles to pass through the intersection safely. However, shortly after the installation and use, the end of the lamp was announced due to an accidental gas explosion.

In 1914, the first electric traffic lights were installed in Cleveland, Ohio.

In 1917, an interconnected signal system including six intersections was installed in Salt Lake City, which was manually controlled.

In 1922, a signal light system was installed in Houston, Texas, which controlled 12 intersections simultaneously by the traffic signal control center. The system adopted the automatic electric timing control function.

In 1925, the manually controlled three-color signal light first appeared at the Piccadilly intersection in London, England. In the following year, the British developed their own automatic control signal controller, which has been continuously popularized in cities in Britain and other European countries.

The early traffic signal system has played a good role in safely dredging the vehicles at intersections. However, with the rapid development of the urban traffic system, this traffic signal system is no longer competent for more and more complex traffic control tasks.

In 1928, traffic engineers successfully developed a traffic signal controller that can store different signal timing schemes to meet the traffic demands in different time periods every day. This traffic signal controller obviously took a big step forward, and the control efficiency was greatly improved compared with the previous traffic signal controller which can only implement a single fixed timing scheme. However, this kind of signal controller still cannot adapt to the actual situation of random change of the traffic flow.

From 1928 to 1930, the inductive traffic signal controller using pressure detectors began to be used in the USA. At that time, the traffic signal controller was only applied to a single intersection. The most significant advantage of this traffic signal controller is that it can flexibly adjust the required signal time according to the arrival of vehicles at the intersection.

Since the advent of the automatic control signal controller for a single intersection, people began to explore the theory and practice of linkage traffic signal control. In 1928, American traffic engineers proposed a flexible push-forward timing system (timing coordinated control), which was soon popularized in American cities.

In 1952, a traffic control system controlled by analog computer was developed and installed in Denver, Colorado, USA. The concept of single point induction control to the control of the urban road network is applied in this traffic control system. The proposed system did not adopt fixed timing control mode. The traffic signal timing was adjusted, and the traffic control was realized according to the traffic flow demand data sampled by the traffic detectors. From 1952 to 1962, at least 100 urban traffic control systems of this type were installed in the USA.

After 1960, researches of a wide range of signal linkage and coordination control system were carried out all over the world. In 1960, an experimental study was carried out in Toronto, Canada, and a centralized control system using digital computer was built. This control system was very popular, and then it was used in Toronto on a large scale. In 1963, 20 intersections were controlled by computers. In the year of 1973, a traffic control system controlling 885 intersections had been developed. During this period, the rapid development of computer technology (including software hardware technologies) provided the technical guarantee for the development of urban road traffic control system. After the first regional urban road traffic control system in Toronto was put into use, the urban road traffic control system has been rapidly applied in large and medium-sized cities in Britain, the USA and other developed countries. After that, urban traffic control has developed vigorously.

9.2.3 The Classification of Urban Road Traffic Control

The development of urban road traffic control is a process of continuous practice. In practice, many different types of traffic control modes and control systems have been proposed and developed, and urban road traffic control has been classified from the aspects of control category, control range and control mode, respectively. In order to have a clear understanding of urban road traffic control, the various types of urban traffic control are summarized in this section.

According to the laws and regulations and technologies used in urban road traffic control, it can be divided into three types: traffic restriction-based control, traffic signal-based control and transmitting intelligence information-based traffic control (traffic guidance control).

1. Traffic restriction-based control
   Traffic restriction-based control mainly adopts channelized traffic organization and traffic adjustment measures, and sets up pavement signs, markings, traffic signs (restrictions, instructions, warnings, etc.) and traffic islands (direction island, central island, etc.) on the road to restrict the passage of various traffic flows in time and space, which can be divided into the following aspects.
   (1) Priority rule control for intersections
   Intersection priority rule control refers to the control method of using the sign of stop and give way to control the passage of vehicles entering the intersection in order to ensure the safety, order and smoothness of traffic at the intersection without traffic signal control. This control method is generally applicable to the intersections with low traffic flow or the directions with obvious primary and secondary relationship. Stop or give way signs are set on the entrance of nonpriority traffic flow directions. On the premise of ensuring the passage of vehicles with priority, the controlled vehicles pass through the intersections by stopping or giving way.
   (2) Pavement marking control
   The traffic sign on the road is a kind of safety facility that transmits specific information with graphic symbols and words to manage urban road traffic. Signs are divided into main signs and auxiliary signs, which are generally set on the roadside or above the road. Road traffic signs provide road users with accurate road traffic information, so that the purpose of safety, smoothness, low pollution and energy conservation can be achieved. Road marking is a traffic safety facility composed of various pavement markings, arrows, words, elevation marks, raised road signs and roadside outline signs. The function

of road mark is to control and guide urban traffic, which can be used alone or in combination with traffic signs. To sum up, pavement marking control is a control method of setting markings and signs on the road to control the vehicles. Traffic signs and markings are the most basic and main means of traffic control.

(3) Parking control
Parking control can be divided into on-road parking control and off-road parking control (including parking lot). When a vehicle is driving on the road, we should use various effective means to control the safety and smoothness of the urban traffic. However, any vehicle always has a destination. When it reaches the destination, it must stop. As a driver, it is often to choose a convenient parking place, and the roadside is the first choice. Arbitrary parking of vehicles has a great impact on traffic safety and smoothness. Therefore, the control of arbitrary parking of vehicles has become a main content of traffic control.

2. Traffic signal-based control
Traffic signal-based control is a traffic organization measure to separate traffic flow in time. Traffic signal lights are set at intersections, freeway mainlines and ramps to control traffic flow through signal color display and transformation. Traffic signal-based control is usually used when the traffic demand is large and priority control cannot solve the problem. This control method needs scientific and reasonable traffic channelization and other measures as the basis.
3. Transmitting intelligence information-based traffic control (traffic guidance control)
Transmitting intelligence information-based traffic control is mainly to transmit information about the operation of road traffic to traffic participants by means of information communication (such as broadcasting) and setting variable information signs on the road, so as to remind drivers to pay attention and select appropriate driving speeds and routes, such as traffic broadcasting, variable speed control of expressway. This control method is generally not legally mandatory, but practice has shown that it plays an important role in dredging traffic and ensuring smooth traffic. Advanced and efficient traffic guidance control is an important part of urban traffic control.

According to the control principle of urban traffic control technology, urban traffic control can be divided into three main forms: traffic signal timing control, traffic induction control and adaptive traffic control.

Traffic signal timing control means that one or more control schemes (for different periods) are given according to the investigated traffic flow data, and each traffic signal controller operates in strict accordance with the preset scheme. In the predetermined traffic signal timing scheme, the green signal ratio, signal cycle and green light starting time are relatively fixed. In addition, according to the fluctuation of traffic flow in a day, several time segments can be divided, and corresponding signal timing schemes can be formulated for the average traffic flow in each time segment. Traffic signal timing control method is simple and effective when the traffic flow is small and stable. However, when the traffic is congested or the traffic flow is unstable, the effect of timing control is often poor.

Traffic induction control is to formulate a control scheme based on the arrival of vehicles detected by the traffic sensors set at the intersection and its upstream road sections. It is mainly used at the intersections of trunk road and branch road. However, when the traffic flow through the intersection is small and irregular, it is easy to produce the phenomenon of unreasonable allocation of green light time in each phase, thus reducing the traffic capacity of the intersection. When the traffic flow is large, traffic induction control is easy to degenerate into traffic signal timing control and lose the induction ability (Zhai et.al. 2011).

Adaptive traffic control means that in a trunk line or an area, the traffic signal control parameters of each intersection are automatically adjusted according to the dynamic and random changes of traffic flow, so that the urban traffic control system can automatically adapt to the random changes of traffic flow.

According to the urban traffic control mode, urban traffic control can be divided into centralized traffic control and decentralized traffic control.

Centralized traffic control refers to the establishment of a central control center in the urban traffic system. All traffic information in the road network is transmitted to the control center and processed by the control center, and all traffic control instructions at all intersections are sent uniformly by the central control center. The advantage of centralized control mode is that it can optimize the performance of the whole urban traffic network as much as possible by establishing the optimization index of the whole urban traffic system. However, the reliability of centralized traffic control is poor. Once the data collection is incomplete or the control center fails, the overall control of the urban traffic system may fail. Therefore, centralized traffic control requires high reliability of data acquisition and transmission equipment and control center processor. In addition, centralized traffic control also requires the control center processor to have high computing power to deal with the control of the entire traffic network.

Decentralized traffic control means that there is no need to establish a central traffic control center, but a traffic control center is set in each control unit (an intersection, a trunk line or a region). The traffic information of different control units flows into different control centers, and the control instructions of each control unit are sent by different control centers. Each traffic control center shall perform its own responsibilities, do not interfere with each other and complete its own control objectives. The advantage of decentralized traffic control is high reliability. Even if some parts of the urban traffic system fail, the impact on the overall control effect is limited. In addition, decentralized traffic control has low requirements for the processor operation speed of each control center and the hardware reliability of the overall urban traffic network. The disadvantage of decentralized traffic control is that it is not easy to optimize the performance of the overall urban road network.

According to the spatial scope of traffic control, urban traffic control can be divided into three categories: single intersection signal control, trunk line signal control and regional (road network level) signal control.

Single intersection signal control refers to the way of traffic signal control for a single intersection or each intersection in a road or a region. In the single intersection signal control, there is no correlation between the signal timing of each intersection, and each intersection operates independently. The main purpose of single intersection control is to improve the traffic efficiency of a single intersection controlled by each traffic signal, regardless of the traffic conditions of other sections and intersections. The main control parameters are the length of the traffic signal cycle and the green signal ratio. Traditional single intersection timing methods mainly include HCM method, ARRB method, TRRL method and so on (Shen 2018). In recent years, with the rapid development of computer technology and the deepening of control theory research, experts and scholars have also proposed a series of single intersection signal parameter optimization methods based on simulated annealing, neural network and other intelligent algorithms (Gu and Wang 1998; Wang et al. 2018).

Trunk line signal control refers to the signal linkage and coordination control of several consecutive adjacent intersections on a road as a whole. The main purpose of trunk line control is to reduce unnecessary parking and queuing delays and maintain continuous traffic flow on trunk roads. In addition to signal period and green signal ratio, the parameters of trunk signal control also include another important parameter—the phase difference. The phase difference refers to the constant time difference of the starting time of the green light at each intersection. By scientifically adjusting the phase difference, vehicles can meet the green light at a certain speed at each intersection, reducing parking times and fuel consumption. At present, the main methods for calculating the phase difference of trunk signal control include numerical solution (Lu et al. 2010), graphic method (Lin et al. 2007), MAXBAND strategy (Gazis 2006), etc. Among them, the numerical solution is to calculate the minimum green time offset by assuming the location of the ideal intersection, so as to obtain the phase difference between the intersections. The graphic method mainly determines the phase difference of the trunk line coordinated control system through the time distance diagram. MAXBAND strategy is to transform the urban traffic trunk line coordination control problem into an optimization problem with the widest green wave band as the control objective. By establishing a linear programming model, the mixed integer linear programming method is used to solve the corresponding control parameters.

Regional (road network level) traffic signal control refers to the coordinated control of traffic signals by taking the intersections controlled by multiple traffic signals in a region or the entire urban road network as a whole. The purpose of regional signal control is to optimize the traffic performance of the whole controlled region. Several typical urban traffic control systems at the regional level are as follows.

1. TRANSYT (traffic network study tool) system (Robertson 1969)
   TRANSYT system was designed and developed by D. I. Robertson and other scholars and engineers of Transport and Road Research Laboratory (TRRL) in 1968. It is a signal timing optimization system program in the whole controlled region. The system was gradually adopted by many countries in the world, which greatly promoted the development of urban road traffic signal control system. The

static mode is adopted in this system. The green signal ratio and phase difference are taken as the control parameters, and the mountain climbing method is used as the optimization method.

2. SCOOT (split, cycle, offset optimization technique) system (Hunt et al. 1982) Since 1973, TRRL engineers began to study the second generation of regional urban road traffic signal control system and successfully developed SCOOT system in 1979. Different from the first generation of regional road traffic signal control system, SCOOT system is an automatic control system based on data feedback. The basic parameters of the signal timing scheme in real time can be adjusted continuously according to the actual traffic conditions of the urban traffic network, so that the best control effect can be obtained as far as possible, instead of using several fixed signal timing schemes obtained through off-line operation as the first-generation urban regional road traffic control system.
3. SCAT (Sydney Coordinated Adaptive Traffic System) system (Lowrie 1982) SCAT system is a regional traffic control system successfully developed by the Roads and Traffic Authority of New South Wales (RTA-NSW) in the late 1970s and has been installed in Sydney and other cities since 1980.

A complete SCAT system consists of three layers: the upper layer is the central monitoring center, which mainly completes the tasks of the management system. The middle level is the regional control center, mainly responsible for strategic control tasks. The lower level is the intersection signal controller, which mainly shares the tactical control task. If the number of controlled intersections is small, the upper layer can be omitted, and only the regional control center and each intersection signal controller can form a two-layer traffic control system.

The above regional signal control systems all take traffic delay as the control objective. However, in practice, traffic delay is difficult to measure and estimate accurately. In addition, the indicator of traffic delay is often not intuitive for urban traffic managers and drivers. For this reason, experts and scholars have also proposed a variety of other regional traffic signal control methods. Girianna and Benekohal (2004) used a simple genetic algorithm with multiple epochs to coordinate and control intersection signals in the region. Liu and Wang (2002) used the multi-agent method to study the coordinated control of regional traffic flow. Beard (1978) established a mixed integer linear programming model to solve the optimal dynamic traffic flow distribution and traffic signal optimal control problems of the urban traffic system.

Although the aforementioned regional traffic signal control methods are considered from the overall level of the region to optimize the traffic conditions in the whole region, all the intersections in the region are taken in these methods as the research object, and they have high implementation cost and calculation burden. In addition, the existing regional signal control methods still have two main shortcomings. First, when the traffic is saturated, the control efficiency is low. Second, the traffic control methods based on optimization also face great difficulties in real time and practical feasibility.

Besides, most of the existing urban traffic control methods are proposed based on the traditional urban traffic flow theory and the modern model-based control methods,

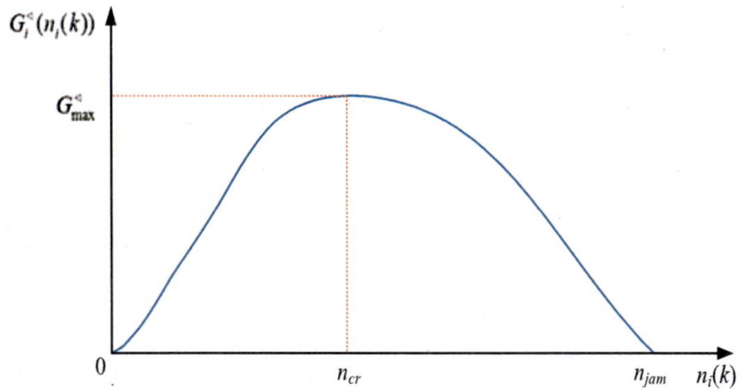

Fig. 9.1 Example of an MFD

where the accurate urban traffic system model is necessity. However, urban traffic model is often very complex, and it is difficult to be accurately established. Besides, the corresponding performance of the control methods may be not good enough, if traffic model is not very precise. Moreover, in reality, traffic congestion in large-scale urban road networks is always distributed unevenly. The urban road networks are often not partitioned reasonably in the above traffic control methods, which will reduce the control effect when the vehicle distribution is not uniform.

In order to address the problem of the heterogeneity of urban road networks, the urban road network needs to be properly partitioned. Some partitioning methods are presented in (Ji and Geroliminis 2012; Saeedmanesh and Geroliminis 2016; An et al. 2018; Lopez et.al. 2017; Saeedmanesh and Geroliminis 2017). After being well partitioned, a macroscopic characteristic called macroscopic fundamental diagram (MFD) is appeared in homogeneous regions. MFD indicates that there is a unimodal and low-scatter relationship between the number of traveling vehicles and the trip completion flow in homogeneous urban road networks or regions, which is an inherent property of urban traffic system, see Fig. 9.1. The original idea of MFD was provided in Godfrey (1969), and similar approaches can also be found in (Herman and Prigogine 1979; Daganzo 2007). The existence of MFD was verified with dynamic features in (Geroliminis and Daganzo 2007, 2008), and a derivation of MFD via a model of urban traffic flow at the intersection level can be found in Helbing (2009). The above research results show that MFD has the following characteristics: (i) some homogeneous urban regions with suitable size approximately exhibit an MFD relating vehicle accumulation to space-mean flow, (ii) there is a robust linear relation between the regions space-mean flow and its trip completion flow (rate that vehicles reaching their destinations, including leaving the region and finishing their trips in it), and (iii) although MFD is effected by the infrastructure and control strategies of the regions (see Ramezani et al. 2015; Zhang et al. 2013), it is not sensitive to traffic demand. Property (i) is applicable for modeling purpose because it is no longer necessary to know the detailed dynamics of vehicles in each road section and intersection in the process of modeling the urban traffic system.

Property (ii) is important for monitoring purpose since we can measure the average traffic flow in the urban road network that is easy to detect to obtain the trip completion flow, which is not easy to be measured. Property (iii) is helpful for control purpose because the urban traffic management department can reasonably control the urban traffic system without the need for a detailed O-D table.

According to Property (iii) of MFD, it can be used for macroscopic perimeter control. Perimeter control methods for small-scale urban road network composed of a single region are provided in (Keyvan-Ekbatani et al. 2012; Keyvan-Ekbatani et.al 2016; Keyvan-Ekbatani et al. 2015; Haddad 2017; Haddad and Shraiber 2014). Stability analysis of perimeter control for medium-scale two-region urban traffic networks is treated in (Haddad and Geroliminis 2012; Gayah et al. 2014), and some perimeter control strategies for two-region urban traffic systems are presented in (Geroliminis et al. 2013; Haddad 2017). A model-based feedback regulator with online adaptive optimization for multi-region system is presented in Kouvelas et al. (2017), and some other perimeter control strategies for large-scale multi-region urban traffic systems are shown in (Aboudolas and Geroliminis 2013; Keyvan-Ekbatani et al. 2015; Haddad and Mirkin 2017; Hajiahmadi et al. 2015; Haddad and Mirkin 2016). The perimeter control methods mentioned above are all designed based on the model of the urban traffic system, which require the urban traffic model must be known accurately. Besides, actually vehicles not always fully compliant with route guidance, which further causes the decline of urban traffic control effects. Moreover, MFD of heterogeneous regions is not well defined, especially under congested condition. It has high scatter and hysteresis phenomena during the process of formation and dissipation of congestion, see (Buisson and Ladier 2009; Mazloumian and Geroliminis 2010; Gayah and Daganzo 2011; Geroliminis and Sun 2011a, b; Saberi and Mahmassani 2012). In reality, absolutely homogeneous regions are hard to be partitioned. Heterogeneity of regions leads to MFD errors, which may further result in the discount effect of the model-based perimeter control methods. Furthermore, using the above existing perimeter control approaches are all need to know all the urban traffic information such as, route choice ratios,regional O-D tables and time-varying traffic demands, etc., which are difficult to obtain. Finally, the perimeter control strategies for urban traffic systems mentioned above are mostly centralized control methods, and the problem of internal coupling of MIMO nonlinear systems is difficult to deal with.

Considering the fact that thanks to the rapid development of traffic detection technology, massive traffic data can be generated every day in the urban traffic systems, it is of great significance to use these urban traffic data rather than urban traffic models to solve the problems faced in urban traffic systems, such as traffic control (Chi and Hou 2010; Wang 2010), etc., and traffic flow prediction (Abadi et al. 2015; Lv et al. 2015; Hou and Li 2016). Using urban traffic data to design the perimeter control strategies can avoid all the shortcomings and deficiencies of the model-based perimeter control methods, such as modeling difficulty, modeling inaccuracy, robustness lack; thus, it can improve the urban traffic control performance.

9.3 One-Step Model Free Adaptive Predictive Learning Perimeter Control

In this section, a novel data-driven control method called model free adaptive predictive learning control (MFAPLC) with input and output constraints (IOC-MFAPLC) is proposed in order to address the problem of perimeter control for medium-scale two-region urban traffic systems. In this chapter and Chap. 10, a series of urban traffic perimeter control methods based on MFAPLC will be proposed. MFAPLC is a typical PRC method, which can control the urban road network by using the characteristics of the periodic operation of the urban traffic system. A medium-scale two-region urban road network is shown in Fig. 9.2. There are two control inputs and two outputs of the urban traffic system in the medium-scale two-region urban traffic system . The control inputs of the proposed urban road network include the two perimeter control ratios, and the outputs of the urban traffic system consist of the vehicle accumulations in the two regions. In the proposed IOC-MFAPLC perimeter control strategy, a data-driven one-step prediction mechanism is adopted, and the constraints of vehicle accumulations in each region and perimeter control ratios between the two regions are both considered in the designing of the perimeter control method and the solution process of perimeter control ratios.

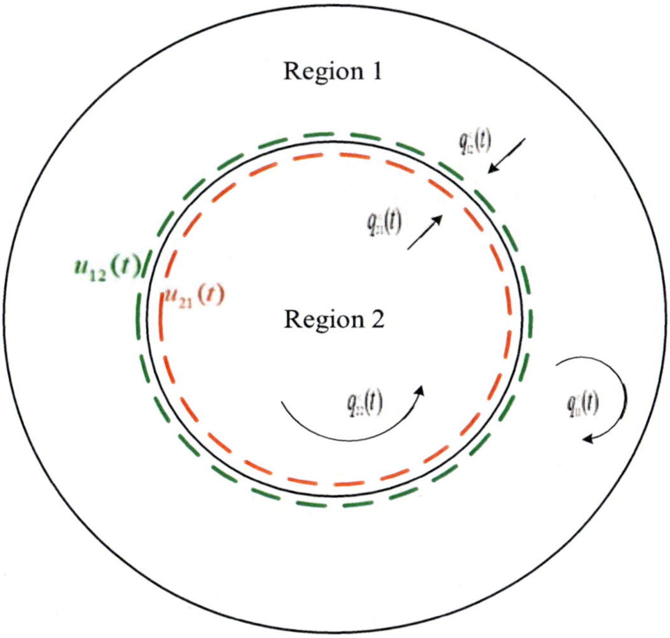

Fig. 9.2 A medium-scale urban road network partitioned into 2 regions

Table 9.1 List of variables

Notation	Description
$i, j = 1, 2$	Label of regions
$n_i(t)$	The total number of vehicles traveling in region i at time t
$n_{ij}(t)$	Number of vehicles in region i with destination to region j at time t
$n_{ii}(t)$	Number of vehicles in region i with destination in itself at time t
n_{cr}	The critical accumulation of each region
n_{jam}	The jammed accumulation of each region
$q_{ij}^{\triangleleft}(t)$	Exogenous traffic demands generated in region i with destination to region j at time t
$q_{ii}^{\triangleleft}(t)$	Endogenous traffic demands generated in region i at time t
$G_i^{\triangleleft}(n_i(t))$	Trip completion flow for region i at time t
$M_{ij}^{\triangleleft}(t)$	Transfer flow from region i to region j at time t
$M_{ii}^{\triangleleft}(t)$	Internal flow in region i at time t
$u_{ij}(t)$	Perimeter control ratio from region i to region j at time t
$u_{\min}$	The maximum value of $u_{ij}(t)$
$u_{\max}$	The minimum value of $u_{ij}(t)$

The main contribution of the work of this section is that a new data-driven IOC-MFAPLC method is proposed to deal with the problem of perimeter control for medium-scale two-region urban traffic systems. Besides, this perimeter control strategy is no longer just "perimeter restriction," but further optimizes the perimeter control according to the traffic conditions of each region. A remarkable advantage of the perimeter control method proposed in this section is that in the design process of the boundary controller, it is no longer necessary to know the details of the urban transportation system and establish the accurate mathematical model of the urban traffic system. On the contrary, the proposed perimeter control strategy can be designed only by using the control input and system output data of the urban traffic system.

9.3.1 Traffic Dynamics for Two-Region Urban Traffic Systems

As mentioned above, the urban traffic network is divided into two internal and external regions. Region 1 is in the periphery of the city, and region 2 is in the city center. In this section, the macroscopic traffic dynamics for the proposed medium-scale two-region urban traffic system is presented, which is introduced as follows.

It can be seen from (Geroliminis et al. 2013) that the dynamic equations for the proposed two-region urban traffic system based on vehicle conservation are as below. The definitions of the variables are listed in Table 9.1.

$$\frac{\mathrm{d}n_{ij}(t)}{\mathrm{d}t} = q_{ij}^{\lhd}(t) - u_{ij}(t)M_{ij}^{\lhd}(t) \tag{9.1}$$

$$\frac{\mathrm{d}n_{ii}(t)}{\mathrm{d}t} = q_{ii}^{\lhd}(t) + u_{ji}(t)M_{ji}^{\lhd}(t) - M_{ii}^{\lhd}(t) \tag{9.2}$$

$$n_i(t) = n_{ii}(t) + n_{ij}(t) \tag{9.3}$$

$$M_{ij}^{\lhd}(t) = \frac{n_{ij}(t)}{n_i(t)} G_i^{\lhd}(n_i(t)) \tag{9.4}$$

$$M_{ii}^{\lhd}(t) = \frac{n_{ii}(t)}{n_i(t)} G_i^{\lhd}(n_i(t)) \tag{9.5}$$

$$G_i^{\lhd}(n_i(t)) = M_{ij}^{\lhd}(t) + M_{ii}^{\lhd}(t) \tag{9.6}$$

where $n_{ij}(t)$ and $n_{ii}(t)$ represent the number of vehicles traveling in region i with destination to region j and i at time t, $i = 1, 2$, $j = 1, 2$, $i \neq j$, respectively. $M_{ij}^{\lhd}(t)$ is the transfer flow from region i to region j, while $M_{ii}^{\lhd}(t)$ is the internal flow from region i with destination to itself. $n_i(t)$ represents the vehicle accumulation in region i, and $G_i^{\lhd}(n_i(t))$ means the trip completion flow, which is a unimodal and low-scatter function of $n_i(t)$, which is depicted in Fig. 9.3. $q_{ii}^{\lhd}(t)$ and $q_{ij}^{\lhd}(t)$ represent endogenous and exogenous traffic demands generated in region i, respectively. As defined in Lin et al. (2007), $u_{ij}(t)$ represents the perimeter control ratio from region i to j, representing the ratio of vehicles that can actually drive from region i to region j, which is relative to the uncontrolled value.

According to the existing works of perimeter control, the perimeter control ratio $u_{ij}(t)$ is always less than 1. However, in other words, all the existing perimeter control strategies should be called "perimeter restriction." In the work of this section, the perimeter control ratio $u_{ij}(t)$ is allowed to be greater than 1, in order to make full use of the road space in the unobstructed regions.

The above state equations of the urban traffic system are differential form for continuous system. In reality, because there are a large number of traffic lights in the urban traffic networks, the urban traffic system has become a discrete-time system. In the following, (9.1)–(9.6) will be discretized by the first-order Euler method.

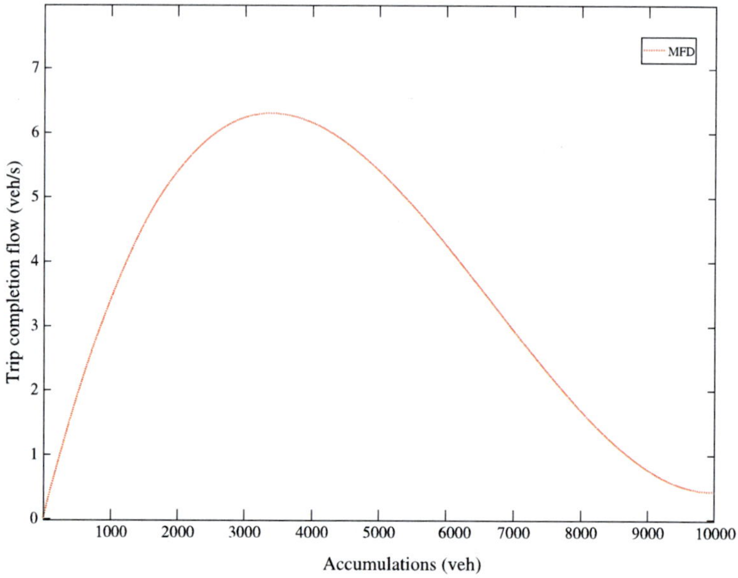

Fig. 9.3 MFD of each region

$$n_{ij}(k+1) = n_{ij}(k) + T^{\triangleleft}(q_{ij}^{\triangleleft}(k) - u_{ij}(k)M_{ij}^{\triangleleft}(k)) \tag{9.7}$$

$$n_{ii}(k+1) = n_{ii}(k) + T^{\triangleleft}(q_{ii}^{\triangleleft}(k) + u_{ji}(k)M_{ji}^{\triangleleft}(k) - M_{ii}^{\triangleleft}(k)) \tag{9.8}$$

$$n_i(k) = n_{ii}(k) + n_{ij}(k) \tag{9.9}$$

$$M_{ij}^{\triangleleft}(k) = \frac{n_{ij}(k)}{n_i(k)} G_i^{\triangleleft}(n_i(k)) \tag{9.10}$$

$$M_{ii}^{\triangleleft}(k) = \frac{n_{ii}(k)}{n_i(k)} G_i^{\triangleleft}(n_i(k)) \tag{9.11}$$

$$G_i^{\triangleleft}(n_i(k)) = M_{ij}^{\triangleleft}(k) + M_{ii}^{\triangleleft}(k) \tag{9.12}$$

where $k = 1, 2, \ldots, K^{\triangleleft}$ represents the time instant, and $T^{\triangleleft}$ is the sampling time, $i = 1, 2,\ j = 1, 2,\ i \neq j$. In the actual urban traffic system control operation, the sampling time generally selected is as one of the common multiples of the signal period of all the boundary intersections in order to ensure the synchronism and integrity of sampled traffic data.

9.3.2 Methodology

As mentioned above, the medium-scale two-region urban road network system is a kind of MIMO nonlinear system, in which there are two perimeter control inputs and two system outputs. In this work, IOC-MFAPLC method based on the dynamic linearization technique (DLT) is utilized to address the two-region perimeter control problem. DLT has three forms (Hou and Jin 2013): the compact form dynamic linearization (CFDL), the partial form dynamic linearization (PFDL), and the full form dynamic linearization (FFDL) techniques. The time-varying dynamical relationship between the variation of the output of the urban traffic system at next time step and that of the perimeter control input at the current time step is considered in CFDL approach. Further, the impacts on the variation of output of the urban traffic system at the next time step imposed by the change of perimeter control input within a fixed length moving time window at the current time are all taken into consideration in PFDL technique. Furthermore, all the effects on the output of the urban traffic system increment at the next time step imposed by both the variation of the perimeter control input and the output of the urban traffic system within input-related and output-related fixed length moving time windows at current time step are fully taken into account in FFDL method. It will be more complex and will take more computation time if PFDL or FFDL and for the sake of simplicity is utilized for the urban traffic system. Thus, without loss of generality and for the sake of simplicity, CFDL data model is used in this section. The CFDL technique for the urban traffic system and one-step model free adaptive predictive learning perimeter controller designing process are presented respectively as below.

A. Dynamic Linearization for Medium-Scale Two-Region Urban Traffic System
The aforementioned medium-scale two-region urban road network system is a two-input and two-output discrete-time nonlinear system. For the convenience of reading, the CFDL technique for general MIMO nonlinear systems is presented below.

Consider the nonlinear system with two control inputs and two system outputs as following:

$$\mathbf{y}(k+1) = \mathbf{f}(\mathbf{y}(k), \ldots, \mathbf{y}(k - n_y^{\triangleleft}), \mathbf{u}(k), \ldots, \mathbf{u}(k - n_u^{\triangleleft})) \tag{9.13}$$

where $\mathbf{y}(k) \in \Re^2$ and $\mathbf{u}(k) \in \Re^2$ are the system output and control input vectors of the system at time step k, respectively. $n_y^{\triangleleft}$ and $n_u^{\triangleleft}$ are two unknown integers, and $\mathbf{f}(\ldots) = [f_1(\ldots), f_2(\ldots)]^T \in \prod_{n_y^{\triangleleft}+n_u^{\triangleleft}+2} \Re^2 \mapsto \Re^2$ is an unknown nonlinear function vector.

In the aforementioned medium-scale two-region urban road network system, the control input and the system output vectors are defined as $u(k) = [u_{12}(k), u_{21}(k)]^T$ and $y(k) = [n_1(k), n_2(k)]^T$, respectively. In other words, the control input and output of the proposed urban traffic system consist of the perimeter control ratios between the regions and the vehicle accumulations in the two regions, respectively. It can be obtained from (9.7) to (9.9) that

$$n_i(k+1) = n_i(k) + T^{\triangleleft}(q_{ij}^{\triangleleft}(k) + q_{ii}^{\triangleleft}(k) - u_{ij}(k)M_{ij}^{\triangleleft}(k) \\ + u_{ji}(k)M_{ji}^{\triangleleft}(k) - M_{ii}^{\triangleleft}(k)), \ i, j = 1, 2, \ i \neq j. \tag{9.14}$$

Comparing (9.13) and (9.14), we can see that $n_y^{\triangleleft} = 0$ and $n_u^{\triangleleft} = 0$ in the proposed medium-scale urban traffic system.

In order to establish CFDL data model and utilize one-step IOC-MFAPLC strategy, the following two assumptions are proposed below.

Assumption 9.1 The partial derivatives of (9.13) with respective to each element of $\mathbf{u}(k)$ are continuous.

Assumption 9.2 The medium-scale two-region urban traffic system (9.13) is generalized Lipschitz, i.e. $\|\mathbf{y}(k_1+1) - \mathbf{y}(k_2+1)\| \leq b^{\triangleleft}\|\mathbf{u}(k_1) - \mathbf{u}(k_2)\|$ for any $k_1 \neq k_2, k_1, k_2 > 0$, and $\mathbf{u}(k_1) \neq \mathbf{u}(k_2)$.

Remark 9.1 From a practical viewpoint, Assumptions 9.1 and 9.2 mentioned above imposed on the medium-scale two-region urban road network system and other controlled systems are acceptable and reasonable. Assumption 9.1 can be easily and obviously verified from (9.14), and it is a typical assumption for nonlinear systems. Assumption 9.2 is a physical constraint for real urban road network systems, and it illustrates that finite change of perimeter control ratios cannot cause infinite change of the number of vehicles traveling in the urban road network. Moreover, it is reasonable for the real systems from the energy point of view.

Theorem 9.1 : *Consider the nonlinear system (9.13) satisfying Assumptions 9.1 and 9.2, there must exist a time-varying matrix $\mathbf{\Phi}^{\triangleleft}(k) \in \Re^{2\times 2}$ named pseudo-Jacobian matrix (PJM), so that the dynamics of system (9.13) can be transformed into the following form called CFDL data model:*

$$\Delta\mathbf{y}(k+1) = \mathbf{\Phi}^{\triangleleft}(k)\Delta\mathbf{u}(k) \tag{9.15}$$

where $\Delta\mathbf{y}(k+1) = \mathbf{y}(k+1) - \mathbf{y}(k)$, $\Delta\mathbf{u}(k) = \mathbf{u}(k) - \mathbf{u}(k-1)$, *and* $\mathbf{\Phi}^{\triangleleft}(k) = \begin{bmatrix} \phi_{11}^{\triangleleft}(k) & \phi_{12}^{\triangleleft}(k) \\ \phi_{21}^{\triangleleft}(k) & \phi_{22}^{\triangleleft}(k) \end{bmatrix} \in \Re^{2\times 2}$ *is bounded for all the time step k .*

Proof
Proof See (Hou and Jin 2013).

B. One-Step IOC-MFAPLC Perimeter Controller Designing
After CFDL for the aforementioned medium-scale two-region urban traffic system, the one-step IOC-MFAPLC strategy for perimeter control of the urban road network can be designed. The existence of PJM $\mathbf{\Phi}^{\triangleleft}(k)$ is guaranteed by Theorem 9.1, and the learning law of $\mathbf{\Phi}^{\triangleleft}(k)$ is shown as follows.

Consider the cost function of PJM as below:

$$J(\mathbf{\Phi}^{\triangleleft}(k)) = \left\|\mathbf{y}(k) - \mathbf{y}(k-1) - \mathbf{\Phi}^{\triangleleft}(k)\Delta\mathbf{u}(k-1)\right\|^2 + \mu^{\triangleleft}\left\|\mathbf{\Phi}^{\triangleleft}(k) - \hat{\mathbf{\Phi}}(k-1)\right\|^2 \tag{9.16}$$

where $\mu^{\triangleleft} > 0$ is a weighting factor to avoid the situation that the excessive change of the estimation of the PJM. The modified projection algorithm can be utilized to obtain the learning law of $\mathbf{\Phi}^{\triangleleft}(k)$ by minimizing (9.16):

$$\hat{\mathbf{\Phi}}^{\triangleleft}(k) = \hat{\mathbf{\Phi}}^{\triangleleft}(k-1) + \frac{\eta^{\triangleleft}(\Delta\mathbf{y}(k) - \hat{\mathbf{\Phi}}^{\triangleleft}(k-1)\Delta\mathbf{u}(k-1))\Delta\mathbf{u}^T(k-1)}{\mu^{\triangleleft} + \|\Delta\mathbf{u}(k-1)\|^2} \tag{9.17}$$

$$\begin{aligned}&\hat{\phi}_{ii}^{\triangleleft}(k) = \hat{\phi}_{ii}^{\triangleleft}(0), if\ \left|\hat{\phi}_{ii}^{\triangleleft}(k)\right| < \varepsilon^{\triangleleft}\ or\ \left|\hat{\phi}_{ii}^{\triangleleft}(k)\right| > \theta^{\triangleleft}\\ &or\ sign(\hat{\phi}_{ii}^{\triangleleft}(k)) \neq sign(\hat{\phi}_{ii}^{\triangleleft}(0)), i = 1, 2.\end{aligned} \tag{9.18}$$

$$\begin{aligned}&\hat{\phi}_{ij}^{\triangleleft}(k) = \hat{\phi}_{ij}^{\triangleleft}(0), if\ \left|\hat{\phi}_{ij}^{\triangleleft}(k)\right| > \xi^{\triangleleft}\\ &or\ sign(\hat{\phi}_{ij}^{\triangleleft}(k)) \neq sign(\hat{\phi}_{ij}^{\triangleleft}(0)),\ i, j = 1, 2,\ i \neq j.\end{aligned} \tag{9.19}$$

where $\hat{\mathbf{\Phi}}^{\triangleleft}(k)$ is the estimation value of $\mathbf{\Phi}^{\triangleleft}(k)$, and $\hat{\phi}_{ij}^{\triangleleft}(0)$ is the initial value of $\hat{\phi}_{ij}^{\triangleleft}(k)$, $i, j = 1, 2$. $\eta^{\triangleleft} \in (0, 2]$ is a weighting factor, $\varepsilon^{\triangleleft}$ is a small positive constant, $\theta^{\triangleleft}$ and $\xi^{\triangleleft}$ are positive constants.

In order to obtain the perimeter control ratios between the two regions, the following one-step prediction cost function with respect to $\Delta u(k)$ is introduced:

$$J(\Delta\mathbf{u}(k)) = \left\|\mathbf{y}_d^{\triangleleft}(k+1) - \mathbf{y}(k+1)\right\|^2 + \alpha^{\triangleleft}\lambda^{\triangleleft}\|\Delta\mathbf{u}(k)\|^2 \tag{9.20}$$

where $\mathbf{y}_d^{\triangleleft}(k+1) \in \Re^2$ is the desired number of vehicles traveling in the two regions, $\lambda^{\triangleleft} > 0$ is a weighting factor to restrain the perimeter control ratios' changes, and $\alpha^{\triangleleft}$ is a positive factor which is introduced to make the order of magnitude of the two items to the right of the equal sign in (9.20) are in the same level.

It is worth noting that the number of vehicles traveling in each region cannot be greater than $\mathbf{n}_{jam}$ or less than $\mathbf{0}$, i.e.

$$\mathbf{0} \leq \mathbf{y}(k+1) \leq \mathbf{n}_{jam} \tag{9.21}$$

where $\mathbf{n}_{jam} = [n_{jam}, n_{jam}]^T$ and $\mathbf{0} = [0, 0]^T$. Meanwhile, the perimeter control ratios should under the following constraints according to the reality of urban traffic control:

$$\mathbf{u}_{\min} \leq \mathbf{u}(k) \leq \mathbf{u}_{\max} \tag{9.22}$$

where $\mathbf{u}_{\max} = [u_{\max}, u_{\max}]^T$, and $\mathbf{u}_{\min} = [u_{\min}, u_{\min}]^T$. Combining (9.15), (9.21) and (9.22), it can be concluded that the optimization variable vector $\Delta\mathbf{u}(k)$ is constrained by a series of inequalities as follows:

$$\begin{bmatrix} \boldsymbol{\Phi}^{\triangleleft}(k) & & & \\ & -\boldsymbol{\Phi}^{\triangleleft}(k) & & \\ & & \boldsymbol{I} & \\ & & & -\boldsymbol{I} \end{bmatrix} \begin{bmatrix} \Delta\mathbf{u}(k) \\ \Delta\mathbf{u}(k) \\ \Delta\mathbf{u}(k) \\ \Delta\mathbf{u}(k) \end{bmatrix} \leq \begin{bmatrix} \mathbf{n}_{jam} - \mathbf{y}(k) \\ \mathbf{0} - \mathbf{y}(k) \\ \mathbf{u}_{\max} - \mathbf{u}(k-1) \\ \mathbf{u}(k-1) - \mathbf{u}_{\min} \end{bmatrix} \tag{9.23}$$

where $\boldsymbol{I} = \begin{bmatrix} 1 & 0 \\ 0 & 1 \end{bmatrix}$ is identity matrix. The optimization problem (9.20) with constraints of (9.23) is a convex nonlinear programming problem, which can be solved by sequential quadratic programming (SQP), interior point method, etc. Here, it is solved via fmincon toolbox in MATLAB. After $\Delta\mathbf{u}(k)$ is calculated from (9.20)–(9.23), the perimeter control ratios $\mathbf{u}(k)$ can be obtained as below:

$$\mathbf{u}(k) = \mathbf{u}(k-1) + \Delta\mathbf{u}(k) \tag{9.24}$$

Besides, in the work of this section, the maximum perimeter control ratios $\mathbf{u}_{\max}$ is set to be greater than $\mathbf{1}$. That is, it is no longer only perimeter restriction in the proposed perimeter control strategy. The advantage of this setting is that it can make more reasonable and full use of the road space resources in the unblocked regions.

9.3.3 Numerical Simulation Results

In this section, the performance of the proposed one-step IOC-MFAPLC perimeter control method for medium-sized two-region urban traffic system is tested via numerical simulation. In this case study, it is assumed that an urban road network is divided into two homogeneous regions, and the well-defined MFD of each region is the same, which is shown as below:

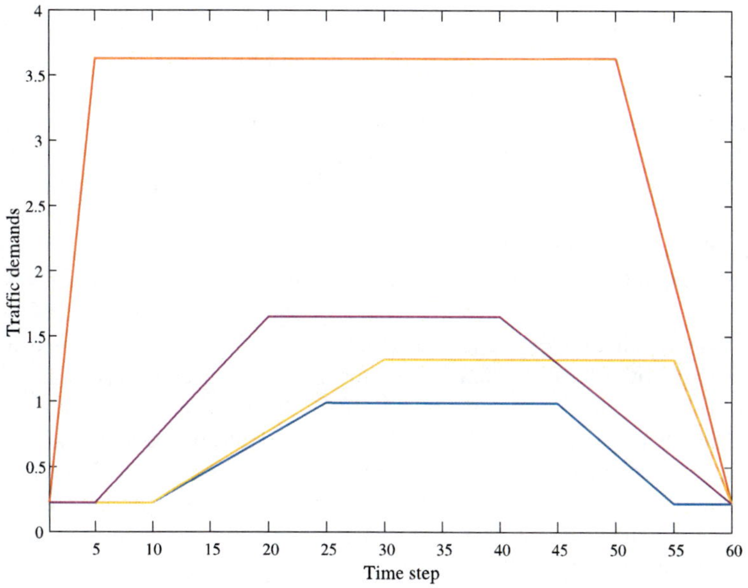

Fig. 9.4 Traffic demands, in which the blue solid line denotes $q_{11}^{\triangleleft}$, the brown solid line denotes $q_{12}^{\triangleleft}$, the yellow solid line denotes $q_{21}^{\triangleleft}$, and the purple solid line denotes $q_{22}^{\triangleleft}$

$$G_i^{\triangleleft}(n_i(k)) = 4.133 \times 10^{-11} n_i(k)^3 - 8.282 \times 10^{-7} n_i(k)^2 + 0.0042 n_i(k) \tag{9.25}$$

The MFD of the two regions is depicted in Fig. 9.3. It is assumed the same as the one obtained in the downtown of Yokohama (Geroliminis et al. 2013). The jammed accumulation of each region $n_{jam} = 10,000(veh)$, and the corresponding trip completion flow $G_i^{\triangleleft}(n_{jam}) = 0.43(veh/s)$. The critical accumulation of each region $n_{cr} = 3400(veh)$, and the maximum trip completion flow $G_i^{\triangleleft}(n_{cr}) = 6.3(veh/s)$.

In practice, the urban traffic system has obvious periodicity and repeatability. For example, from Monday to Friday in every week (i.e., working days), there are two travel peaks: morning peak and evening peak. On weekends in every week and holidays, people's travel behavior is often similar. The case study simulates real traffic situation in morning peak hour of one weekday, and the same periodic urban traffic control can also be carried out on other working days. The simulation time in this chapter and the next chapter is set based on this principle. The simulation time lasts 240 min, in other words, 4 h, and the traffic signal period of all the intersections in the road network is set as 120 s, while the sampling and control cycle is chosen as double of the traffic signal period, i.e., 4 min. Hence, the simulation duration is divided into 60 time steps. The change process of traffic demand between the two regions and within each region is shown in Fig. 9.4. It can be clearly seen that the process of the morning peak from generation to dissipation. Since the traffic signal period is 120 s, it is assumed that there are two phases in each boundary intersections, and the initial green time of both phases are 60 s. In view of the actual traffic situation, the minimum and maximum green time of each phase are 12 s and 108 s, respectively. Thus, it gives that $u_{\min} = 0.2$ and $u_{\max} = 1.8$ according to the definition of perimeter control ratio.

In this case study, the initial number of vehicles traveling in each region are: $n_1(0) = 6400$, $n_{11}(0) = 3200$, $n_{12}(0) = 3200$, $n_2(0) = 5200$, $n_{21}(0) = 2600$, and $n_{22}(0) = 2600$. Both of the two regions are suffering morning peak under initially congested condition. The purpose of setting such initial congestion conditions is to verify the applicability and effectiveness of the proposed IOC-MFAPLC perimeter control method. The desired vehicle accumulations in each region are selected as $0.97 \cdot n_{cr}$, in order to maximize the traffic throughput and ensure the robustness of the urban road network system. The initial perimeter control ratios between the two regions are both set as 1, which is the same as fixed time control (FTC).

In addition, in order to compare with IOC-MFAPLC method, fixed time control (FTC) and bang-bang control (BBC) strategies are also applied to the same medium-scale two-region urban road network under the same condition. FTC is a benchmark for comparing perimeter control effects. Under this strategy, both of the perimeter control ratios between the two regions are constant to the initial value of 1 all the time, regardless of the traffic conditions in the two regions, i.e., $u_{ij}(k) = 1$, $k = 1, 2, \ldots, K^{\triangleleft}$ under FTC method.

BBC (Daganzo 2007; Aboudolas and Geroliminis 2013), is a commonly used perimeter control strategy. The main idea of BBC method is that if a region is congested, the perimeter control ratio entering it will be minimized. On the contrary, if a region is unblocked, the corresponding perimeter control ratio entering it will be maximized, i.e.,

$$u_{ij}(k) = \begin{cases} u_{\max}, & if\ n_j(k) \le n_{cr} \\ u_{\min}, & if\ n_j(k) > n_{cr} \end{cases} \tag{9.26}$$

The simulation results of all the above perimeter control methods are described in Figs. 9.5, 9.6 and 9.7, respectively. The result of FTC method is shown in Fig. 9.5. It can be seen from the figure that region 1 fell into jammed state quickly and the traffic congestion in region 1 maintained and lasted till the end of the simulation, while the road space in region 2 was wasted due to there were few vehicles traveling in it.

Figure 9.6 describes the result of BBC strategy. It can be found that under this BBC approach, region 1 was no longer congested in the two hours, but traffic congestion still occurred in the later period. Meanwhile, region 1 was still jammed, which is similar to FTC.

The result of IOC-MFAPLC is depicted in Fig. 9.7. It is shown that both of the two regions were no longer congested. In the initial stage of the simulation, because the congestion in region 1 was more serious than that in region 2, $u_{12}(k)$ increased and $u_{21}(k)$ decreased. After a short time, both regions were no longer congested, and until the end of the simulation, both of the two regions were still unblocked. Under the one-step IOC-MFAPLC perimeter control strategy, with the optimization of perimeter control ratios under constraints at each time step, both of the two regions remained unblocked.

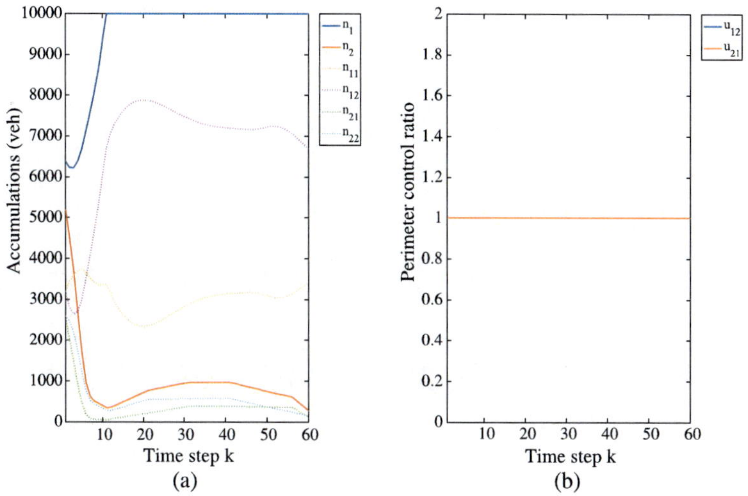

Fig. 9.5 Result of FTC: (**a**) evolution of the number of vehicles traveling in each region; (**b**) perimeter control ratio of FTC

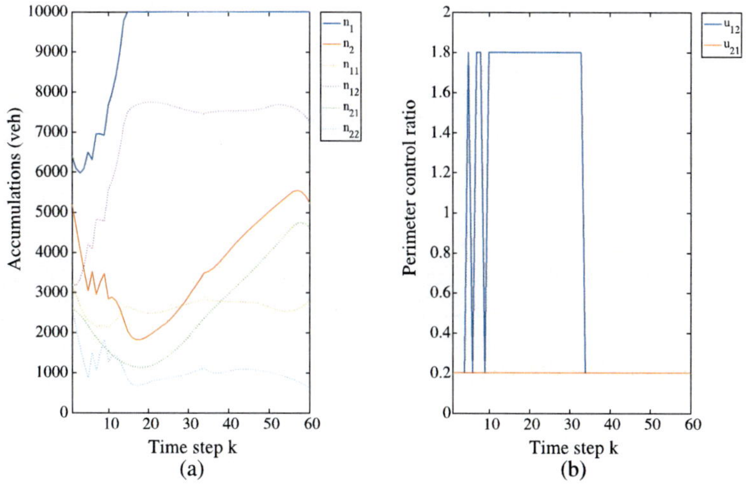

Fig. 9.6 Result of BBC: (**a**) evolution of the number of vehicles traveling in each region; (**b**) perimeter control ratio of BBC

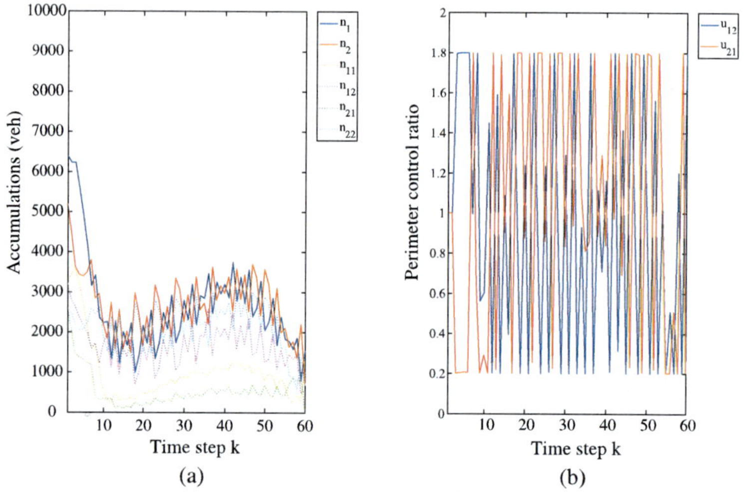

Fig. 9.7 Result of IOC-MFAPLC: (**a**) evolution of the number of vehicles traveling in each region; (**b**) perimeter control ratio of IOC-MFAPLC

Table 9.2 TTT for the urban road network

Strategy	TTT (veh)
FTC	3.0038×10^4
BBC	3.1199×10^4
IOC-MFAPLC	9.3149×10^4

In addition, in order to evaluate the perimeter control performance of the above different perimeter control strategies, the total traffic throughput (TTT) of the entire medium-scale two-region urban road network system is introduced here, and it can be calculated below:

$$TTT = T^{\triangleleft} \sum_{k=1}^{K^{\triangleleft}} \sum_{i=1}^{2} M_{ii}^{\triangleleft}(k) \tag{9.27}$$

The result of TTT under different perimeter control strategies is compared in Table 9.2, where the improvement is relative to FTC and it is strategy, expressed in percentages. It can be seen from Table 9.2 that the performance of the proposed one-step IOC-MFAPLC strategy is much better than FTC, while BBC approach works a little better than FTC. Hence, the one-step IOC-MFAPLC method for perimeter control can improve the performance of the urban traffic network significantly.

9.4 Multi-step Model Free Adaptive Predictive Learning Perimeter Control

In the work of this section, a new data-driven perimeter control method called multi-step model free adaptive predictive learning control with constraints (multi-step cMFAPLC) is presented. The merits of MFAPLC and MPC are combined in the multi-step cMFAPLC strategy. On one hand, the perimeter control input sequence on the control horizon can be obtained by only using the perimeter control input and the system output data of the urban road network. On the other hand, the system output sequence on the prediction horizon could be predicted without the model information of the urban road network. Multi-step MFAPLC for SISO systems is proposed in Hou and Jin (2013). In this work, it is extended to the MIMO form, and the input/output constraints are also considered.

The main contribution of this work is that a novel data-driven multi-step cMFAPLC perimeter control strategy is proposed for medium-scale two-region urban traffic systems. It combines the excellences of MFAPLC and MPC, and the constraints of the urban traffic system are also considered. Under this strategy, the difficult and complex MFD-based traffic modeling process is avoided. Instead, only the perimeter control input and the system output data are needed in the process of perimeter controller design. Meanwhile, the high uncertainties can be handled.

9.4.1 Methodology

As shown in (9.7)–(9.12), the medium-scale two-region urban road network system is a MIMO nonlinear system, which has two inputs and two outputs. In this section, cMFAPLC method based on the compact form CFDL data model (Hou and Jin 2013) is utilized to address the perimeter control problem of the two-region urban traffic system. Similar to Sect. 9.2.2, the proposed scheme also consists of two parts: dynamic linearization and perimeter controller design, which are introduced in the following respectively.

A. Dynamic Linearization for the Two-Region Urban Traffic System
Since the medium-scale two-region urban road network is a MIMO nonlinear system, an equivalent description of the unknown nonlinear system can be provided by DLT. In Sect. 9.2.2, the two-input two-output system is dynamically linearized. In this section, the CFDL data modeling technique is extended to general MIMO systems. For the convenience of reading, the general form of CFDL for MIMO nonlinear systems is presented below.

Consider the following MIMO nonlinear system with m inputs and m outputs:

$$\mathbf{y}(k+1) = f(\mathbf{y}(k), \ldots, \mathbf{y}(k - n_y^{\triangleleft}), \mathbf{u}(k), \ldots, \mathbf{u}(k - n_u^{\triangleleft})) \tag{9.28}$$

where $\mathbf{u}(k) \in \Re^m$ and $\mathbf{y}(k) \in \Re^m$ are the control input and the output vectors of the system at time step k, respectively. $n_y^{\triangleleft}$ and $n_u^{\triangleleft}$ are two unknown integers, and $\mathbf{f}(\ldots) = [f_1, \ldots, f_m(\ldots)]^T \in \prod_{n_y^{\triangleleft}+n_u^{\triangleleft}+2} \Re^m \mapsto \Re^m$ is an unknown nonlinear function vector.

The proposed medium-scale two-region urban road network has two perimeter control inputs and two system outputs, i.e., m=2. The inputs of the urban road network are the perimeter control ratios between the two regions, while the outputs are the number of vehicles driving in the regions. So the input and output vectors are defined as $\mathbf{u}(k) = [u_{12}(k), u_{21}(k)]^T$ and $\mathbf{y}(k) = [n_1(k), n_2(k)]^T$, respectively. Combining (9.7)–(9.12), one has

$$\begin{aligned} n_i(k+1) = n_i(k) + T^{\triangleleft}(q_{ij}^{\triangleleft}(k) + q_{ii}^{\triangleleft}(k) - u_{ij}(k)M_{ij}^{\triangleleft}(k) \\ +u_{ji}(k)M_{ji}^{\triangleleft}(k) - M_{ii}^{\triangleleft}(k)),\ i, j = 1, 2,\ i \neq j. \end{aligned} \tag{9.29}$$

Comparing (9.28) and (9.29), it can be seen that $n_y^{\triangleleft}$ and $n_u^{\triangleleft}$ are all equal to 0 in the medium-scale urban road network.

In order to utilize the cMFAPLC scheme for the two-region urban traffic system, the following assumptions and theorem are proposed below:

Assumption 9.3 The partial derivatives of $\mathbf{f}(\ldots)$ with respective to every element of $\mathbf{u}(k)$ are continuous.

Assumption 9.4 The system (9.28) is generalized Lipschitz, i.e., for all $k_1 \neq k_2, k_1, k_2 > 0$, and $\mathbf{u}(k_1) \neq \mathbf{u}(k_2)$, one has $\|\mathbf{y}(k_1+1) - \mathbf{y}(k_2+1)\| \leq b^{\triangleleft}\|\mathbf{u}(k_1) - \mathbf{u}(k_2)\|$, where $b^{\triangleleft}$ is a positive constant.

Remark 9.2 In reality, the above assumptions imposed on the controlled system such as the urban traffic system are reasonable and acceptable. Assumption 9.1 is easy to be verified from the dynamics of the two-region urban traffic system (9.29), and it is a typical assumption for controller design for nonlinear systems. Assumption 9.2 is a physical constraint by the inherent characteristic of urban traffic system, i.e., finite change of perimeter control ratios cannot lead to infinite change of the number of vehicles in the regions. Meanwhile, it is a physical constraint of the real system from the energy point of view.

Theorem 9.2 *Consider the nonlinear system (9.28) satisfying Assumptions 9.3 and 9.4, there must exist a time-varying matrix pseudo-Jacobian matrix (PJM) denoted by $\mathbf{\Phi}^{\triangleleft}(k) \in \Re^{m\times m}$, so that the system (9.28) can be transformed into the following equivalent CFDL data model:*

$$\mathbf{y}(k+1) = \mathbf{y}(k) + \mathbf{\Phi}^{\triangleleft}(k)\Delta\mathbf{u}(k) \tag{9.30}$$

where $\Delta\mathbf{u}(k) = \mathbf{u}(k) - \mathbf{u}(k-1)$, and $\mathbf{\Phi}^{\triangleleft}(k) = \begin{pmatrix} \phi_{11}^{\triangleleft}(k) & \cdots & \phi_{1m}^{\triangleleft}(k) \\ \vdots & \ddots & \vdots \\ \phi_{m1}^{\triangleleft}(k) & \cdots & \phi_{mm}^{\triangleleft}(k) \end{pmatrix} \in \Re^{m\times m}$ is bounded for all the time step k.

Proof
Proof See (Hou and Jin 2013).

B. Multi-Step cMFAPLC Perimeter Controller Designing
After the CFDL data model for the medium-scale urban road network, the multi-step cMFAPLC perimeter controller will be designed as below.

The objective function of PJM at the current time step k is proposed below:

$$J(\mathbf{\Phi}^{\lhd}(k)) = \left\|\mathbf{y}(k) - \mathbf{y}(k-1) - \mathbf{\Phi}^{\lhd}(k)\Delta\mathbf{y}(k-1)\right\|^2 + \mu^{\lhd}\left\|\mathbf{\Phi}^{\lhd}(k) - \hat{\mathbf{\Phi}}^{\lhd}(k-1)\right\|^2 \tag{9.31}$$

where $\mu^{\lhd} > 0$ is a weighting factor to penalize excessive change of the estimation of the PJM. The following modified projection algorithm can be applied to obtain the learning law of $\phi^{\lhd}(k)$ by minimizing (9.31):

$$\hat{\mathbf{\Phi}}^{\lhd}(k) = \hat{\mathbf{\Phi}}^{\lhd}(k-1) + \frac{\eta^{\lhd}(\Delta\mathbf{y}(k) - \hat{\mathbf{\Phi}}^{\lhd}(k-1)\Delta\mathbf{u}(k-1))\Delta\mathbf{u}^T(k-1)}{\mu^{\lhd} + \|\Delta\mathbf{u}(k-1)\|^2} \tag{9.32}$$

$$\begin{aligned}&\hat{\phi}_{ii}^{\lhd}(k) = \hat{\phi}_{ii}^{\lhd}(0),\ if\left|\hat{\phi}_{ii}^{\lhd}(k)\right| < \varepsilon^{\lhd}\\ &or\left|\hat{\phi}_{ii}^{\lhd}(k)\right| > M^{\lhd}\ or\ sign(\hat{\phi}_{ii}^{\lhd}(k)) \neq sign(\hat{\phi}_{ii}^{\lhd}(0)), i = 1, 2.\end{aligned} \tag{9.33}$$

$$\begin{aligned}&\hat{\phi}_{ij}^{\lhd}(k) = \hat{\phi}_{ij}^{\lhd}(0),\ if\left|\hat{\phi}_{ij}^{\lhd}(k)\right| > \zeta^{\lhd}\\ &or\ sign(\hat{\phi}_{ij}^{\lhd}(k)) \neq sign(\hat{\phi}_{ij}^{\lhd}(0)), i, j = 1, 2,\ \ i \neq j.\end{aligned} \tag{9.34}$$

where $\hat{\phi}^{\lhd}(k)$ is the estimation value of $\phi^{\lhd}(k)$, and $\hat{\phi}_{ij}^{\lhd}(0)$is the initial value of $\hat{\phi}_{ij}^{\lhd}(k)$,$i, j = 1, 2$. $\eta^{\lhd} \in (0, 2]$ is a weighting factor, $\varepsilon^{\lhd}$ is a small positive constant, $M^{\lhd}$ and $\zeta^{\lhd}$ are positive constants.

Let $K_y^{\lhd}$ and $K_u^{\lhd}$ represent the prediction and control horizons, respectively, and it can be obtained from (9.30) that

$$\mathbf{Y}(k+1) = \mathbf{E}(k)\mathbf{y}(k) + \mathbf{A}^{\lhd}(k)\Delta\mathbf{U}(k) \tag{9.35}$$

where the variables are defined as below:

$$\begin{cases}\mathbf{Y}(k+1) = [\mathbf{y}(k+1)^T, \ldots, \mathbf{y}(k+K_y^{\lhd})^T]^T \in \Re^{2K_y^{\lhd}}\\ \Delta\mathbf{U}(k) = [\Delta\mathbf{u}(k)^T, \ldots, \Delta\mathbf{u}(k+K_u^{\lhd}-1)^T]^T \in \Re^{2K_u^{\lhd}}\\ \mathbf{E}(k) = [I_{2\times 2}, I_{2\times 2}, \ldots, I_{2\times 2}]^T \in \Re^{2K_y^{\lhd}}\\ \mathbf{A}^{\lhd}(k) = \begin{bmatrix}\mathbf{\Phi}^{\lhd}(k) & 0 & 0 & 0\\ \mathbf{\Phi}^{\lhd}(k) & \mathbf{\Phi}^{\lhd}(k+1) & 0 & 0\\ \vdots & \vdots & \ddots & \vdots\\ \mathbf{\Phi}^{\lhd}(k) & \mathbf{\Phi}^{\lhd}(k+1) & \cdots & \mathbf{\Phi}^{\lhd}(k+K_u^{\lhd}-1)\\ \vdots & \vdots & \cdots & \vdots\\ \mathbf{\Phi}^{\lhd}(k) & \mathbf{\Phi}^{\lhd}(k+1) & \cdots & \mathbf{\Phi}^{\lhd}(k+K_u^{\lhd}-1)\end{bmatrix} \in \Re^{2K_y^{\lhd}\times 2K_u^{\lhd}}\end{cases}$$

At each time step k, the following autoregressive (AR) model is utilized to predict the PJM of the next $K_u^{\triangleleft}$ time steps:

$$\hat{\mathbf{\Phi}}^{\triangleleft}(k+j) = \boldsymbol{\theta}_1^{\triangleleft}(k)\hat{\mathbf{\Phi}}^{\triangleleft}(k+j-1) + \boldsymbol{\theta}_2^{\triangleleft}(k)\hat{\mathbf{\Phi}}^{\triangleleft}(k+j-2) + \ldots + \boldsymbol{\theta}_{n_p^{\triangleleft}}^{\triangleleft}(k)\hat{\mathbf{\Phi}}^{\triangleleft}(k+j-n_p^{\triangleleft}),\ j = 1, 2, \ldots, K_u^{\triangleleft} \quad (9.36)$$

$$\hat{\phi}_{pp}^{\triangleleft}(k+j) = \hat{\phi}_{pp}^{\triangleleft}(0),\ \ if \left|\hat{\phi}_{pp}^{\triangleleft}(k+j)\right| < \varepsilon^{\triangleleft}\ or\ \left|\hat{\phi}_{pp}^{\triangleleft}(k+j)\right| > M^{\triangleleft}$$
$$or\ sign(\hat{\phi}_{pp}^{\triangleleft}(k+j)) \neq sign(\hat{\phi}_{pp}^{\triangleleft}(0)),\ p = 1, 2. \quad (9.37)$$

$$\hat{\phi}_{pq}^{\triangleleft}(k+j) = \hat{\phi}_{pq}^{\triangleleft}(0),\ \ if \left|\hat{\phi}_{pq}^{\triangleleft}(k+j)\right| > \zeta^{\triangleleft}$$
$$or\ sign(\hat{\phi}_{pq}^{\triangleleft}(k+j)) \neq sign(\hat{\phi}_{pq}^{\triangleleft}(0)),\ p, q = 1, 2,\ \ p \neq q. \quad (9.38)$$

where $\boldsymbol{\theta}_i^{\triangleleft}(k) \in \Re^{2\times 2}$, i=1, 2, $\cdots$, $n_p^{\triangleleft}$, is the coefficient matrix, and $n_p^{\triangleleft}$ is a proper model order. $n_p^{\triangleleft}$ is usually set to be 2–7 according to Han (1984). In this section, $n_p^{\triangleleft}$=2. Denote $\mathbf{\Theta}^{\triangleleft}(k) = [\boldsymbol{\theta}_1^{\triangleleft}(k), \ldots, \boldsymbol{\theta}_{n_p^{\triangleleft}}^{\triangleleft}(k)]$ and $\hat{\mathbf{\Phi}}^{\triangleleft}(k+j-1) = [\hat{\mathbf{\Phi}}^{\triangleleft T}(k+j-1), \ldots, \hat{\mathbf{\Phi}}^{\triangleleft T}(k+j-n_p^{\triangleleft})]^T$, the PJM prediction model (9.36) can be transformed into the following form:

$$\mathbf{\Phi}^{\triangleleft}(k+j) = \mathbf{\Theta}^{\triangleleft}(k)\hat{\mathbf{\Phi}}^{\triangleleft}(k+j-1) \quad (9.39)$$

where $\mathbf{\Theta}^{\triangleleft}(k)$ can be calculated below:

$$\mathbf{\Theta}^{\triangleleft}(k) = \mathbf{\Theta}^{\triangleleft}(k-1) + \frac{(\hat{\mathbf{\Phi}}^{\triangleleft}(k) - \mathbf{\Theta}^{\triangleleft}(k-1)\hat{\mathbf{\Phi}}^{\triangleleft}(k-1))}{\delta^{\triangleleft} + \left\|\hat{\mathbf{\Phi}}^{\triangleleft}(k-1)\right\|^2}\hat{\mathbf{\Phi}}^{\triangleleft T}(k-1) \quad (9.40)$$

where $\delta^{\triangleleft} > 0$ is a weighting factor to penalize excessive change of the estimation of $\mathbf{\Theta}^{\triangleleft}(k)$.

The multi-step prediction cost function for obtaining $\Delta\mathbf{U}(\mathbf{k})$ is as below:

$$J(\Delta\mathbf{U}(k)) = \sum_{i=1}^{K_y^{\triangleleft}} \left\|\mathbf{y}(k+i) - \mathbf{y}_d^{\triangleleft}(k+i)\right\|^2 + \xi^{\triangleleft}\lambda^{\triangleleft} \sum_{j=0}^{K_u^{\triangleleft}-1} \|\Delta\mathbf{u}(k+j)\|^2 \quad (9.41)$$

where $\mathbf{y}_{\mathbf{d}}^{\triangleleft}(k+i) \in \Re^2$ is the desired vehicle accumulations of the regions, $\lambda^{\triangleleft} > 0$ is a weighting factor to restrain the change of perimeter control inputs, and $\xi^{\triangleleft}$ is a positive factor which is introduced to make the order of magnitude of the two items in (9.41) are in the same level.

It is noteworthy that both the inputs and the outputs have constraints. On the one hand, the vehicle accumulations in each region cannot be less than **0** or greater than $n_i^{\lhd jam}$, i.e.

$$\mathbf{0} \leq \mathbf{y}(k+i) \leq \mathbf{n}_{jam}, \quad i = 1, \ldots, K_y^{\lhd} \tag{9.42}$$

where $\mathbf{0} = [0, 0]^T$, and $\mathbf{n}_{jam} = [n_1^{jam}, n_2^{jam}]^T$. On the other hand, the perimeter control inputs are under the following constraints:

$$\mathbf{u}_{\min} \leq \mathbf{u}(k+j) \leq \mathbf{u}_{\max}(k+j), \; j = 0, \cdots, K_u^{\lhd} - 1. \tag{9.43}$$

where $\mathbf{u}_{\max}(k+j) = [u_{12,\max}(k+j), u_{21,\max}(k+j)]^T$, and $\mathbf{u}_{\min} = [u_{\min}, u_{\min}]^T$. Combining (9.35), (9.42) and (9.43), one has

$$\begin{bmatrix} \boldsymbol{A}^{\lhd}(k) & & & \\ & -\boldsymbol{A}^{\lhd}(k) & & \\ & & \boldsymbol{I} & \\ & & & -\boldsymbol{I} \end{bmatrix} \begin{bmatrix} \Delta\mathbf{U}(k) \\ \Delta\mathbf{U}(k) \\ \Delta\mathbf{U}(k) \\ \Delta\mathbf{U}(k) \end{bmatrix} \leq \begin{bmatrix} \mathbf{N}_{jam} - \mathbf{E}(k)\mathbf{y}(k) \\ \mathbf{E}(k)\mathbf{y}(k) \\ \mathbf{U}_{\max}(k) - \mathbf{U}(k-1) \\ \mathbf{U}(k-1) - \mathbf{U}_{\min} \end{bmatrix} \tag{9.44}$$

where $\mathbf{U}(k-1) = [\mathbf{u}(k-1)^T, \mathbf{u}(k)^T, \ldots, \mathbf{u}(k+K_u^{\lhd}-2)^T]^T \in \Re^{2K_u^{\lhd}}$, $\mathbf{N}_{jam} = [n_{jam}^T, \cdots, n_{jam}^T]^T \in \Re^{2K_y^{\lhd}}$, and $\mathbf{U}_{\max}(k) = [\mathbf{u}_{\max}(k)^T, \cdots, \mathbf{u}_{\max}(k+K_u^{\lhd}-1)^T]^T \in \Re^{2K_u^{\lhd}}$.

The optimization problem (9.41) with constraints of (9.44) is a convex quadratic programming (QP) problem, which can be solved in many ways, SQP, interior point method, etc. After is calculated from (9.41)–(9.44), one has $\Delta\mathbf{u}(k) = \mathbf{g}^{\lhd T}\Delta\mathbf{U}(k)$, where $\mathbf{g}^{\lhd} = [1, 1, 0, \cdots, 0]^T$. Thus, the perimeter control inputs $\mathbf{u}(k)$ is obtained as follow:

$$\mathbf{u}(k) = \mathbf{u}(k-1) + \mathbf{g}^{\lhd T}\Delta\mathbf{U}(k) \tag{9.45}$$

The multi-step cMFAPLC perimeter control scheme is constructed of (9.32)–(9.45). The reset mechanism (9.33)–(9.34) and (9.37)–(9.38) are designed to make the ability for tracking time-varying parameters of the PJM estimation algorithm stronger.

9.4.2 Numerical Simulation Results

In this section, the performance of the proposed multi-step cMFAPLC strategy for the medium-scale two-region urban traffic network system is tested via numerical simulation. As mentioned above, an urban road network is partitioned into two regions.

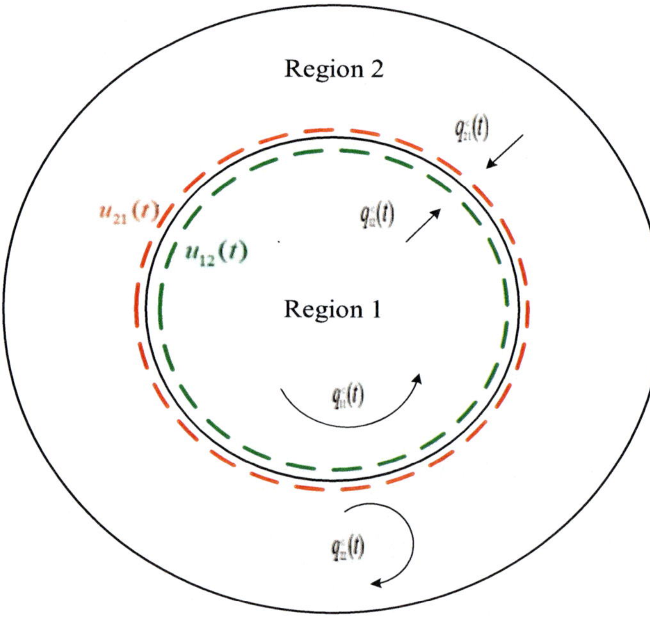

Fig. 9.8 The urban road network partitioned into 2 regions

Region 1 locates at the city center, while region 2 is on the periphery, see Fig. 9.8. As depicted in Fig. 9.9, the MFD of region 2 is the same as the one in the downtown of Yokohama (Hunt et al. 1982), while the MFD of region 1 is 1.2 times as the one in region 2, i.e., $a_1^{\triangleleft} = 4.959 \times 10^{-7}$, $b_1^{\triangleleft} = 9.938 \times 10^{-7}$, $c_1^{\triangleleft} = 5.04 \times 10^{-3}$, $a_2^{\triangleleft} = 4.133 \times 10^{-7}$, $b_2^{\triangleleft} = 8.282 \times 10^{-7}$, and $c_2^{\triangleleft} = 4.2 \times 10^{-3}$, respectively. The jammed and critical accumulations of the two regions are the same, which are n_{jam}=10,000 (veh) and n_{cr}=3400 (veh), respectively. The boundary capacity parameters are $\alpha^{\triangleleft} = 0.64$, $C_{12}^{\triangleleft\max} = 3.2$ (veh/s) and $C_{21}^{\triangleleft\max} = 3.84$ (veh/s), respectively.

The initial number of vehicles are: $n_1(0) = 6400, n_{11}(0) = 3200, n_{12}(0) = 3200$, $n_2(0) = 6000, n_{21}(0) = 3000$, and $n_{22}(0) = 3000$. Both of the regions are suffering morning peak under the initially congested condition. The desired vehicle accumulations in each region are chosen as $0.95n_{cr}$ in order to make the the traffic throughput maximized and ensure the robustness of the urban road network system. The initial, maximum and minimum green time ratios are $\omega^{\triangleleft} = 0.5$, $\omega_{\max}^{\triangleleft} = 0.9$ and $\omega_{\min}^{\triangleleft} = 0.1$, respectively. Thus, $u_{ij,\max}^{g} = 1.8$ and $u_{ij,\min}^{g} = 0.2$ according to the definition of perimeter control ratio. Other controller parameters are: $\xi^{\triangleleft} = 1000$, $\lambda^{\triangleleft} = 3$, $\eta^{\triangleleft} = \mu^{\triangleleft} = \delta^{\triangleleft} = 1$, respectively.

The measurement noises in the demands, accumulations and MFDs obey normal and uniform distribution, respectively, which are shown as below:

$$\tilde{q}_{ij}^{\triangleleft}(k) = q_{ij}^{\triangleleft}(k)(1 + \mathcal{N}^{\triangleleft}(0, \sigma_{q_{ij}^{\triangleleft}}^2)) \tag{9.46}$$

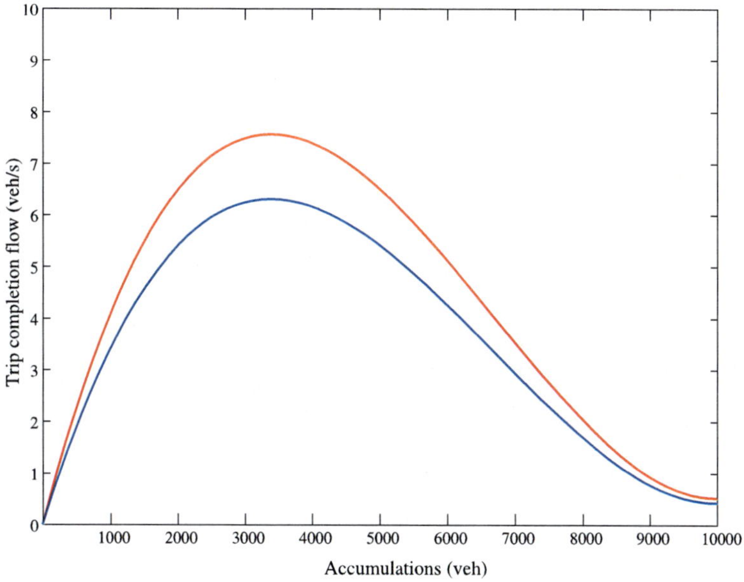

Fig. 9.9 MFD of each region, in which the red solid line denotes $G_1^{\triangleleft}$, and the blue solid line denotes $G_2^{\triangleleft}$

$$\tilde{n}_i(k) = n_i(k)\left(1 + \mathcal{N}^{\triangleleft}\left(0, \sigma_{n_i}^2\right)\right) \tag{9.47}$$

$$\tilde{G}_i^{\triangleleft}(n_i(k)) = G_i^{\triangleleft}(n_i(k))\left(1 + \mathcal{U}^{\triangleleft}\left(-\alpha_{G_i^{\triangleleft}}^{\triangleleft},\ \alpha_{G_i^{\triangleleft}}^{\triangleleft}\right)\right) \tag{9.48}$$

where the error parameters are $\sigma_{q_{ij}^{\triangleleft}}^{\triangleleft 2} = 0.1$, $\sigma_{n_i}^{\triangleleft 2} = 0.1$, and $\alpha_{G_i^{\triangleleft}}^{\triangleleft} = 0.3$, respectively. The simulation duration lasts 4 h including the morning peak hour, and the signal cycle length is 120 s, while the control cycle length is $T^{\triangleleft}$=240 s, which is double of the signal cycle length for control convenience. Hence, the 4 h of simulation time is divided into 60 time steps. The time-varying traffic demands are the same as that shown in Fig. 9.4. The simulation is based on MATLAB.

Meanwhile, in order to compare with the proposed multi-step cMFAPLC strategy, other two strategies (no control (NC) and MPC) are also tested in the same urban traffic system under the same condition. In NC strategy, the perimeter control ratios are equal to 1 all the time (Aboudolas and Geroliminis 2013), and it is the same as FTC in Sect. 9.3.3. MPC is commonly used for perimeter control. The urban traffic model in Section. 9.4.1 is only used to design the MPC perimeter controller. The objective function and the constraints of MPC strategy are the same as the ones in Geroliminis et al. (2013). The predictive horizon and control horizon of MPC and cMFAPLC are both chosen as $K_p^{\triangleleft} = 5$ and $K_c^{\triangleleft} = 2$.

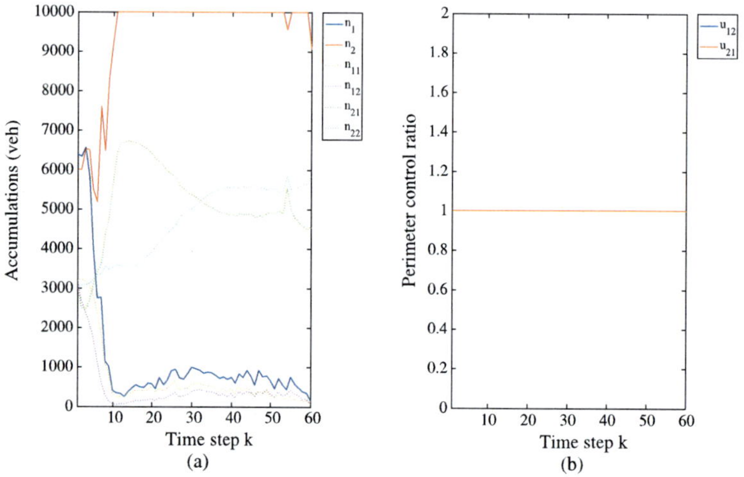

Fig. 9.10 Result of NC: (**a**) evolution of the number of vehicles traveling in each region; (**b**) perimeter control ratio of NC

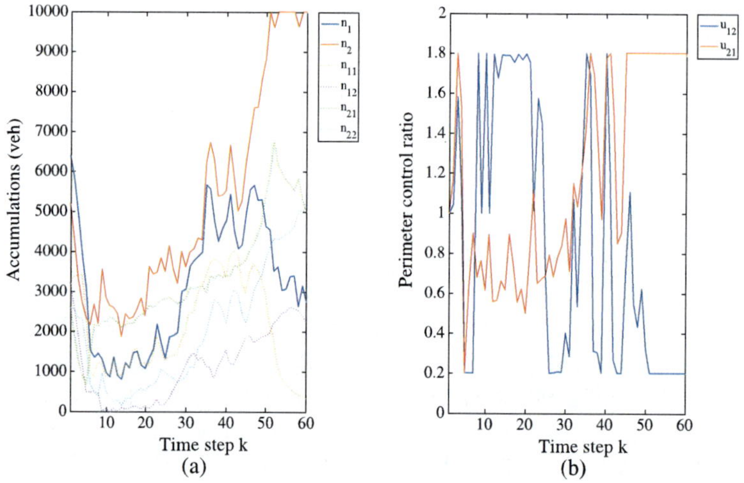

Fig. 9.11 Result of MPC: (**a**) evolution of the number of vehicles traveling in each region; (**b**) perimeter control ratio of MPC

The simulation results are depicted in Figs. 9.10, 9.11 and 9.12. Figure 9.10 shows the NC result. It can be seen from the figures that region 2 fell into jammed state after a short half time and the congestion maintained till the end, while the road space in region 1 was wasted because there were very few vehicles driving in it.

Figure 9.11 shows the MPC result. It can be seen from Fig. 9.11a that the region 1 was no longer jammed, but region 2 still jammed at the end of the simulation.

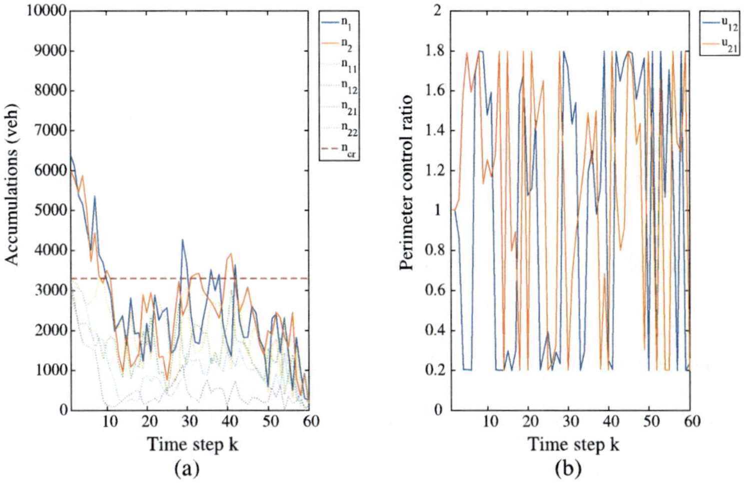

Fig. 9.12 Result of cMFAPLC: (**a**) evolution of the number of vehicles traveling in each region; (**b**) perimeter control ratio of cMFAPLC

Table 9.3 TTS for the urban road network

Strategy	TTS (s)
NC	1.5255×10^8
MPC	1.2032×10^8
cMFAPLC	7.3715×10^7

The result of multi-step cMFAPLC is depicted in Fig. 9.12. As shown in Fig. 9.12a, the congestion phenomenon in both regions was disappeared. Instead, the number of vehicles in the two regions was both near the desired value, which made the traffic throughput maximized.

In addition, the total time spent (TTS) of the medium-scale urban traffic network is introduced to evaluate the perimeter control performance of the above strategies, which is calculated as $TTS = T^{\triangleleft} \sum_{k=1}^{K^{\triangleleft}} \sum_{i=1}^{2} n_i(k)$. The results are compared in Table 9.3, where the improvement is relative to NC, expressed in percentages. As shown in Table 9.3, the proposed multi-step cMFAPLC method worked better than NC and MPC.

9.5 Conclusion

Two new data-driven control method called one-step IOC-MFAPLC and multi-step cMFAPLC are introduced to address the problem of perimeter control for medium-scale two-region urban road network system in this chapter. A remarkable advantage of the proposed strategies is that only the input/output data of the system is utilized to design the perimeter controller, and the urban traffic model is no longer needed. Meanwhile, the input/output constraints are considered to make it more practical. The simulation results show that the proposed two strategies are superior to some other perimeter control methods under model errors.

References

Abadi A, Rajabioun T, Ioannou P (2015) Traffic flow prediction for road transportation networks with limited traffic data. IEEE Trans Intell Transp Syst 16(2):653–662

Aboudolas K, Geroliminis N (2013) Perimeter and boundary flow control in multi-reservoir heterogeneous networks. Transp Res Part B: Methodol 55(9):265–281

An K, Chiu YC, Hu X et al (2018) A network partitioning algorithmic approach for macroscopic fundamental diagram-based hierarchical traffic network management. IEEE Trans Intell Transp Syst 19(4):1130–1139

Beard C, Ziliaskopoulos A (2006) System optimal signal optimization formulation. Transp Res Rec 1:102–112

Buisson C, Ladier C (2009) Exploring the impact of homogeneity of traffic measurements on the existence of macroscopic fundamental diagrams. Transp Res Rec 137(2124):127–136

Chi RH, Hou ZS (2010) A model-free periodic adaptive control for freeway traffic density via ramp metering. Acta Automatica Sinica 36:1029–1033

Daganzo CF (2007) Urban gridlock: macroscopic modeling and mitigation approaches. Transp Res Part B: Methodol 41(1):49–62

Gayah VV, Daganzo CF (2011) Clockwise hysteresis loops in the macroscopic fundamental diagram: an effect of network instability. Transp Res Part B: Methodol 45(4):643–655

Gayah VV, Gao X, Nagle AS (2014) On the impacts of locally adaptive signal control on urban network stability and the macroscopic fundamental diagram. Transp Res Part B: Methodol 70(1):255–268

Gazis DC (2006) Traffic theory. Springer Science Business Media

Geroliminis N, Daganzo CF (2008) Existence of urban-scale macroscopic fundamental diagrams: some experimental findings. Transp Res Part B: Methodol 42(9):759–770

Geroliminis N, Sun J (2011) Hysteresis phenomena of a macroscopic fundamental diagram in freeway networks. Transp Res Part A: Policy Pract 45(9):966–979

Geroliminis N, Sun J (2011) Properties of a well-defined macroscopic fundamental diagram for urban traffic. Transp Res Part B: Methodol 45(3):605–617

Geroliminis N, Haddad J, Ramezani M (2013) Optimal perimeter control for two urban regions with macroscopic fundamental diagrams: a model predictive approach. IEEE Trans Intell Transp Syst 14(1):348–359

Geroliminis N, Daganzo CF (2007) Macroscopic modeling of traffic in cities. In: 86th proceedings of the human factors and ergonomics society annual meeting, Washington, DC

Girianna M, Benekohal RF (2004) Using genetic algorithms to design signal coordination for oversaturated networks. J Intell Transp Syst 8(2):117–129

Godfrey JW (1969) The mechanism of a road network. Traffic Eng Control 11(7):323–327

Gu HZ, Wang W (1998) A global optimization simulated annealing algorithm for intersection signal timing. J Southeast Univ 28(3):69–72

Haddad J (2017) Optimal coupled and decoupled perimeter control in one-region cities. Control Eng Pract 61:134–148

Haddad J (2017) Optimal perimeter control synthesis for two urban regions with aggregate boundary queue dynamics. Transp Res Part B: Methodol 96:1–25

Haddad J, Geroliminis N (2012) On the stability of traffic perimeter control in two-region urban cities. Transp Res Part B: Methodol 46(9):1159–1176
Haddad J, Mirkin B (2016) Adaptive perimeter traffic control of urban road networks based on MFD model with time delays. Int J Robust Nonlinear Control 26(6):1267–1285
Haddad J, Mirkin B (2017) Coordinated distributed adaptive perimeter control for large-scale urban road networks. Transp Res Part C: Emerging Technol 77:495–515
Haddad J, Shraiber A (2014) Robust perimeter control design for an urban region. Transp Res Part B: Methodol 68:315–332
Hajiahmadi M, Haddad J, Schutter BD et al (2015) Optimal hybrid perimeter and switching plans control for urban traffic networks. IEEE Trans Intell Transp Syst 23(2):464–478
Han Z (1984) On the identification of time-varying parameters in dynamic systems. Acta Automatica Sinica 10(4):330–337
Helbing D (2009) Derivation of a fundamental diagram for urban traffic flow. Eur Phys J B: Condensed Matter Complex Syst 70(2):229–241
Herman R, Prigogine I (1979) A two-fluid approach to town traffic. Science 204(4389):148–151
Hou ZS, Jin ST (2013) Model free adaptive control: theory and applications. CRC Press, Florida
Hou ZS, Li XY (2016) Repeatability and similarity of freeway traffic flow and long-term prediction under big data. IEEE Trans Intell Transp Syst 17(6):1786–1796
Hunt P, Robertson D, Bretherton R et al (1982) The SCOOT on-line traffic signal optimisation technique. Traffic Eng Control 23(4):190–192
Ji Y, Geroliminis N (2012) On the spatial partitioning of urban transportation networks. Transp Res Part B: Methodol 46(10):1639–1656
Keyvan-Ekbatani M, Kouvelas A, Papamichail I et al (2012) Exploiting the fundamental diagram of urban networks for feedback-based gating. Transp Res Part B: Methodol 46:1393–1403
Keyvan-Ekbatani M, Papageorgiou M, Knoop VL (2015) Controller design for gating traffic control in presence of time-delay in urban road networks. Transp Rese Part C: Emerging Technol 59:308–322
Keyvan-Ekbatani M, Yildirimoglu M, Geroliminis N et al (2015) Multiple concentric gating traffic control in large-scale urban networks. IEEE Trans Intell Transp Syst 16(4):2141–2154
Keyvan-Ekbatani M, Carlson RC, Knoop VL et al (2016) Queuing under perimeter control: analysis and control strategy. In: IEEE 19th international conference on Intelligent Transportation Systems (ITSC), Rio de Janeiro, pp 1502–1507
Kouvelas A, Saeedmanesh M, Geroliminis N (2017) Enhancing model-based feedback perimeter control with data-driven online adaptive optimization. Transp Res Part B: Methodol 96:26–45
Lin XH, Xu JM, Lu K et al (2007) A design method of two way green wave of each phase for entrance. J Transp Inf Safety 25(5):8–12
Liu XM, Wang FY (2002) Study of city area traffic coordination control on the basis of agent. In: The IEEE 5th international conference on intelligent transportation systems, pp 758-761
Lopez C, Krishnakumari P, Leclercq L et al (2017) Spatio-Temporal partitioning of transportation network using travel time data. Transportation research record. J Transp Res Board 2623: 98–107
Lowrie P (1982) SCATS: the Sydney co-ordinated adaptive traffic system: principles, methodology, algorithms. Proc Int Conf Road Traffic Signalling 4:67–70
Lu K, Xu JM, Li YS (2010) Algebraic method of arterial road coordinate control for bidirectional green wave under signal design mode of one-phase-one-approach. China J Highway Transp 23(3):95–101
Lv Y, Duan Y, Kang W et al (2015) Traffic flow prediction with big data: a deep learning approach. IEEE Trans Intell Transp Syst 16(2):865–873
Mazloumian A, Geroliminis N, Helbing D (1928) The spatial variability of vehicle densities as determinant of urban network capacity. Philos Trans Royal Soc Londn Ser A: Math Phys Eng Sci 368:4627–4647
Ramezani M, Haddad J, Geroliminis N (2015) Dynamics of heterogeneity in urban networks: aggregated traffic modeling and hierarchical control. Transp Res Part B: Methodol 74:1–19
Robertson DI (1969) TRANSYT: a traffic network study tool

Saberi M, Mahmassani HS (2012) Exploring properties of networkwide flow-density relations in a freeway network. J Transp Res Board 2012: 1–21 (Washington, DC)
Saeedmanesh M, Geroliminis N (2016) Clustering of heterogeneous networks with directional flows based on Snake similarities. Transp Res Part B: Methodol 91:250–269
Saeedmanesh M, Geroliminis N (2017) Dynamic clustering and propagation of congestion in heterogeneously congested urban traffic networks. Transp Res Part B: Methodol 105:193–211
Shen L (2018) Boundary control based on macroscopic fundamental diagram. M. D. thesis, Zhejiang University, Hangzhou, China (In Chinese)
Wang FY (2010) Parallel control and management for intelligent transportation systems: concepts, architectures, and applications. IEEE Trans Intell Transp Syst 11(3):630–638
Wang Y, Yao ZH, Jiang YS et al (2018) The dual-phase signal timing optimization model based on adaptive genetic algorithm. Ind Eng J 21(5):72
Zhai R, Zhou T, Liu G (2011) Principle and application of road traffic control. People's Public Security University of China Press (In Chinese)
Zhang L, Garoni TM, Gier JD (2013) A comparative study of macroscopic fundamental diagrams of arterial road networks governed by adaptive traffic signal systems. Transp Res Part B: Methodol 49:1–23

Chapter 10
Large-Scale Multi-region Urban Road Networks

10.1 Introduction

For the small-scale single-region and medium-sized two-region urban transportation systems, vehicles drive within the controlled region or only between the inner and outer controlled areas, and there is no other path between each region. Therefore, there is no need to consider the problem of vehicle route guidance between the regions. However, for large-scale multi-region urban traffic systems (MRUTS), when the destination of vehicles in a certain region is in its nonadjacent regions, vehicles need to pass through other regions to reach the destination, which requires interregional route guidance for vehicles while implementing perimeter control among various regions, so as to make full use of the traffic resources in each region. It also improves the service level of the whole urban traffic network. In order to balance the traffic load in each region and reduce traffic congestion, the perimeter control among the regions and the regional route guidance for vehicles need to be both solved. In this chapter, a decentralized one-step model free adaptive predictive learning control method and a centralized multi-step model free adaptive predictive learning control strategy are proposed for route guidance and perimeter control for the MRUTS.

The main contributions of this chapter are as follows.

- Two novel data-driven model free adaptive predictive learning control (MFAPLC) method is first applied to perimeter control and route guidance for the large-scale MRUTS. Only the perimeter control and route guidance input and the system output data of the urban traffic system is used to design the perimeter control and route guidance strategies to handle the problem of model mismatching.
- In the proposed strategies, the O-D tables, absolutely homogeneous regions and well-defined MFDs, are no longer needed. Instead, it is sufficient to simply divide the large-scale urban traffic network according to geometric characteristics or some other factors, since the precise model is no longer needed.

© The Author(s), under exclusive license to Springer Nature Singapore Pte Ltd. 2023

Q. Yu et al., *Predictive Learning Control for Unknown Nonaffine Nonlinear Systems*, Intelligent Control and Learning Systems 8,
https://doi.org/10.1007/978-981-19-8857-8_10

- The CPU time of the proposed methods is much smaller than the traffic signal cycle, so that they can be implemented in real time.

This chapter is organized as follows. Section 10.2 introduces one-step model free adaptive predictive learning perimeter control method. Section 10.3 shows the muti-step model free adaptive learning route guidance and perimeter control strategy. Some conclusions are given in Sect. 10.4.

10.2 One-Step Model Free Adaptive Predictive Learning Perimeter Control

In this section, a data-driven approach called MFAPLC is utilized for perimeter control of large-scale multi-region urban traffic systems. The one-step predictive control mechanism is used to obtain the optimal perimeter control ratios among the regions in the proposed MFAPLC scheme. Taking into consideration that the large-scale MRUTS is a MIMO nonlinear system, the decentralized estimation and decentralized MFAPLC (DED-MFAPLC) strategy (Hou and Jin 2013) can be used for perimeter control to handle the measurable interactions.

The main contributions of the work in this section are: (i) A novel data-driven MFAPLC method is first applied to perimeter control for the large-scale MRUTS. Only the perimeter control input and the system output data of the urban traffic system is used to design the perimeter control strategy to handle the problem of model mismatching. (ii) In the proposed method, the O-D tables, absolutely homogeneous regions, and well-defined MFDs are no longer needed. Instead, it is sufficient to simply divide the large-scale urban traffic network according to geometric characteristics or some other factors, since the precise model is no longer needed. (iii) The internal coupling of MIMO nonlinear systems can be addressed by subsystem decomposition, and the measurable interactions between the subsystems can be handled by the one-step DED-MFAPLC scheme. (iv) Route choice is combined with the proposed perimeter control strategy in order to further enhance the efficiency of the urban road network system, while the precise route choice ratios are also not needed in the process of perimeter controller design. (v) The CPU time of this scheme is much smaller than the traffic signal cycle, so that it can be implemented in real time.

10.2.1 Dynamics for the Large-Scale Multi-region Urban Road Network

In this section, it is assumed that there is a heterogeneous urban road network divided into N homogenous regions. A network consisting of five regions is diagrammed in Fig. 10.1 as an example, and that each region has a well-defined MFD with trip completion flow $G_i^{\triangleright}(n_i(k))$. The discrete form of dynamics after the first-order Eulerian

Table 10.1 List of variables

Notation	Description
i, j, h, g	Symbols representing regions and subsystems
N_i	The set of regions neighboring region i
V_{ij}	The set of regions that vehicles from region i to region j can go through next immediately
$y_i(k)$	The total number of vehicles traveling in region i at time step k
$y_{ij}(k)$	Number of vehicles in region i traveling to region j at time step k
$y_{ij}^{f}(k)$	Number of vehicles in region i with final destination to region j at time step k
$y_{ii}(k)$	Number of vehicles in region i with destination to itself at time step k
$y_{ihj}(k)$	Number of vehicles in region i traveling to the next immediate region h with destination to region j at time step k
y_i^{cr}	The critical accumulation of region i
y_i^{jam}	The jammed accumulation of region i
$q_{ij}^{\triangleright}(k)$	Exogenous traffic demands generated in region i with destination to region j at time step k
$q_{ii}^{\triangleright}(k)$	Endogenous traffic demands in region i at time step k
$M_{ij}^{\triangleright}(k)$	Transfer flow from region i to region j at time step k
$M_{ij}^{\triangleright f}(k)$	Transfer flow from region i to region j with final destination to region j at time step k
$M_{ihj}^{\triangleright}(k)$	Transfer flow from region i to region h with final destination to region j at time step k
$M_{ii}^{\triangleright}(k)$	The internal flow from region i with destination in itself at time step k
$G_i^{\triangleright}(n_i(k))$	Trip completion flow for region i at time step k
$u_{ij}(k)$	Perimeter control ratio from region i to region j at time step k
$b_{ihj}^{\triangleright}(k)$	The ratio of vehicles in region i choosing to go through the next immediate region h to reach their destination region j at time step k
z_{hj}	The interaction of subsystem h acting on subsystem j
y_j	The output of subsystem j

discretization is directly given in the following to save space. The reason why we use the discrete-time model to describe the dynamics of the MRUTS is that the discrete-time one is more suitable for traffic control in practice due to the actual traffic control system is always controlled by a computer. The urban traffic model is only utilized to generate the urban traffic data instead of perimeter control strategy design. The definition of the main variables used in this section is shown in Table 10.1.

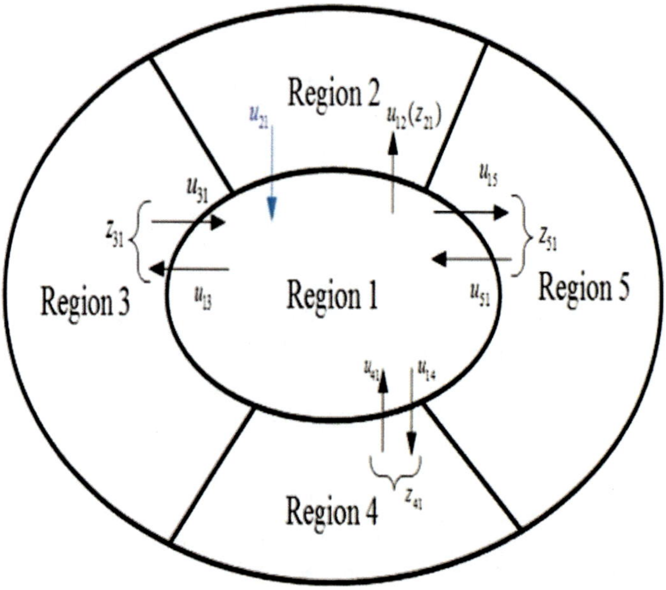

Fig. 10.1 Urban road network partitioned into five regions

A. Dynamics of multi-Region urban traffic networks
For the case that region i and j are adjacent to each other gives

$$n_{ij}^{f}(k+1) = n_{ij}^{f}(k) + T^{\triangleright}(q_{ij}^{\triangleright}(k) + \sum_{\substack{g \in N_i, g \notin N_j \\ i \in V_{gj}}} u_{gi}(k) M_{gij}^{\triangleright}(k) - \sum_{h \in V_{ij}} u_{ih}(k) M_{ihj}^{\triangleright f}(k)), j \in N_i \tag{10.1}$$

$$n_{ij}(k) = n_{ij}^{f}(k) + \sum_{\substack{h \notin N_i \\ j \in V_{ih}}} n_{ijh}(k), j \in N_i \tag{10.2}$$

where $T^{\triangleright}$ is the sampling period, N_i and V_{ih} represent the set of regions adjacent to region i and the set of regions where the vehicles can pass immediately when driving from region i to h that is not adjacent to it, respectively. The exogenous traffic demands generated in region i with destination to region j and endogenous traffic demands generated in region i are denoted by $q_{ij}^{\triangleright}(k)$ and $q_{ii}^{\triangleright}(k)$, respectively. The perimeter control $u_{ij}(k)$ is defined as the ratio of vehicles that can actually drive from region i to region j, and it is relative to the proportion of uncontrolled value (Haddad and Geroliminis 2012; Geroliminis et al. 2013). A more detailed distinction is proposed for state variables $n_{ij}(k)$ and $n^f_{ij}(k)$, and the same for the corresponding transfer flow below:

$$M_{ij}^{\triangleright}(k) = \frac{n_{ij}(k)}{n_i(k)} G_i^{\triangleright}(n_i(k)), j \in N_i \tag{10.3}$$

$$M_{ij}^{\triangleright f}(k) = \frac{n^f_{ij}(k)}{n_i(k)} G_i^{\triangleright}(n_i(k)), j \in N_i \tag{10.4}$$

$$M_{ijh}^{\triangleright}(k) = \frac{n_{ijh}(k)}{n_i(k)} G_i^{\triangleright}(n_i(k)), j \in N_i, h \notin N_i, j \in V_{ih} \tag{10.5}$$

$$M_{ii}^{\triangleright}(k) = \frac{n_{ii}(k)}{n_i(k)} G_i^{\triangleright}(n_i(k)) \tag{10.6}$$

$$M_{ij}^{\triangleright}(k) = M_{ij}^{\triangleright f}(k) + \sum_{\substack{h \notin N_i \\ j \in V_{ih}}} M_{ijh}^{\triangleright}(k), j \in N_i \tag{10.7}$$

It is worth noting that the transfer flow $M_{ij}^{\triangleright}(k)$ and $M_{ij}^{\triangleright f}(k)$ are only for adjacent regions. Define that $M_{ij}^{\triangleright}(k)$ and $M_{ij}^{\triangleright f}(k)$ are both equal to zero for $j \notin N_i$. Under perimeter control, $u_{ij}(k) \cdot M_{ij}^{\triangleright}(k)$, $u_{ij}(k) \cdot M_{ij}^{\triangleright f}(k)$, and $u_{ij}(k) \cdot M_{ijh}^{\triangleright}(k)$ are the actual transfer flows. However, it is worth noting that there is a boundary capacity between every two adjacent regions, see Fig. 10.2. The boundary capacity $C_{ij}^{\triangleright}((n_j(k))$ is considered as below:

$$C_{ij}^{\triangleright}(n_j(k)) = \begin{cases} C_{ij}^{\triangleright \max} & , \text{if } 0 \le n_j(k) \le \alpha^{\triangleright} n_j^{\text{jam}} \\ \frac{C_{ij}^{\triangleright \min} - C_{ij}^{\triangleright \max}}{(1-\alpha^{\triangleright}) \cdot n_j^{\text{jam}}} n_j(k) + \frac{C_{ij}^{\triangleright \max} - \alpha^{\triangleright} C_{ij}^{\triangleright \min}}{1-\alpha^{\triangleright}}, & if\ \alpha^{\triangleright} n_j^{\text{jam}} \le n_j(k) \le n_j^{\text{jam}} \end{cases} \tag{10.8}$$

where $C_{ij}^{\triangleright \max}$ and $C_{ij}^{\triangleright \min}$ are the maximum and the minimum value of the boundary capacity from region i to region j, respectively, and $0 < \alpha^{\triangleright} < 1$. $C_{ij}^{\triangleright \min}$ is equal to zero in Ji and Geroliminis (2012), Daganzo (2007). However, in reality, however, there still has a very small trip completion flow (i.e., $G_j^{\triangleright}(n_j^{\text{jam}})$) under this circumstance. Therefore, $C_{ij}^{\triangleright \min}$ is modified to a portion of $G_j^{\triangleright}(n_j^{\text{jam}})$ here, which is more realistic.

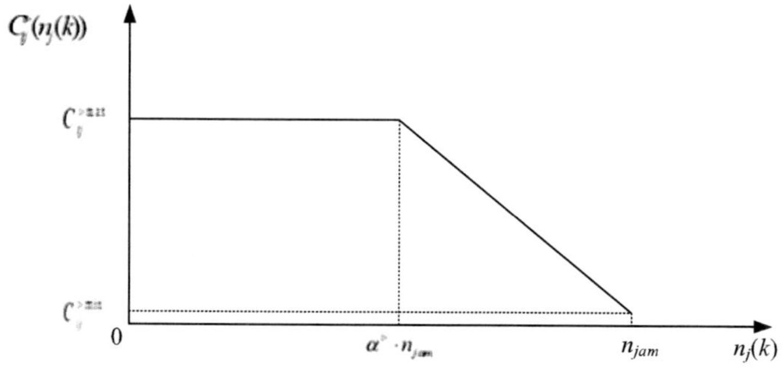

Fig. 10.2 Boundary capacity

In consideration of boundary capacity, one has $u_{ij}(k) \le \frac{C_{ij}^{\triangleright}(n_j(k))}{M_{ij}^{\triangleright}(k)}$. In addition, $u_{ij}(k)$ is also limited by green time ratio. Assume that the initial green time ratio of boundary intersections is $\omega^{\triangleright}$, and the maximum and minimum green time ratios are $\omega_{\min}^{\triangleright}$ and $\omega_{\max}^{\triangleright}$, respectively. Thus, $u_{ij,\max}^{g} = \frac{\omega_{\max}^{\triangleright}}{\omega^{\triangleright}}$ and $u_{ij,\min}^{g} = \frac{\omega_{\min}^{\triangleright}}{\omega^{\triangleright}}$ are the theoretical approximation of maximum and minimum perimeter control ratio, respectively. Therefore, the range of perimeter control input is presented below:

$$u_{ij,\min}^{g} \le u_{ij}(k) \le \min\left\{u_{ij,\max}^{g}, \frac{C_{ij}^{\triangleright}(n_j(k))}{M_{ij}^{\triangleright}(k)}\right\} \tag{10.9}$$

For another case that regions i and j are not adjacent to each other, the following equations are established:

$$\begin{aligned} n_{ij}^{f}(k+1) = n_{ij}^{f}(k) + T^{\triangleright}(q_{ij}^{\triangleright}(k) + &\sum_{\substack{g \in N_i, g \notin N_j \\ i \in V_{gj}}} u_{gi}(k) M_{gij}^{\triangleright}(k) \\ &- \sum_{h \in V_{ij}} u_{ih}(k) M_{ihj}^{\triangleright}(k)), j \notin N_i \end{aligned} \tag{10.10}$$

$$n_{ij}^{f}(k) = \sum_{h \in V_{ij}} n_{ihj}(k), \ j \notin N_i \tag{10.11}$$

$$n_{ij}(k) = n_{ij}^{f}(k), \ j \notin N_i \tag{10.12}$$

$$n_{ihj}(k) = b_{ihj}^{\triangleright}(k) n_{ij}(k), j \notin N_i, h \in V_{ij} \tag{10.13}$$

where $b_{ihj}^{\triangleright}(k)$ is the ratio of vehicles in region i choosing to go through the next immediate region h to reach their destination region j, and obviously $\sum_{h\in V_{ij}} b_{ihj}^{\triangleright}(k)=1$.

For region i, the number of vehicles in region i with destination to itself and the corresponding dynamic equation are shown as follows:

$$n_{ii}(k) = n_{ii}^{f}(k) \tag{10.14}$$

$$n_{ii}(k+1) = n_{ii}(k) + T^{\triangleright}(q_{ii}^{\triangleright}(k) + \sum_{j\neq i, j\in N_i} u_{ji}(k)M_{ji}^{\triangleright f}(k) - M_{ii}^{\triangleright}(k)) \tag{10.15}$$

The total number of vehicles traveling in region i is calculated as below:

$$n_i(k) = n_{ii}(k) + \sum_{j\neq i} n_{ij}^{f}(k) \tag{10.16}$$

B. Route Choice (RC)

In this section, RC is integrated to characterize drivers reaction to traffic conditions and further improve the service level of the urban road network system, which is shown as below.

For region $j \in N_i$, vehicles can enter region j from region i directly. It is suggested that $b_{ijj}^{\triangleright}(k) = 1, b_{ihj}^{\triangleright}(k) = 0, j \in N_i$. That is, all vehicles with destinations in adjacent regions can directly drive into the destination region without passing through other regions.

For region $j \notin N_i$, vehicles from region i to region j need to go through other regions. A logit model integrating with Dijkstras algorithm for K-shortest paths strategy (similar to Ramezani et al. (2015), Sirmatel and Geroliminis (2016)) is utilized to calculate the route choice ratio, denoted by. The shortest average travel time for the route from region i to region j via the next immediate region h, denoted by $b_{ihj}^{\triangleright \text{logit}}(k)$, can be obtained by $t_{ihj}(k)$ Dijkstras algorithm (Dijkstra 1959). Similar to Ben-Akiva and Bierlaire (1999), the following logic strategy for RC is used to calculate $b_{ihj}^{\triangleright \text{logit}}(k)$:

$$b_{ihj}^{\triangleright \text{logit}}(k) = \frac{e^{\frac{\tau^{\triangleright}}{t_{ihj}(k)}}}{\sum_{h\in V_{ij}} e^{\frac{\tau^{\triangleright}}{t_{ihj}(k)}}} \tag{10.17}$$

where $\tau^{\triangleright} > 0$ is a scale parameter. In fact, the routes which take a lot of time and need to go too far away are often not selected by drivers. Therefore, the first K-shortest routes are generally provided for drivers to choose from via (10.17).

Meanwhile, vehicles have their initial tendency of RC. Therefore, the RC strategy is presented as below:

$$b_{ihj}^{\triangleright}(k) = \beta^{\triangleright} b_{ihj}^{\triangleright \mathrm{init}} + (1 - \beta^{\triangleright}) b_{ihj}^{\triangleright \mathrm{logit}}(k),\ j \notin N_i,\ h \in V_{ij} \tag{10.18}$$

where $b_{ihj}^{\triangleright \mathrm{init}}$ is the initial route choice ratio, which is determined by drivers driving habits. $0 \leq \beta^{\triangleright} \leq 1$ is the proportion of drivers who adhere to the initial route selection.

10.2.2 Methodology Framework

As depicted in the macroscopic urban traffic dynamics (10.1)–(10.16), the large-scale MRUTS is a complex interconnected system, and it is strong-coupled and will definitely leading to too complex control problems if the model-based control method is used. In this section, one-step DED-MFAPLC for complex interconnected system is applied to the perimeter control for the nonlinear large-scale MRUTS.

As mentioned above, the urban traffic network is partitioned into N regions, so the whole system has N outputs, which are the accumulations in each region. There is a two-way boundary between every two adjacent regions in the urban road network, and there have two perimeter control inputs in each boundary. Therefore, the urban traffic system is a MIMO nonlinear system, which is complex interconnected. The one-step DED-MFAPLC scheme includes three parts: subsystem decomposition, dynamic linearization, and perimeter controller design, which are presented below, respectively.

A. Subsystem decomposition

In this work, the multi-region urban traffic system can be decomposed into N MISO subsystems, see Fig. 10.3. Each region represents a subsystem, whose perimeter control inputs are u_{ij}, $i \in N_j$, and the output is n_j. Due to the complex connectivity of large-scale MRUTS, there are interconnected affects between the subsystems after decomposition. For subsystem j, when one related perimeter control input u_{ij} is selected as the current input of the subsystem, the other perimeter control inputs related to subsystem j are regarded as the interconnected affects.

The interreaction of subsystem h on subsystem j ($h \in N_j$) consists of two parts: u_{jh} leads to the reduction of the output of subsystem j, while u_{hj} results in the increase of it. The interconnected influence of other subsystems on subsystem 1 is shown in Fig. 10.1 as an example. u_{21} is regarded as the current control input of subsystem 1 in the proposed example, and other related perimeter control inputs are the interreactions of other subsystems acting on subsystem 1. The same is true for other subsystems.

For all the regions $h \in N_j$, the subsystem h has interconnected influence on subsystem j, which is denoted by z_{hj}. z_{hj} represents the increment of vehicles contributed by region h to region j, i.e.,

$$z_{hj}(k) = T^{\triangleright}(u_{hj}(k-1)M_{hj}^{\triangleright}(k-1) - u_{jh}(k-1)M_{jh}^{\triangleright}(k-1)) \tag{10.19}$$

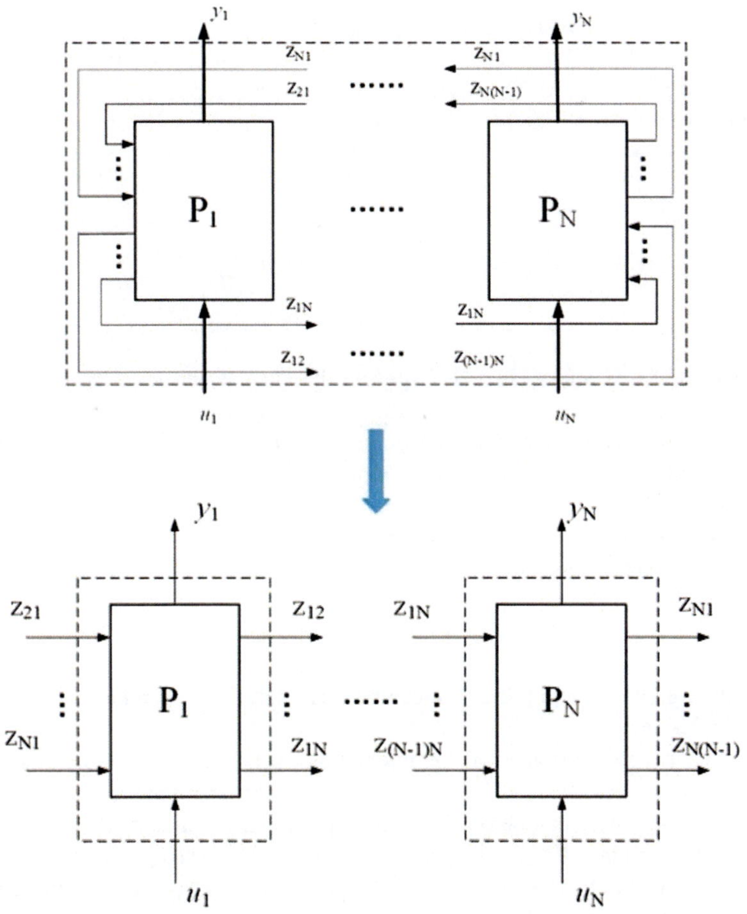

Fig. 10.3 Decomposition of a complex interconnected urban traffic system

$$z_{ij}(k) = -Tu_{ji}(k-1)M_{ji}^{\triangleright}(k-1) \tag{10.20}$$

The reason why there is a sign "−" in (10.20) is that z_{ij} means the number of vehicles flowing out of region j to region i, and it will lead to the reduction of n_j. Since u_{ij} is considered as the current perimeter control input of subsystem j, the interconnected effect of subsystem i on subsystem j is only composed of u_{ji}, see (10.20) and Fig. 10.1. Furthermore, note that $z_{hj}(k) = 0$ if the region $h \notin N_j$. The interactions (10.19)–(10.20) are measurable. After decomposition, dynamic linearization for each subsystem is presented below.

B. Dynamic Linearization for Subsystems

In this work, the CFDL data modeling technique is applied to the perimeter control for the large-scale MRUTS. Assume that there is a complex interconnected MIMO nonlinear system composed of N subsystems, and the jth subsystem is portrayed as below:

$$\begin{aligned} y_j(k+1) = & f_j(y_j(k), \ldots, y_j(k - n_{y_j}^{\triangleright}), u_{ij}(k), \ldots, u_{ij}(k - n_{u_{ij}}^{\triangleright}), \\ & z_{1j}(k), \ldots, z_{1j}(k - n_{z_{1j}}^{\triangleright}), \ldots, z_{hj}(k), \ldots, \\ & z_{hj}(k - n_{z_{hj}}^{\triangleright}), \ldots, z_{Nj}(k), \ldots, z_{Nj}(k - n_{z_{Nj}}^{\triangleright})), \\ & i, j, h = 1, 2, \ldots, N, \quad i, h \in N_j \end{aligned} \tag{10.21}$$

where $u_{ij}(k) \in \Re$ and $y_j(k) \in \Re$ represent the current perimeter control input and the system output of subsystem j, respectively. $z_{hj}(k) \in \Re$, $h \neq j$, is the measurable interactions of subsystem h acting on subsystem j. $n_{y_j}^{\triangleright}$, $n_{u_{ij}}^{\triangleright}$ and $n_{z_{hj}}^{\triangleright}$ are unknown integers, and $f_j(...)$ is an unknown nonlinear function describing the urban traffic dynamics of subsystem j.

In the proposed MRUTS, the interactions are same as (10.19)–(10.20). It can be obtained from (10.1)–(10.16) that

$$y_j(k+1) = f(y_j(k), u_{ij}(k), z_{hj}(k)) \; i, h, j = 1, \ldots, N \tag{10.22}$$

Comparing (10.21) and (10.22), it can be seen that $n_{y_j}^{\triangleright}$, $n_{u_{ij}}^{\triangleright}$ and $n_{z_{hj}}^{\triangleright}$ are all equal to zero in this section.

In general circumstances, if the interactions $z_{hj}(k)$ are measurable, the complex interconnected system (10.21) can be disassembled into N independent subsystems, and each subsystem can be regarded as a multiple-input and single-output (MISO) system. Combining (10.1)–(10.20), the form of (10.21) or (10.22) for each subsystem can be obtained. In this framework, the perimeter control ratio is chosen as the current control input of the subsystem j, while other perimeter control ratios relative to subsystem j are considered as the interactions of other subsystems acting on subsystem j, see (10.19)–(10.20).

Define the augmented control input vector of subsystem j with respect to $\mathbf{u}_{ij}$ as $\mathbf{u}_{ij}(k) = [u_{ij}(k), z_{1j}(k), \ldots, z_{hj}(k), \ldots, z_{Nj}(k)]^{\mathrm{T}}, h \neq j, i = 1, \ldots, N, j = 1, \ldots, N$, and $h = 1, \ldots, N$, $z_{hj}(k) = 0$ if $h \notin N_j$. The jth subsystem (10.21) can be rewritten as

$$y_j(k+1) = f_j(y_j(k), \ldots, y_j(k - n_{y_j}^{\triangleright}), \mathbf{u}_{ij}(k), \ldots, \mathbf{u}_{ij}(k - n_{u_{ij}}^{\triangleright})) \tag{10.23}$$

where $n_{u_{ij}}^{\triangleright} = \max\{n_{u_{ij}}^{\triangleright}, n_{z_{1j}}^{\triangleright}, \ldots, n_{z_{hj}}^{\triangleright}, \ldots, n_{z_{Nj}}^{\triangleright}\}, h \neq j, i = 1, \ldots, N, j{=}1, \ldots, N$, and $h = 1, \ldots, N$. The augmented control input vector of each subsystem j is composed of the current perimeter control and the interactions from other subsystems. In order to utilize the DED-MFAPLC scheme for the complex interconnected multi-region urban traffic system, the following assumptions and theorem are proposed:

Assumption 10.1 *The partial derivatives of $f_j(\ldots)$ with respect to each component of the control vector $\mathbf{u}_{ij}(k)$ are continuous.*

Assumption 10.2 *Each subsystem (10.23) is generalized Lipschitz, i.e.,* $\|y_j(k_1+1)-y_j(k_2+1)\| \le b^{\triangleright} \left\|\mathbf{u}_{ij}(k_1)-\mathbf{u}_{ij}(k_2)\right\|$ *for all* $k_1 \neq k_2$, $k_1, k_2 > 0$ *and* $\mathbf{u}_{ij}(k_1) \neq \mathbf{u}_{ij}(k_2)$, *where* $y_j(k_p+1) = f(y_j(k_p), \ldots, y_j(k_p-n_{y_j}), \mathbf{u}_{ij}(k_p), \ldots, \mathbf{u}_{ij}(k_p-n_{u_{ij}}))$, $p = 1, 2$, *and* $b^{\triangleright}$ *is a positive constant.*

Remark 10.1 In reality, the above assumptions imposed on the controlled system such as urban traffic system are reasonable and acceptable. Assumption 10.1 is easy to be verified from the dynamics of the multi-region urban traffic system (10.1)–(10.22), and it is a typical assumption for controller design for nonlinear systems. Assumption 10.2 is a physical constraint by the inherent nature of urban traffic system, i.e., finite change of vehicle flow does not lead to infinite change of the number of vehicles in a region. Meanwhile, it is a physical constraint of the real system from the energy point of view.

Theorem 10.1 *Consider the nonlinear subsystem j, satisfying the Assumptions 10.1 and 10.2 and* $\left\|\Delta\mathbf{u}_{ij}(k)\right\| \neq 0$, *there must exist a pseudo gradient (PG) denoted by* $\mathbf{\Phi}_{ij}^{\triangleright}(k)$, *such that the subsystem (10.23) can be transformed into the following equivalent CFDL data model:*

$$\Delta y_j(k+1) = \mathbf{\Phi}_{ij}^{\triangleright \mathrm{T}}(k)\Delta\mathbf{u}_{ij}(k) \tag{10.24}$$

where $\mathbf{\Phi}_{ij}^{\triangleright}(k) = [\phi_{ij}^{\triangleright}(k), \varphi_{1j}^{\triangleright}(k), \ldots, \varphi_{hj}^{\triangleright}(k), \ldots, \varphi_{Nj}^{\triangleright}(k)]^{\mathrm{T}}$, $h \neq j$, $i, j, h = 1, \ldots, N$ *is the PG of the subsystem* $i, j, h = 1, \ldots, N$ *with respect to* u_{ij}, $\Delta y_j(k+1) = y_j(k+1) - y_j(k)$, *and* $\Delta\mathbf{u}_{ij}(k) = \mathbf{u}_{ij}(k) - \mathbf{u}_{ij}(k-1)$.

Proof See (Hou and Jin, 2013).

Remark 10.2 It can be seen from (10.24) that the change of the jth subsystems output $\Delta y_j(k+1)$ is not only related to the change of the control input of subsystem j, but also bound up with the change of the interactions from other subsystems, which are depicted by $\phi_{ij}^{\triangleright}(k)\Delta u_{ij}(k)$ and $\sum_{h=1,h\neq j}^{N} \varphi_{hj}^{\triangleright}(k) \cdot \Delta z_{hj}(k)$, respectively. The latter item fully illustrates the impact of other subsystems acting on subsystem j.

C. One-Step DED-MFAPLC Perimeter Controller Design

After CFDL for subsystem j, the perimeter control of one-step DED-MFAPLC scheme will be designed as below.

The objective function of PG is proposed below:

$$\begin{aligned} J_j(\mathbf{\Phi}_{ij}^{\triangleright}(k)) = &\left|y_j(k) - y_j(k-1) - \mathbf{\Phi}_{ij}^{\triangleright \mathrm{T}}(k)\Delta\mathbf{u}_{ij}(k-1)\right|^2 \\ &+\mu^{\triangleright}\left\|\mathbf{\Phi}_{ij}^{\triangleright}(k) - \hat{\mathbf{\Phi}}_{ij}^{\triangleright}(k-1)\right\|^2 \end{aligned} \tag{10.25}$$

where $\mu^{\triangleright} > 0$ is a weighting factor to avoid too fast changes in PG estimates. The learning law of PG is estimated by minimizing (10.25) via the modified projection algorithm as below:

$$\hat{\mathbf{\Phi}}_{ij}^{\triangleright}(k) = \hat{\mathbf{\Phi}}_{ij}^{\triangleright}(k-1) + \frac{\eta^{\triangleright}\Delta\mathbf{u}_{ij}(k-1)(y_j(k) - y_j(k-1) - \hat{\mathbf{\Phi}}_{ij}^{\triangleright\mathrm{T}}(k-1)\Delta\mathbf{u}_{ij}(k-1))}{\mu^{\triangleright} + \left\|\Delta\mathbf{u}_{ij}(k-1)\right\|^2} \tag{10.26}$$

$$\hat{\phi}_{ij}^{\triangleright}(k) = \hat{\phi}_{ij}^{\triangleright}(0), \text{ if } \left|\hat{\phi}_{ij}^{\triangleright}(k)\right| < \varepsilon^{\triangleright} \text{ or } sign(\hat{\phi}_{ij}^{\triangleright}(k)) \neq sign(\hat{\phi}_{ij}^{\triangleright}(0)) \tag{10.27}$$

$$\hat{\varphi}_{hj}^{\triangleright}(k) = \hat{\varphi}_{hj}^{\triangleright}(0), \text{ if } \left|\hat{\varphi}_{hj}^{\triangleright}(k)\right| < \varepsilon^{\triangleright} \text{ or } sign(\hat{\varphi}_{hj}^{\triangleright}(k)) \neq sign(\hat{\varphi}_{hj}^{\triangleright}(0)) \tag{10.28}$$

where $\hat{\phi}_{ij}^{\triangleright}(0)$ and $\hat{\varphi}_{hj}^{\triangleright}(0)$ are the initial values of $\hat{\phi}_{ij}^{\triangleright}(k)$ and $\hat{\varphi}_{hj}^{\triangleright}(k)$, $h \neq j$, $i, j, h = 1, \ldots, N$, respectively, and $\hat{\mathbf{\Phi}}_{ij}^{\triangleright}(k)$ is the estimation of $\mathbf{\Phi}_{ij}^{\triangleright}(k)$.

Consider the one-step objective function for perimeter control input below:

$$J_j(u_{ij}(k)) = \left|y_d^{\triangleright}(k+1) - y_j(k+1)\right|^2 + \zeta^{\triangleright}\lambda_j^{\triangleright}\left|u_{ij}(k) - u_{ij}(k-1)\right|^2 \tag{10.29}$$

where $y_d^{\triangleright}(k+1)$ is the desired output signal of subsystem j, which represents the expected number of vehicles in region j. $\lambda_j^{\triangleright} > 0$ is a weighting constant to penalize excessive changes of the control input signal; i.e., the perimeter control ratios should not change too fast. $\zeta^{\triangleright}$ is a factor that makes the order of magnitude of the two items in (10.29) which are in the same level.

The perimeter control ratio $u_{ij}(k)$ is obtained by substituting (10.24) into (10.29), then differentiating it with respect to $u_{ij}(k)$, and letting it zero as below:

$$u_{ij}(k) = u_{ij}(k-1) + \frac{\rho^{\triangleright}\hat{\phi}_{ij}^{\triangleright}(k)(y_d^{\triangleright}(k+1) - y_j(k) - \sum_{h=1,h\neq j}^{N}\hat{\varphi}_{hj}^{\triangleright}(k)\Delta z_{hj}(k))}{\zeta^{\triangleright}\lambda_j^{\triangleright} + \hat{\phi}_{ij}^{\triangleright 2}(k)} \tag{10.30}$$

$$u_{ij}(k) = \begin{cases} u_{ij,\min}^{g}, & \text{if } u_{ij}(k) < u_{ij,\min}^{g} \\ u_{ij}(k), & \text{if } u_{ij,\min}^{g} \leq u_{ij}(k) \\ & \leq \min\{u_{ij,\max}^{g}, C_{ij}^{\triangleright}\frac{((n_j(t))}{M_{ij}^{\triangleright}}(t)\} \\ \min\{u_{ij,\max}^{g}, C_{ij}^{\triangleright}\frac{((n_j(t))}{M_{ij}^{\triangleright}}(t)\}, & \\ \text{if } u_{ij}(k) > \min\{u_{ij,\max}^{g}, C_{ij}^{\triangleright}\frac{((n_j(t))}{M_{ij}^{\triangleright}}(t)\} & \end{cases} \tag{10.31}$$

where $\eta^{\triangleright} \in (0, 2]$ and $\rho^{\triangleright} \in (0, 1]$ are step-size constants, and $\varepsilon^{\triangleright}$ is a positive constant (10.31) has the same meaning with (10.9).

The one-step DED-MFAPLC scheme is constructed of (10.26)–(10.28) and (10.30)–(10.31). The reset mechanism (10.27)–(10.28) are designed to make the ability for tracking time-varying parameters of the learning law (10.26) stronger.

The outstanding features of the proposed one-step DED-MFAPLC scheme are as follows:

(1) Since the MIMO nonlinear MRUTS can be decomposed to N subsystems, and each subsystem is a MISO nonlinear system, when one controllable input is considered as the current control input of a certain subsystem, the other inputs are considered as the interconnected actions from other subsystems, which are measurable. The interactions between the subsystems are compensated via the last item in the numerator of the second item on the right side of the equal sign in (10.30). Thus, the internal coupling of the MRUTS can be handled by the one-step DED-MFAPLC strategy. By establishing CFDL data model with measurable interactions and designing MFAPLC scheme for each subsystem, the one-step DED-MFAPLC scheme for the whole multi-region urban traffic system is realized.
(2) The DED-MFAPLC strategy is independent of the mathematical model of the large-scale MRUTS. Instead, only the input/output data of the urban road network is needed in perimeter controller designing process. Besides, there is no unmodeled dynamics in the one-step DED-MFAPLC scheme strategy, and the structure and the orders of mathematical model of the controlled MRUTS are both no longer required to be known. Furthermore, the one-step DED-MFAPLC method does not need any training process or satisfy the persistence excitation conditions, which are usually necessary for other adaptive control methods and the neural networks based approaches.
(3) The existing linearization methods have more or less various limitations. For example, the high-order terms are ignored in Taylors linearization, more model information of the controlled system is required in piecewise linearization, the precise mathematical model and the measurement of full states are needed in feedback linearization, and it is difficult to handle the parameter uncertainties or disturbances, etc. Different from the previously mentioned linearization methods, an equivalent description of the original MIMO urban traffic system is provided by CFDL method, in which only the input and output data of the system is needed, while the system dynamics is no longer required.
(4) The existence of the PG of the large-scale MRUTS is guaranteed by rigorous mathematical analysis, see Theorem 10.1. Meanwhile, it is simple to obtain the CFDL data model, since the PG is very easy to be estimated by using the system measurement data, without heavy computational burden. Finally, since the PG is a slowly time-varying parameter, which is insensitive to the changes in the system variables, such as accumulations, demands, etc., the one-step DED-MFAPLC method has strong robustness.

10.2.3 Simulation Results

In this research, a simulation-based case study is presented to test the performance of the proposed one-step DED-MFAPLC strategy for the multi-region perimeter control problem. In addition, some other commonly used perimeter control methods (FTC and MPC) are also applied to the multi-region urban traffic system under the same condition to compare with DED-MFAPLC strategy. The same route choice strategy is drawn into all the presented perimeter control methods to compare the perimeter control performance.

Two cases under different degree of noise are simulated in this section. In the first case, the low measurement noises in demands and accumulations and low errors in MFDs are added to test the perimeter control effect with high accuracy of the urban traffic model. In the other case, the high noises are integrated to study the control effect under severe traffic model mismatch.

A. Parameter and Traffic Demand Setting

The urban traffic network partitioned into five homogenous regions (i.e., N=5, see Fig. 10.1) with MFDs as (10.32) is considered. The trip completion flow $G_i^{\triangleright}(n_i(k))$ is approximately formulated as a third-order function of $n_i(k)$, see Fig. 10.4, i.e.,

$$G_i^{\triangleright}(n_i(k)) = \begin{cases} (1.7852 \times 10^{-7} n_i(k)^3 - 3.5022 \times 10^{-3} n_i(k)^2 \\ \qquad +18.1094 n_i(k))/3600, \ i = 1. \\ (1.4877 \times 10^{-7} n_i(k)^3 - 2.9185 \times 10^{-3} n_i(k)^2 \\ \qquad +15.0912 n_i(k))/3600, \ i = 2, 3, 4, 5. \end{cases} \tag{10.32}$$

where $n_i^{\text{cr}} = 3400$ (veh), $G_i^{\triangleright}(n_i^{\text{cr}}) = 6.3$ (veh/s), $n_i^{\text{jam}} = 10000$ (veh), and $G_i^{\triangleright}(n_i^{\text{jam}}) = 0.43$ (veh/s), i=2, 3, 4, 5, while for region 1, one has $G_1^{\triangleright}(n_1^{\text{cr}}) = 7.56$ (veh/s) and $G_1^{\triangleright}(n_1^{\text{jam}}) = 0.52$ (veh/s), and the other MFD parameters are the same as the ones in the MFD of region 2–5. The parameters of boundary capacity are $\alpha^{\triangleright} = 0.64$, $C_{ij}^{\triangleright\max} = 3.2$ (veh/s) (Ji and Geroliminis, 2012), and $C_{ij}^{\triangleright\min} = 0.15$ (veh/s) for $j = 2, 3, 4, 5$ and $i \in N_j$, while for region 1, the corresponding parameters are $C_{i1}^{\triangleright\max} = 3.84$ (veh/s) and $C_{i1}^{\triangleright\min} = 0.18$ (veh/s), $i \neq 1$, respectively. The simulation duration is set as 4 h, and the signal cycle length of the boundary intersections is set as 2 min, while the sample and control cycle length is $T^{\triangleright}$=4min. Therefore, the 4 h of simulation duration is divided into 60 time steps (i.e., $k^{\text{end}} = 60$). The time-varying traffic demands are shown in Fig. 10.5 to simulate the real traffic situation in morning peak hour.

It is assumed that all the regions are initially congested; i.e., the initial number of vehicles are: $n_1(0) = 5000$, $n_2(0) = 4000$, $n_3(0) = 4000$, $n_4(0) = 4000$, and $n_5(0) = 4000$, respectively. The initial conditions are set up to test whether the DED-MFAPLC method proposed in this work can solve the problem of traffic congestion under such unfavorable conditions that the multi-region urban traffic system went through the early peak under original congestion state. The initial green time ratio is $\omega^{\triangleright} = 0.5$, while the maximum and minimum green time ratio are $\omega_{\max}^{\triangleright} = 0.9$ and $\omega_{\min}^{\triangleright} = 0.1$, respectively. According to the definition of perimeter control ratio, one has $u_{ij,\max}^{g} = 1.8$ and $u_{ij,\min}^{g} = 0.2$.

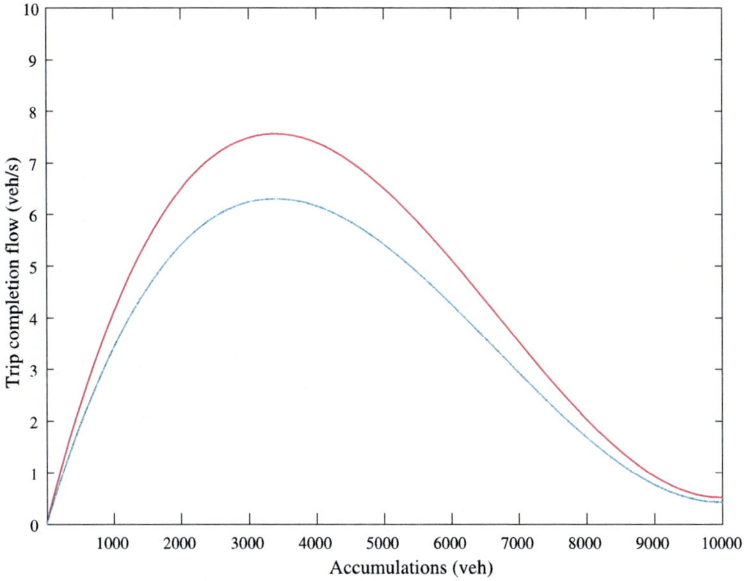

Fig. 10.4 MFD of each region, in which the magenta solid line denotes $G_1^{\triangleright}$, the brown dashed line denotes $G_2^{\triangleright}$, the blue dotted line denotes $G_3^{\triangleright}$, the green chain line denotes $G_4^{\triangleright}$, and the cyan chain line denotes $G_5^{\triangleright}$

The errors in MFDs and the measurement noises in demands and accumulations are added as follows:

$$\tilde{n}_i(k) = n_i(k)(1 + \mathcal{N}(0, \sigma_{n_i}^{\triangleright 2})) \tag{10.33}$$

$$\tilde{q}_{ij}^{\triangleright}(k) = q_{ij}^{\triangleright}(k)(1 + \mathcal{N}(0, \sigma_{q_{ij}^{\triangleright}}^{\triangleright 2})) \tag{10.34}$$

$$\tilde{G}_i^{\triangleright}(n_i(k)) = G_i^{\triangleright}(n_i(k))(1 + U(-\alpha_{G_i^{\triangleright}}^{\triangleright}, \ \alpha_{G_i^{\triangleright}})) \tag{10.35}$$

where in the first case of low noise, the error parameters are chosen as $\sigma_{n_i}^{\triangleright 2} = 0.1$, $\sigma_{q_{ij}^{\triangleright}}^{\triangleright 2} = 0.1$, and $\alpha_{G_i^{\triangleright}}^{\triangleright} = 0.1$, respectively, while in the second case of high noise, the above parameters are, respectively, chosen as $\sigma_{n_i}^{\triangleright 2} = 0.1$, $\sigma_{q_{ij}^{\triangleright}}^{\triangleright 2} = 0.1$, and $\alpha_{G_i^{\triangleright}}^{\triangleright} = 0.35$.

In the proposed one-step DED-MFAPLC strategy, the desired output of each subsystem i is chosen as $0.98 \cdot n_i^{\text{cr}}$ in order to make the trip completion flow near the maximum value without congestion. Other parameters are: $\varsigma^{\triangleright} = 1000$, $\lambda_i^{\triangleright} = 10$, $\eta^{\triangleright} = 1$, $\rho^{\triangleright} = 1$, and $\mu^{\triangleright} = 1$, respectively.

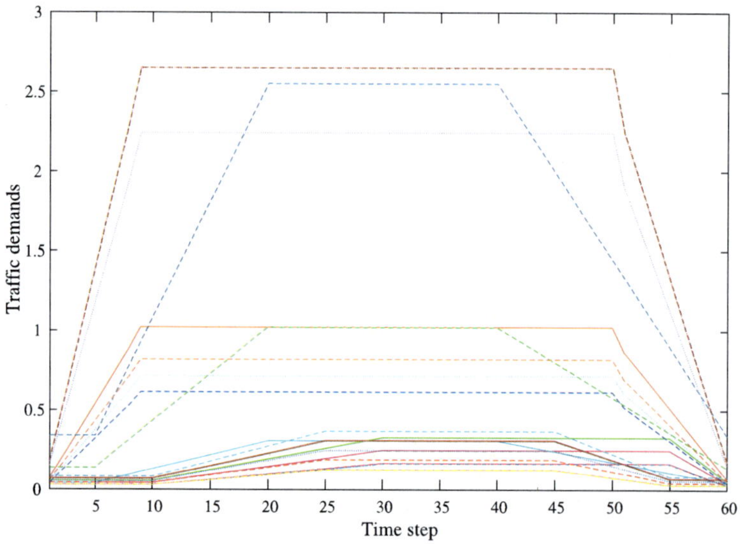

Fig. 10.5 Traffic demands among the five regions, in which the red solid line denotes $q_{11}^{\triangleright}$, the magenta solid line denotes $q_{12}^{\triangleright}$, the blue solid line denotes $q_{13}^{\triangleright}$, the green solid line denotes $q_{14}^{\triangleright}$, the cyan solid line denotes $q_{15}^{\triangleright}$, the orange solid line denotes $q_{21}^{\triangleright}$, the brown solid line denotes $q_{22}^{\triangleright}$, the pink solid line denotes $q_{23}^{\triangleright}$, the gray blue dashed line denotes $q_{24}^{\triangleright}$, the magenta dashed line denotes $q_{25}^{\triangleright}$, the blue dashed line denotes $q_{31}^{\triangleright}$, the green dashed line denotes $q_{32}^{\triangleright}$, the cyan dashed line denotes $q_{33}^{\triangleright}$, the yellow dashed line denotes $q_{34}^{\triangleright}$, the brown dashed line denotes $q_{35}^{\triangleright}$, the orange dashed line denotes $q_{41}^{\triangleright}$, the red dotted line denotes $q_{42}^{\triangleright}$, the magenta dotted line denotes $q_{43}^{\triangleright}$, the blue dotted line denotes $q_{44}^{\triangleright}$, the green dotted line denotes $q_{45}^{\triangleright}$, the cyan dotted line denotes $q_{51}^{\triangleright}$, the yellow dotted line denotes $q_{52}^{\triangleright}$, the purple dotted line denotes $q_{53}^{\triangleright}$, the pink dotted line denotes $q_{54}^{\triangleright}$, and the red dashed line denotes $q_{55}^{\triangleright}$

B. Simulation Analysis

1. Simulation results with small model mismatch

The simulation results in the case of small model mismatch are presented in Figs. 10.6, 10.7, 10.8, 10.9, 10.10 and 10.11. The evolution of accumulations in the five regions under FTC strategy is depicted in Fig. 10.6a. As can be seen from the figure, region 1 fell into jammed state soon, and subsequently, another three regions were in a state of congestion. The congestion continued to the end, even with the reduction of traffic demands after morning peak. As time went on, the corresponding trip completion flow for region 1 dropped rapidly depending on the accumulation, and the same is true for regions 2, 3, and 4 soon after, see Fig. 10.7a. The perimeter control inputs for every two adjacent regions were constant as the initial value 1 all the time, which can be seen from Fig. 10.6b, and the route choice ratios are shown in Fig. 10.7b.

Figures 10.8 and 10.9 depict the result of MPC strategy. Different from FTC strategy, the performance of the urban road network system has been greatly improved. It can be seen from Fig. 10.8a that all the regions were running smoothly and they

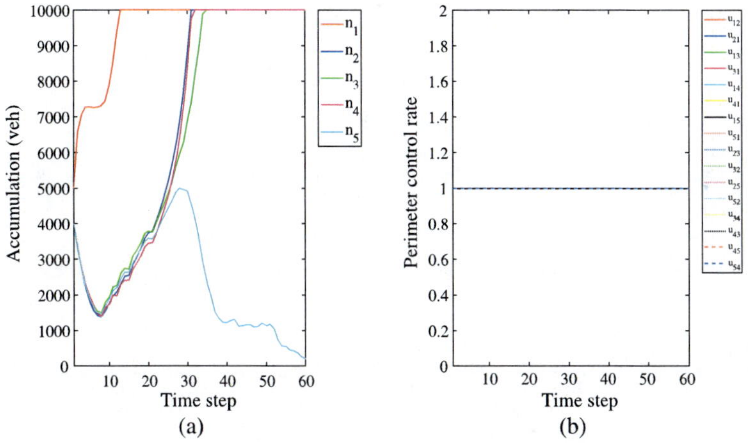

Fig. 10.6 Result of FTC strategy with low measurement noises (part one): (**a**) evolution of the number of vehicles traveling in each region; (**b**) perimeter control ratios of FTC

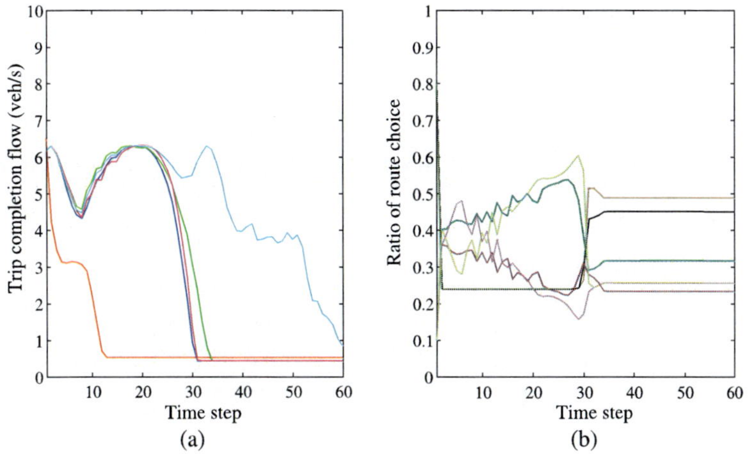

Fig. 10.7 Result of FTC strategy with low measurement noises (part two): (**a**) trip completion flow for each region; (**b**) ratio of route choice. Where in sub-figure (**a**), the red solid line denotes $G_1^{\triangleright}$, the blue solid line denotes $G_2^{\triangleright}$, the green solid line denotes $G_3^{\triangleright}$, the magenta solid line denotes $G_4^{\triangleright}$, the cyan solid line denotes $G_5^{\triangleright}$. In sub-figure (**b**), the red solid line denotes $b_{214}^{\triangleright}$, the blue solid line denotes $b_{234}^{\triangleright}$, the green solid line denotes $b_{254}^{\triangleright}$, the magenta solid line denotes $b_{315}^{\triangleright}$, the cyan solid line denotes $b_{325}^{\triangleright}$, the yellow solid line denotes $b_{345}^{\triangleright}$, the black solid line denotes $b_{412}^{\triangleright}$, the red dotted line denotes $b_{432}^{\triangleright}$, the blue dotted line denotes $b_{452}^{\triangleright}$, the green dotted line denotes $b_{513}^{\triangleright}$, the magenta dotted line denotes $b_{523}^{\triangleright}$, and the cyan dotted line denotes $b_{543}^{\triangleright}$

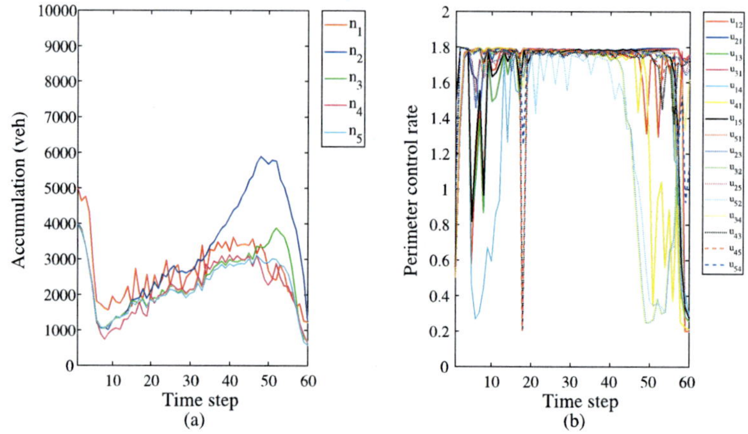

Fig. 10.8 Result of MPC strategy with low measurement noises (part one): (**a**) evolution of the number of vehicles traveling in each region; (**b**) perimeter control ratios of MPC

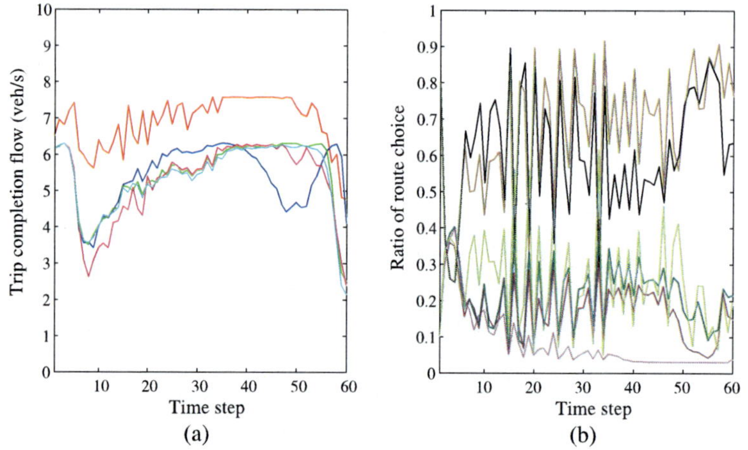

Fig. 10.9 Result of MPC strategy with low measurement noises (part two): (**a**) trip completion flow for each region; (**b**) ratio of route choice. Where in sub-figure (**a**), the red solid line denotes $G_1^{\triangleright}$, the blue solid line denotes $G_2^{\triangleright}$, the green solid line denotes $G_3^{\triangleright}$, the magenta solid line denotes $G_4^{\triangleright}$, the cyan solid line denotes $G_5^{\triangleright}$. In sub-figure (**b**), the red solid line denotes $b_{214}^{\triangleright}$, the blue solid line denotes $b_{234}^{\triangleright}$, the green solid line denotes $b_{254}^{\triangleright}$, the magenta solid line denotes $b_{315}^{\triangleright}$, the cyan solid line denotes $b_{325}^{\triangleright}$, the yellow solid line denotes $b_{345}^{\triangleright}$, the black solid line denotes $b_{412}^{\triangleright}$, the red dotted line denotes $b_{432}^{\triangleright}$, the blue dotted line denotes $b_{452}^{\triangleright}$, the green dotted line denotes $b_{513}^{\triangleright}$, the magenta dotted line denotes $b_{523}^{\triangleright}$, and the cyan dotted line denotes $b_{543}^{\triangleright}$

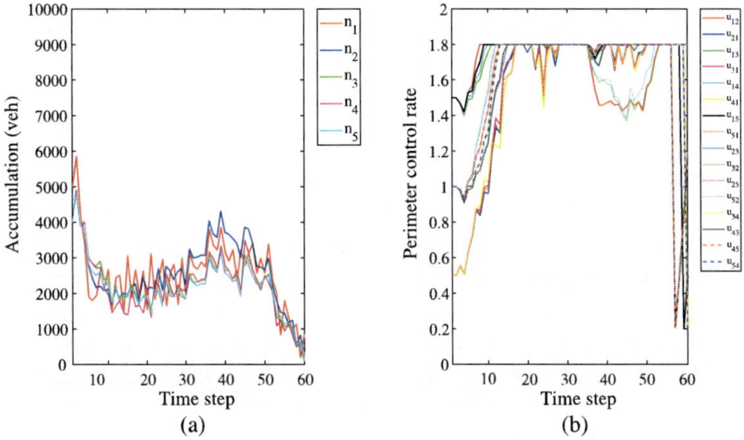

Fig. 10.10 Result of DED-MFAPLC strategy with low measurement noises (part one): (**a**) evolution of the number of vehicles traveling in each region; (**b**) perimeter control ratios of DED-MFAPLC

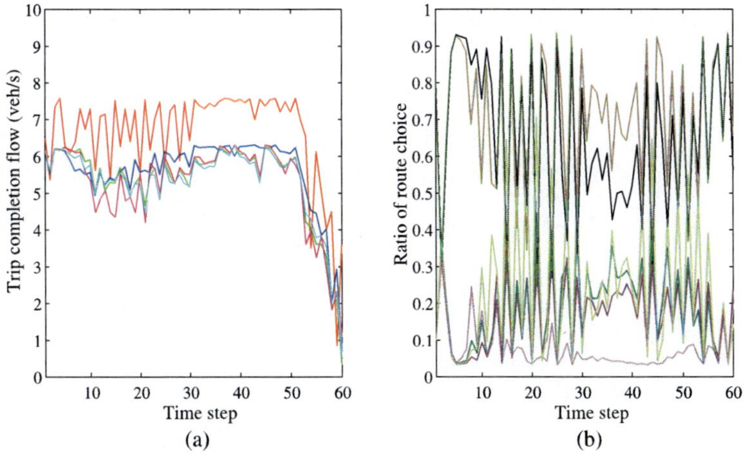

Fig. 10.11 Result of DED-MFAPLC strategy with low measurement noises (part two): (**a**) trip completion flow for each region; (**b**) ratio of route choice. Where in sub-figure (**a**), the red solid line denotes $G_1^{\triangleright}$, the blue solid line denotes $G_2^{\triangleright}$, the green solid line denotes $G_3^{\triangleright}$, the magenta solid line denotes $G_4^{\triangleright}$, the cyan solid line denotes $G_5^{\triangleright}$. In sub-figure (**b**), the red solid line denotes $b_{214}^{\triangleright}$, the blue solid line denotes $b_{234}^{\triangleright}$, the green solid line denotes $b_{254}^{\triangleright}$, the magenta solid line denotes $b_{315}^{\triangleright}$, the cyan solid line denotes $b_{325}^{\triangleright}$, the yellow solid line denotes $b_{345}^{\triangleright}$, the black solid line denotes $b_{412}^{\triangleright}$, the red dotted line denotes $b_{432}^{\triangleright}$, the blue dotted line denotes $b_{452}^{\triangleright}$, the green dotted line denotes $b_{513}^{\triangleright}$, the magenta dotted line denotes $b_{523}^{\triangleright}$, and the cyan dotted line denotes $b_{543}^{\triangleright}$

were all in unimpeded states, except for the short-term slight congestion in region 2. The trip completion flow of in each region remained at a relatively high level, see Fig. 10.9a. The result indicates that although there is a certain model mismatch and measurement noise, the perimeter control effect of MPC is quite good.

The result of one-step DED-MFAPLC strategy is depicted in Figs. 10.10 and 10.11. The desired output of each subsystem is chosen as $0.97n_{\text{cr}}$, in order to make the trip completion flow near the maximum value without congestion. As shown in Fig. 10.10a, all the five regions were no longer congested all the way, and the perimeter control input changed gently, see Fig. 10.10b. Under DED-MFAPLC strategy for perimeter control, all the regions were no longer congested despite the high traffic demand, and the accumulations in the regions were near the desired value during the morning peak, which makes the trip completion flows maximized, see Fig. 10.11a. The performance of one-step DED-MFAPLC and MPC is similar under small model mismatch condition.

After simulation for all the perimeter control methods mentioned in this section, the total network throughput (TNT) for the entire urban road network, the total time spent (TTS) and the average travel time (ATT) for all vehicles are used to evaluate the effectiveness of the perimeter control methods. The above evaluation criteria are defined as below:

$$\text{TNT} = T^{\triangleright} \sum_{k=1}^{k^{\text{end}}} \sum_{i=1}^{N} M_{ii}^{\triangleright}(k) \tag{10.36}$$

$$\text{TTS} = T^{\triangleright} \sum_{k=1}^{k^{\text{end}}} \sum_{i=1}^{N} n_i(k) \tag{10.37}$$

$$\text{ATT} = \frac{\text{TTS}}{\text{TNT}} \tag{10.38}$$

Besides, the total CPU time (TCT) which means the total time for calculating perimeter control input during the simulation is also considered as an auxiliary evaluation indicator to evaluate the performance of the above perimeter control strategies. Under the same initial condition and traffic demands, the more TNT, and the less TTS, ATT, and TCT, the better performance the urban traffic system has. The simulation results with mild uncertainty are presented in Table 10.2.

The result shows that under the condition of low measurement noises and MFD errors, the urban traffic system under MPC and one-step DED-MFAPLC perimeter control strategies were both operated better than FTC, and the jammed phenomenon were all disappeared. Meanwhile, the control effects of MPC and one-step DED-MFAPLC are similar, but the total CPU time for one-step DED-MFAPLC is much less than MPC.

Table 10.2 Evaluation results of different strategies with low noise

Strategy	Total network throughput	Total time spent	Average travel time	Total CPU time
	TNT (veh)	TTS (s)	ATT (s)	TCT (s)
FTC	5.8372×10^4	4.7802×10^8	8.1892×10^3	0.23
MPC	2.1486×10^4	1.8164×10^8	845.39	171.75
DED-MFAPLC	2.1610×10^4	1.7024×10^8	787.79	0.43

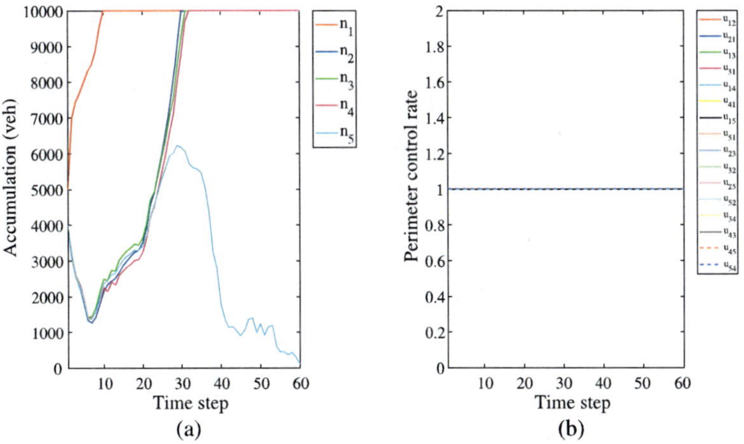

Fig. 10.12 Result of FTC strategy with high measurement noises (part one): (**a**) evolution of the number of vehicles traveling in each region; (**b**) perimeter control ratios of FTC

2. Simulation Results with Severe Model Mismatch

The results of the above strategies with high uncertainty are shown in Figs. 10.12, 10.13, 10.14, 10.15, 10.16 and 10.17 and Table 10.3, respectively. The evolution of accumulations in the five regions under FTC strategy is depicted in Fig. 10.12a. As can be seen from the figure, only one region remained unobstructed during the whole period, while all the other four regions experienced extreme congestion. The congestion continued to the end, even with the reduction of traffic demands after morning peak. As time went on, the corresponding trip completion flow for the jammed regions dropped rapidly depending on the accumulation, see Fig. 10.13a. The perimeter control inputs for every two adjacent regions were constant as the initial value 1 all the time, which can be seen from Fig. 10.12b, and the route choice ratios are shown in Fig. 10.13b.

Figures 10.14 and 10.15 depict the result of MPC strategy. It can be seen from Fig. 10.14a that although three regions kept unblocked till the end, there was still one region fell jammed and another region congested at the end of morning peak. The reason is that MPC requires the system model to be known precisely, so the perimeter control effect of MPC strategy under high noise is not good.

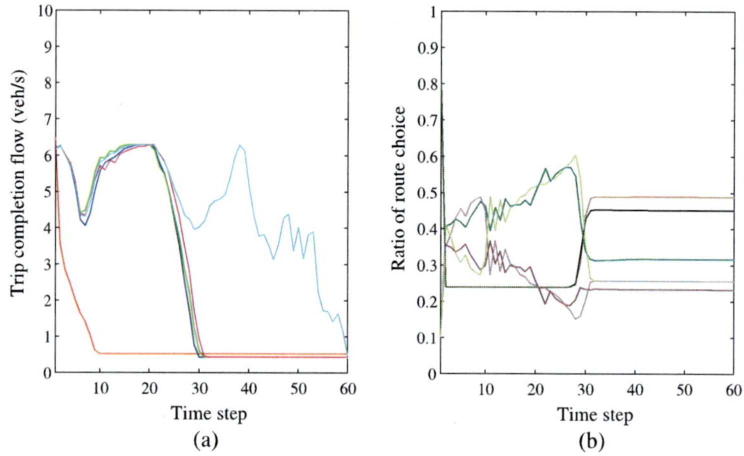

Fig. 10.13 Result of FTC strategy with high measurement noises (part two): (**a**) trip completion flow for each region; (**b**) ratio of route choice. Where in sub-figure (**a**), the red solid line denotes $G_1^{\triangleright}$, the blue solid line denotes $G_2^{\triangleright}$, the green solid line denotes $G_3^{\triangleright}$, the magenta solid line denotes $G_4^{\triangleright}$, the cyan solid line denotes $G_5^{\triangleright}$. In sub-figure (**b**), the red solid line denotes $b_{214}^{\triangleright}$, the blue solid line denotes $b_{234}^{\triangleright}$, the green solid line denotes $b_{254}^{\triangleright}$, the magenta solid line denotes $b_{315}^{\triangleright}$, the cyan solid line denotes $b_{325}^{\triangleright}$, the yellow solid line denotes $b_{345}^{\triangleright}$, the black solid line denotes $b_{412}^{\triangleright}$, the red dotted line denotes $b_{432}^{\triangleright}$, the blue dotted line denotes $b_{452}^{\triangleright}$, the green dotted line denotes $b_{513}^{\triangleright}$, the magenta dotted line denotes $b_{523}^{\triangleright}$, and the cyan dotted line denotes $b_{543}^{\triangleright}$

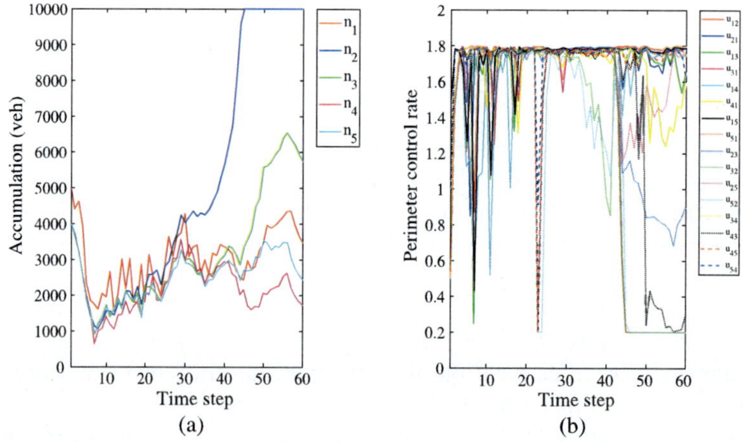

Fig. 10.14 Result of MPC strategy with high measurement noises (part one): (**a**) evolution of the number of vehicles traveling in each region; (**b**) perimeter control ratios of MPC

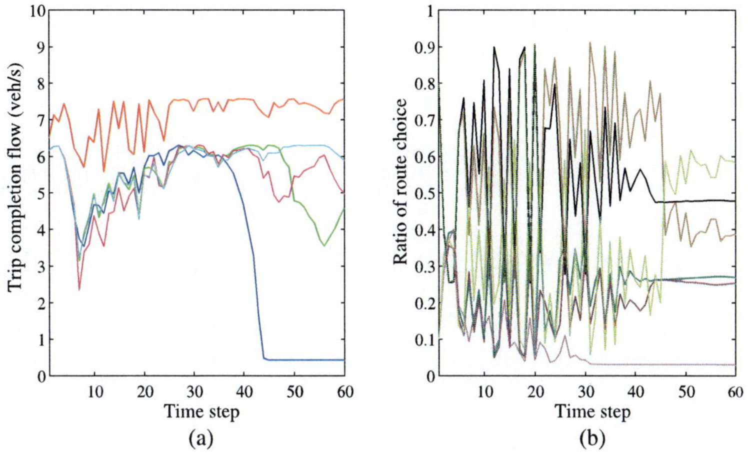

Fig. 10.15 Result of MPC strategy with high measurement noises (part two): (**a**) trip completion flow for each region; (**b**) ratio of route choice. Where in sub-figure (**a**), the red solid line denotes $G_1^{\triangleright}$, the blue solid line denotes $G_2^{\triangleright}$, the green solid line denotes $G_3^{\triangleright}$, the magenta solid line denotes $G_4^{\triangleright}$, the cyan solid line denotes $G_5^{\triangleright}$. In sub-figure (**b**), the red solid line denotes $b_{214}^{\triangleright}$, the magenta solid line denotes $b_{315}^{\triangleright}$, the cyan solid line denotes $b_{325}^{\triangleright}$, the yellow solid line denotes $b_{345}^{\triangleright}$, the black solid line denotes $b_{412}^{\triangleright}$, the red dotted line denotes $b_{432}^{\triangleright}$, the blue dotted line denotes $b_{452}^{\triangleright}$, the green dotted line denotes $b_{513}^{\triangleright}$, the magenta dotted line denotes $b_{523}^{\triangleright}$, and the cyan dotted line denotes $b_{543}^{\triangleright}$

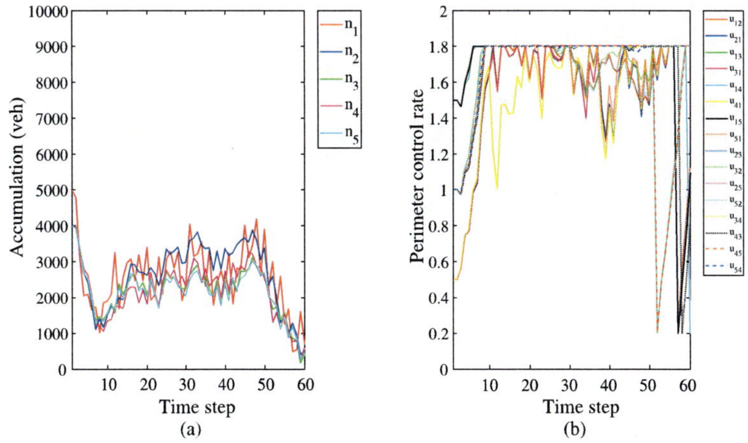

Fig. 10.16 Result of DED-MFAPLC strategy with high measurement noises (part one): (**a**) evolution of the number of vehicles traveling in each region; (**b**) perimeter control ratios of DED-MFAPLC

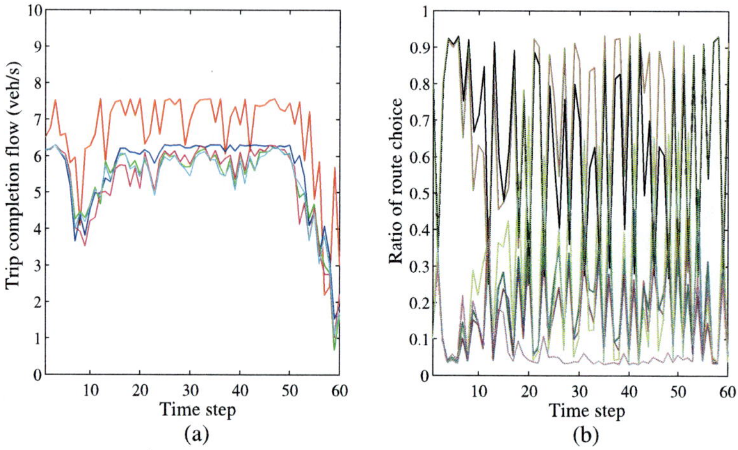

Fig. 10.17 Result of DED-MFAPLC strategy with high measurement noises (part two): (**a**) trip completion flow for each region; (**b**) ratio of route choice. Where in sub-figure (**a**), the red solid line denotes $G_1^{\triangleright}$, the blue solid line denotes $G_2^{\triangleright}$, the green solid line denotes $G_3^{\triangleright}$, the magenta solid line denotes $G_4^{\triangleright}$, the cyan solid line denotes $G_5^{\triangleright}$. In sub-figure (**b**), the red solid line denotes $b_{214}^{\triangleright}$, the magenta solid line denotes $b_{315}^{\triangleright}$, the cyan solid line denotes $b_{325}^{\triangleright}$, the yellow solid line denotes $b_{345}^{\triangleright}$, the black solid line denotes $b_{412}^{\triangleright}$, the red dotted line denotes $b_{432}^{\triangleright}$, the blue dotted line denotes $b_{452}^{\triangleright}$, the green dotted line denotes $b_{513}^{\triangleright}$, the magenta dotted line denotes $b_{523}^{\triangleright}$, and the cyan dotted line denotes $b_{543}^{\triangleright}$

Table 10.3 Evaluation results of different strategies with high noise

Strategy	Total network throughput	Total time spent	Average travel time	Total CPU time
	TNT (veh)	TTS (s)	ATT (s)	TCT (s)
FTC	5.3309×10^4	4.9710×10^8	9.3249×10^3	0.25
MPC	1.8316×10^5	2.3175×10^8	1.2563×10^3	172.89
DED-MFAPLC[a] 1	2.1654×10^5	1.6937×10^8	782.18	0.46
DED-MFAPLC[b]	1.9581×10^5	2.1611×10^8	1.1037×10^3	0.47

[a]DED-MFAPLC without measurement noises in the measurable interactions
[b]DED-MFAPLC with measurement noises in the measurable interactions

The result of one-step DED-MAFPLC strategy is depicted in Figs. 10.16 and 10.17. As shown in Fig. 10.16a, all the five regions were no longer congested all the way, and the perimeter control input changed gently, see Fig. 10.16b. Under one-step DED-MFAPLC strategy for perimeter control, the accumulations in the regions were near the desired value during the morning peak, and the trip completion flow maintained at a relatively high level, see Fig. 10.17a. The performance of the urban road network system has been greatly improved under DED-MFAPLC strategy, and one-step DED-MFAPLC is superior to all other strategies for the multi-region perimeter control problem, as the total network throughput for the entire urban traffic

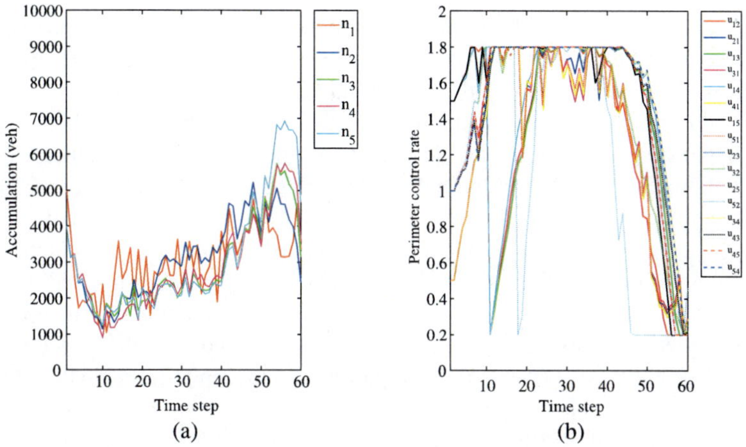

Fig. 10.18 Result of DED-MFAPLC strategy with measurement noises in measurable interactions (part one): (**a**) evolution of the number of vehicles traveling in each region; (**b**) perimeter control ratios of DED-MFAPLC

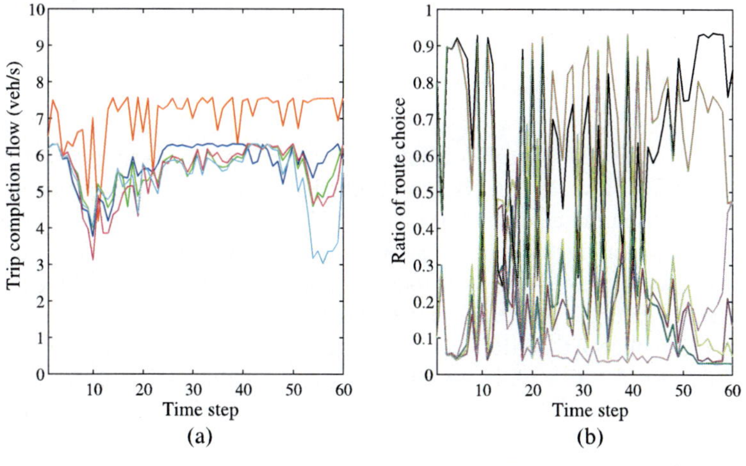

Fig. 10.19 Result of DED-MFAPLC strategy with measurement noises in measurable interactions (part two): (**a**) trip completion flow for each region; (**b**) ratio of route choice. Where in sub-figure (**a**), the red solid line denotes $G_1^{\triangleright}$, the blue solid line denotes $G_2^{\triangleright}$, the green solid line denotes $G_3^{\triangleright}$, the magenta solid line denotes $G_4^{\triangleright}$, the cyan solid line denotes $G_5^{\triangleright}$. In sub-figure (**b**), the red solid line denotes $b_{214}^{\triangleright}$, the blue solid line denotes $b_{234}^{\triangleright}$, the green solid line denotes $b_{254}^{\triangleright}$, the magenta solid line denotes $b_{315}^{\triangleright}$, the cyan solid line denotes $b_{325}^{\triangleright}$, the yellow solid line denotes $b_{345}^{\triangleright}$, the black solid line denotes $b_{412}^{\triangleright}$, the red dotted line denotes $b_{432}^{\triangleright}$, the blue dotted line denotes $b_{452}^{\triangleright}$, the green dotted line denotes $b_{513}^{\triangleright}$, the magenta dotted line denotes $b_{523}^{\triangleright}$, and the cyan dotted line denotes $b_{543}^{\triangleright}$

network is greater than others, while both the total time spent and the average travel time for all vehicles are less than other perimeter control strategies. The problem of high noise can be handled by the proposed DED-MFAPLC method.

In addition, the measurement noises are also added to the measurable interactions under one-step DED-MFAPLC strategy to further study the robustness of it. The simulation results of this additional case are presented in Figs. 10.18 and 10.19 and the last row of Table 10.3, respectively. It can be seen from Fig. 10.18a that although noises are added into the measurable interactions, the number of vehicles in each region are still controlled near expected values, which makes the trip completion flows maximized, see Fig. 10.19a. The perimeter control inputs and the route choice ratios under this case are depicted in Figs. 10.18b and 10.19b, respectively. The last row of Table 10.3 reveals that the proposed one-step DED-MFAPLC perimeter control method is also superior to other strategies in spite of the uncertainties in the measurable interactions. Meanwhile, it is noteworthy that the total CPU time for calculating the control input is much smaller than the signal cycle, so it can be implemented in real time.

10.3 Multi-step Model Free Adaptive Learning Route Guidance and Perimeter Control

In Sect. 10.2, the one-step DED-MFAPLC method is proposed for perimeter control of the MRUTS. However, the constraints of the perimeter control input and the traffic system output in the actual urban transportation system have not been considered. In addition, the application of the above method requires that the interactions (i.e., transfer flows) among different regions are measurable. In this section, the multi-step cMFAPLC strategy is presented, and it is not only used for perimeter control, but also for route guidance of the MRUTS. Compared with the one-step DED-MFAPLC strategy, in the cMFAPLC method, there is a data-driven multi-step prediction mechanism that predicts the the future state of urban transportation system, and the multi-step cMFAPLC strategy realizes the route guidance and perimeter control of large-scale MRUTS through rolling optimization, by which the ability to address the various uncertainties in urban traffic systems can be improved. The main contributions of the work done in this section are as follows:

(1) A new multi-step cMFAPLC strategy is proposed for MIMO systems with different control input and system output dimensions and with physical constraints on both control input and system output to solve route guidance and perimeter control problems of the heterogeneous MRUTS synchronously. Moreover, this method can make the traffic in different regions of the urban road network conform to equilibrium, so as to reduce traffic congestion.
(2) Different from the existing model-based methods (such as MPC and LQI) for route guidance and perimeter control of urban road networks, the multi-step cMFAPLC scheme is a data-driven control method. In the design process of

route guidance and perimeter control method, only the control input and system output data of urban transportation system are needed. That means, the accurate mathematical model of the MRUTS is no longer necessary, and it is also no longer necessary to know the traffic demand that is difficult to obtain. Further, the cMFAPLC strategy can be designed without MFD-based urban traffic models. Thus, the deterioration of urban traffic control effect caused by the difficulty of MFD modeling process of the urban traffic system and the inaccuracy of the urban traffic model can be avoided. In addition, the computational complexity of this method is much less than that of MPC method.

(3) Compared with the existing prototype centralized and decentralized MFAPLC route guidance and perimeter control methods, there is a data-driven prediction mechanism in the proposed cMFAPLC scheme to predict the future traffic conditions only using the control input and system output data of the MRUTS. Besides, perimeter control ratios and route guidance proportions can be realized through rolling optimization process, which enhances the capacity of the proposed scheme to deal with various uncertainties in traffic system.

(4) The practical control input and the system output constraints of the MRUTS have been both taken into account in the designing process of cMFAPLC route guidance and perimeter control strategy, which enable it can solve the practical problems of urban transportation system more effectively.

(5) The multi-step cMFAPLC strategy proposed here is designed for unknown nonlinear nonsquared MIMO nonlinear systems with different control input and the system output dimensions. Moreover, in this method, it is no longer necessary to measure the complex interconnection impact in urban transportation systems, which makes it far less costly and more applicable to deal with the actual problems of the large-scale multi-region urban traffic systems.

10.3.1 Dynamics Model of the MRUTS

The urban traffic dynamics of the large-scale multi-region urban road network is introduced in this section. It is assumed that the MRUTS is divided into N = four regions, which is shown in Fig. 10.20, and it is also assumed that there exists a well-defined MFD in each region shown in Fig. 10.21. The discrete-time macro-level urban traffic dynamics of the MRUTS, after the first-order Euler discretizing method on the continuous time model (Ramezani et al. 2015; Sirmatel and Geroliminis 2016), is introduced as below:

$$
\begin{aligned}
n_{ij}(k+1) = n_{ij}(k) + T^{\triangleright}(q_{ij}^{\triangleright}(k) - \sum_{m \in N_i} u_{im}(k) M_{imj}^{\triangleright}(k) \\
+ \sum_{m \in N_i, m \neq j} u_{mi}(k) M_{mij}^{\triangleright}(k))
\end{aligned}
\tag{10.39}
$$

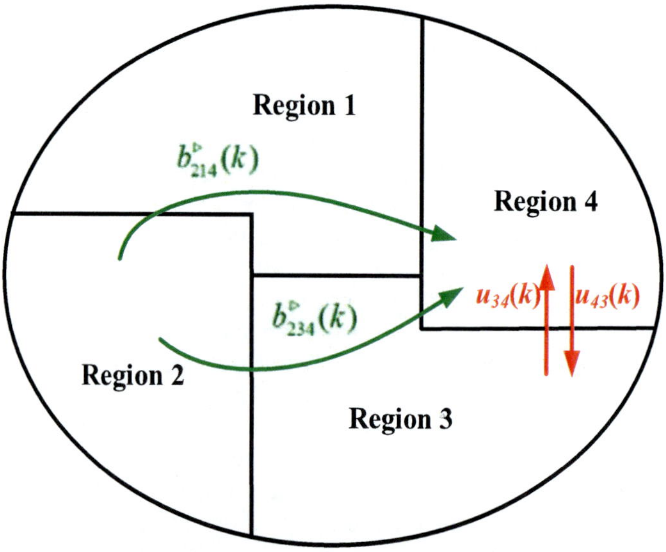

Fig. 10.20 Urban traffic network consisting of four regions

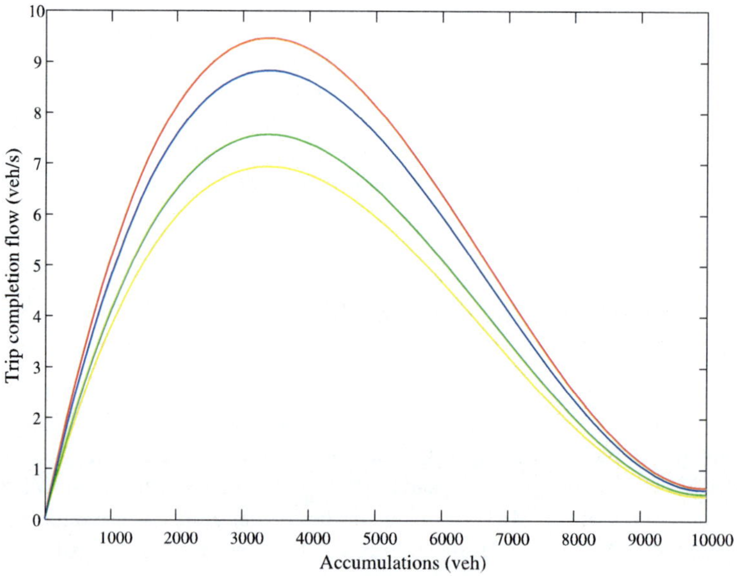

Fig. 10.21 MFD of each region, in which the red solid line denotes $G_1^{\triangleright}$, the green solid line denotes $G_2^{\triangleright}$, the yellow solid line denotes $G_3^{\triangleright}$, and the blue solid line denotes $G_4^{\triangleright}$

$$n_{ii}(k+1) = n_{ii}(k) + T^{\triangleright}(q_{ii}^{\triangleright}(k) - M_{ii}^{\triangleright}(t) + \sum_{m \in N_i} u_{mi}(k) M_{mii}^{\triangleright}(k)) \quad (10.40)$$

$$M_{imj}^{\triangleright}(k) = b_{imj}^{\triangleright}(k) \frac{n_{ij}(k)}{n_i(k)} G_i^{\triangleright}(n_i(k)), \ m \in N_i, \ j \neq i \quad (10.41)$$

$$M_{ii}^{\triangleright}(k) = \frac{n_{ii}(k)}{n_i(k)} G_i^{\triangleright}(n_i(k)) \quad (10.42)$$

$$G_i^{\triangleright}(n_i(k)) = A_i^{\triangleright} n_i^3(k) + B_i^{\triangleright} n_i^2(k) + C_i^{\triangleright} n_i(k) \quad (10.43)$$

$$\begin{aligned} n_{ij}^{og}(k+1) = n_{ij}^{og}(k) + T^{\triangleright}(& \sum_{f \in (N_g^*) \backslash \{i,j\}} u_{gi}(k) M_{gij}^{\triangleright of}(k) \\ & - \sum_{m \in N_i \backslash \{o,g\}} u_{im}(k) M_{imj}^{\triangleright og}(k)), \ g \in N_i \backslash \{j\}, \ i,j \neq o, \ j \neq i \end{aligned} \quad (10.44)$$

$$\begin{aligned} n_{ii}^{og}(k+1) = n_{ii}^{og}(k) + T^{\triangleright}(& \sum_{f \in (N_g^*) \backslash \{i\}} u_{gi}(k) M_{gii}^{\triangleright of}(k) - M_{iii}^{\triangleright og}(k)), \\ & g \in N_i, \ o \neq i \end{aligned} \quad (10.45)$$

$$M_{imj}^{\triangleright og}(k) = b_{imj}^{\triangleright og}(k) \frac{n_{ij}^{og}(k)}{n_i(k)} \cdot G_i^{\triangleright}(n_i(k)) \quad (10.46)$$

$$\begin{aligned} & b_{imj}^{\triangleright og}(k) = \begin{cases} b_{imj}^{\triangleright}(k), & \text{if } m \neq g, \\ 0, & \text{otherwise.} \end{cases} \\ & g \in N_i \backslash \{j\}, \ m \in N_i \backslash \{o\}, \ i,j \neq o, \ j \neq i \end{aligned} \quad (10.47)$$

$$n_{ij}^{ii}(k+1) = n_{ij}^{ii}(k) + T^{\triangleright}(q_{ij}^{\triangleright}(k) - \sum_{m \in N_i} u_{im}(k) M_{imj}^{\triangleright ii}(k)), \ j \neq i \quad (10.48)$$

$$n_{ii}^{ii}(k+1) = n_{ii}^{ii}(k) + T^{\triangleright}(q_{ii}^{\triangleright}(k) - M_{ii}^{\triangleright ii}(k)) \tag{10.49}$$

$$\sum_{o\in\mathcal{N}\setminus\{j\}} \sum_{g\in\mathcal{N}\setminus\{j\}} n_{ij}^{og}(k) = n_{ij}(k) \tag{10.50}$$

$$n_i(k) = \sum_{j\in\mathcal{N}} n_{ij}(k) \tag{10.51}$$

where $n_{ij}^{og}(k)$, $M_{imy}^{\triangleright og}(k)$ and $b_{imj}^{\triangleright og}(k)$ represent the number of vehicles in region i with origin region o, previous region g, and destination region j, the transfer flow from region i to the next immediate region m with origin region o, previous region g, and destination region j, and the ratio of vehicles in region i with origin region o, previous region g, choosing to go through the next immediate region m to reach their destination region j, respectively. Other definitions of variables and symbols in this model are same as that shown in Table 10.1. Different from Sect. 10.2, the trip completion flow in each region i here, i.e., the MFD of region i, $G_i^{\triangleright}(n_i(k))$, is different from each other. The MFD of each region is approximated as a cubic polynomial function of vehicle accumulations, shown in (10.43), where $A_i^{\triangleright}$, $B_i^{\triangleright}$, and $C_i^{\triangleright}$ are the MFD coefficients of region i. For simplicity of writing and subsequent derivation, (10.9) is rewritten into the following form.

$$u_{ij,\min} \le u_{ij}(k) \le u_{ij,\max}(k) \tag{10.52}$$

It is worth mentioning that the model of the MRUTS shown above is only used to generate the traffic data, rather than to design the strategy of route guidance and perimeter control. The multi-step cMFAPLC method for both route guidance and perimeter control is introduced next.

10.3.2 Methodology Framework

The multi-step cMFAPLC route guidance and perimeter control scheme will be designed in this section. It is also composed of two parts: the dynamic linearization data modeling technique for the urban road network system and the designing of the multi-step cMFAPLC route guidance and perimeter control strategy.

A. Dynamic Linearization Data Modeling for MRUTS

Theoretically, the urban traffic dynamics (10.39)–(10.51) of the proposed MRUTS is a complex discrete-time nonlinear MIMO system. Based on this complex MIMO nonlinear system designing the applicable control system is definitely beyond imagery and field application even the dynamics model (10.39)–(10.51) is totally known. In the work of this section, CFDL dynamic linearization data modeling method is used in the urban transportation system. Through this data modeling method, a dynamic linearized data model that can completely and equivalently describe the dynamics of the urban transportation system can be obtained. The data model is designed for the route guidance and perimeter control strategies of the urban traffic system. In Sect. 10.3, CFDL for MIMO systems with the same dimensional input and output is introduced. In this section, it is extended to the more general case where the input and output dimensions of the MIMO system are not equal. Without losing generality, CFDL data modeling technique for general unknown MIMO nonlinear systems will be presented as follow.

In the CFDL data modeling technique, the dynamical relationship between the variation of the urban traffic system output at the future time instant and the route guidance and perimeter control input increments at the current time is well taken into account. In Chap. 9, the cMFAPLC perimeter control for medium-scale two-region urban road network systems is proposed, in which the dimensions of the route guidance and perimeter control input and system output are equal. However, for the MRUTS, the dimensions of control input and system output are not equal, so it is necessary to extend the CFDL data modeling technique to the following more general case. Consider the MIMO nonlinear system with N outputs and M control inputs as below:

$$\mathbf{y}(k+1) = \mathbf{f}(\mathbf{y}(k), \ldots, \mathbf{y}(k-\sigma_y^{\triangleright}), \mathbf{h}(k), \ldots, \mathbf{h}(k-\sigma_h^{\triangleright})) \tag{10.53}$$

where $\mathbf{y}(k) \in \Re^N$ and $\mathbf{h}(k) \in \Re^M$ are the vectors of the output of the system and control input at time interval k, respectively, and $\mathbf{f}(\ldots) = [f_1(\ldots), \ldots, f_N(\ldots)]^T \in \Re^N \times \Re^M \mapsto \Re^N$ is an unknown nonlinear function vector used to describe the dynamics of the controlled system.

The control inputs $\mathbf{h}(k)$ in the MRUTS have of two parts: the perimeter control ratios and the route guidance percentages. That is, $\mathbf{h}(k) = [\mathbf{u}^{\mathrm{T}}(k), \mathbf{b}^{\mathrm{T}}(k)]^{\mathrm{T}}$, where $\mathbf{u}(k) \in \Re^{M_u^{\triangleright}}$ is made up of all the perimeter control ratios $u_{ij}(k), i, j \in \mathcal{N}, j \in N_i$, and $\mathbf{b}(k) \in \Re^{M_b^{\triangleright}}$ include es all the route guidance proportions $b_{imj}^{\triangleright}(k)$, $i, j, m \in \mathcal{N}$, $m \in N_i, j \notin N_i$. $M_u^{\triangleright}$ and $M_b^{\triangleright}$ represent the route guidance and perimeter control inputs dimensions, respectively. For convenience $\mathbf{h}(k) = [h_1(k), \ldots, h_M(k)]^{\mathrm{T}}$ is also used to represent $\mathbf{h}(k) = [\mathbf{u}^{\mathrm{T}}(k), \mathbf{b}^{\mathrm{T}}(k)]^{\mathrm{T}}$ in the following statement. It is noticeable that $N \leq M$ should be satisfied according to the geometric characteristics of the MRUTS. The outputs of the MRUTS consist of the number of vehicles traveling in each region, i.e., $\mathbf{y}(k) = [y_1(k), y_2(k), \ldots, y_N(k)]^{\mathrm{T}}$. It can be found from (10.39)–(10.51) and (10.53) that $\sigma_y^{\triangleright} = 0$ and $\sigma_h^{\triangleright} = 0$ in the MRUTS. There are two assumptions for the CFDL data modeling technique of the dynamics of the MRUTS.

Assumption 10.3 *The partial derivatives of the function* $\mathbf{f}(\ldots)$ *with respect to all the elements of control inputs* $\mathbf{h}(k)$ *are continuous.*

Assumption 10.4 *The system (10.53) is generalized Lipschitz, i.e., for all* $k_1 \neq k_2$, $k_1, k_2 > 0$, *and* $\mathbf{h}(k_1) \neq \mathbf{h}(k_2)$, $\|\mathbf{n}(k_1+1) - \mathbf{n}(k_2+1)\| \leq L^{\triangleright} \|\mathbf{h}(k_1) - \mathbf{h}(k_2)\|$ *always satisfied, where* $L^{\triangleright} > 0$ *is the Lipschitz constant.*

Remark 10.3 In reality, the assumptions aforementioned on the MRUTS are reasonable. Assumption 10.3 can be intuitively verified from (10.1)–(10.51), and it is also a typical assumption for the controller designing of general MIMO nonlinear systems. Assumption 10.4 is an inherent characteristic of the urban traffic network. It indicates that the finite change of route guidance and perimeter control inputs will not lead to the infinite change of the number of vehicles traveling in the road network. In addition, it is also a physical constraint on the actual system based on energy point of view (Hou and Jin, 2011).

Theorem 10.2 *Given the nonlinear MIMO system (10.53) with* $N \leq M$, *if Assumptions 10.3 and 10.4 are satisfied and* $\Delta h(k) \neq 0$, *then there must exist a time-varying matrix* $\mathbf{\Phi}^{\triangleright}(k) \in \Re^{N\times M}$ *named pseudo-Jacobian matrix (PJM), making that the original system (10.53) can be equivalently transformed into the CFDL data model:*

$$\Delta \mathbf{y}(k+1) = \mathbf{\Phi}^{\triangleright}(k)\Delta \mathbf{h}(k) \tag{10.54}$$

where $\Delta \mathbf{h}(k) = \mathbf{h}(k) - \mathbf{h}(k-1)$, *and for any time interval* k, $\mathbf{\Phi}^{\triangleright}(k) = \begin{bmatrix} \phi_{11}^{\triangleright}(k) & \cdots & \phi_{1M}^{\triangleright}(k) \\ \vdots & \ddots & \vdots \\ \phi_{N1}^{\triangleright}(k) & \cdots & \phi_{NM}^{\triangleright}(k) \end{bmatrix} \in \Re^{N\times M}$ *is bounded.*

Proof From the definition of $\Delta \mathbf{y}(k+1)$ and (10.53), one has

$$\begin{aligned}\Delta \mathbf{y}(k+1) = &\mathbf{f}(\mathbf{y}(k), \ldots, \mathbf{y}(k-\sigma_y^{\triangleright}), \mathbf{h}(k), \ldots, \mathbf{h}(k-\sigma_h^{\triangleright}))\\ &-\mathbf{f}(\mathbf{y}(k), \ldots, \mathbf{y}(k-\sigma_y^{\triangleright}), \mathbf{h}(k-1), \mathbf{h}(k-1), \ldots, \mathbf{h}(k-\sigma_h^{\triangleright}))\\ &+\mathbf{f}(\mathbf{y}(k), \ldots, \mathbf{y}(k-\sigma_y^{\triangleright}), \mathbf{h}(k-1), \mathbf{h}(k-1), \ldots, \mathbf{h}(k-\sigma_h^{\triangleright}))\\ &-\mathbf{f}(\mathbf{y}(k-1), \ldots, \mathbf{y}(k-\sigma_y^{\triangleright}-1), \mathbf{h}(k-1), \ldots, \mathbf{h}(k-\sigma_h^{\triangleright}-1))\end{aligned} \tag{10.55}$$

Denote

$$\begin{aligned}\mathbf{\Psi}^{\triangleright}(k) = &\mathbf{f}(y(k), \ldots, \mathbf{y}(k-\sigma_y^{\triangleright}), \mathbf{h}(k-1), \mathbf{h}(k-1), \ldots, \mathbf{h}(k-\sigma_h^{\triangleright}))\\ &-\mathbf{f}(\mathbf{y}(k-1), \ldots, \mathbf{y}(k-\sigma_y^{\triangleright}-1), \mathbf{h}(k-1), \ldots, \mathbf{h}(k-\sigma_h^{\triangleright}-1))\end{aligned} \tag{10.56}$$

According to the differential mean value theorem, (10.55) could be transformed to Eq. (10.57)

$$\Delta \mathbf{y}(k+1) = \frac{\partial \mathbf{f}^*}{\partial \mathbf{h}(k)}\Delta \mathbf{h}(k) + \mathbf{\Psi}^{\triangleright}(k) \tag{10.57}$$

where

$$\frac{\partial \mathbf{f}^*}{\partial \mathbf{h}(k)} = \begin{bmatrix} \frac{\partial f_1^*}{\partial h_1(k)} & \frac{\partial f_1^*}{\partial h_2(k)} & \cdots & \frac{\partial f_1^*}{\partial h_N(k)} \\ \frac{\partial f_2^*}{\partial h_1(k)} & \frac{\partial f_2^*}{\partial h_2(k)} & \cdots & \frac{\partial f_2^*}{\partial h_N(k)} \\ \vdots & \vdots & \ddots & \vdots \\ \frac{\partial f_M^*}{\partial h_1(k)} & \frac{\partial f_M^*}{\partial h_2(k)} & \cdots & \frac{\partial f_M^*}{\partial h_N(k)} \end{bmatrix} \tag{10.58}$$

and $\partial f_i^*/\partial h_j(k)$ means the partial derivative value of f_i with respect to h_j at certain point in $[h_j(k-1), h_j(k)]$ or $[h_j(k), h_j(k-1)]$.

For each fixed time step k, considering the function with respective to $\Delta\mathbf{h}(k)$ below:

$$\boldsymbol{\Psi}(k) = \mathbf{F}^{\triangleright}(k)\Delta\mathbf{h}(k) \tag{10.59}$$

where $\mathbf{F}^{\triangleright}(k)$ has M rows and N columns. Since $\|\Delta\mathbf{h}(k)\| \neq 0$ and $M \leq N$ are satisfied, there must exist at least one solution $\mathbf{F}^{\triangleright *}(k)$ of (10.59). Denote $\boldsymbol{\Phi}^{\triangleright}(k) = \mathbf{F}^{\triangleright *}(k) + \partial\mathbf{f}^*/\partial\mathbf{h}(\mathbf{k})$, thus (10.57) could be rewritten as $\Delta\mathbf{y}(k+1) = \boldsymbol{\Phi}^{\triangleright}(k)\Delta\mathbf{h}(k)$. The boundness of $\boldsymbol{\Phi}^{\triangleright}(k)$ can be guaranteed directly via Assumption 10.4.

B. Design of route guidance and perimeter control Strategy

Based on the CFDL data model of the MRUTS, the multi-step cMFAPLC-based route guidance and perimeter control scheme can be designed, including PJM learning, PJM prediction, and computation of route guidance and perimeter control ratios as follows.

1. PJM learning

The PJM cost function of the MRUTS at each time step k is as below:

$$\begin{aligned} J(\boldsymbol{\Phi}^{\triangleright}(k)) = & \left\|\mathbf{y}(k) - \mathbf{y}(k-1) - \boldsymbol{\Phi}^{\triangleright}(k)\Delta\mathbf{h}(k-1)\right\|^2 \\ & + \mu^{\triangleright}\left\|\boldsymbol{\Phi}^{\triangleright}(k) - \hat{\boldsymbol{\Phi}}^{\triangleright}(k-1)\right\|^2 \end{aligned} \tag{10.60}$$

where $\mu^{\triangleright} > 0$ is a weighting factor that avoid PJM changing too fast. The learning law (10.61) can be obtained to estimate $\boldsymbol{\Phi}^{\triangleright}(k)$, by minimizing (10.60) with respect to $\hat{\boldsymbol{\Phi}}^{\triangleright}(k)$ (Hou and Jin, 2013):

$$\begin{aligned} \hat{\boldsymbol{\Phi}}^{\triangleright}(k) = & (\Delta\mathbf{y}(k) - \hat{\boldsymbol{\Phi}}^{\triangleright}(k-1)\Delta\mathbf{h}(k-1))\Delta\mathbf{h}^{\mathrm{T}}(k-1)(\mu^{\triangleright}\boldsymbol{I} \\ & + \Delta\mathbf{h}(k-1)\Delta\mathbf{h}^{\mathrm{T}}(k-1))^{-1} + \hat{\boldsymbol{\Phi}}^{\triangleright}(k-1) \end{aligned} \tag{10.61}$$

When the control input and the system output dimensions of the controlled system, especially the large-scale multi-region urban transportation system, are too high, the calculation of formula (10.61) will become very difficult due to the existence of inverse operation. Therefore in practice, the learning law (10.61) is not suitable for calculating the PJM of the MRUTS. In fact, we can use the simplified algorithm (10.62)–(10.63) shown below to estimate the PJM of the multi-region urban road network system, in which it is no longer necessary to solve the inverse of high-dimensional matrix, and the robustness of cMFAPLC route guidance and perimeter control strategy is enhanced:

$$\hat{\mathbf{\Phi}}^{\triangleright}(k) = \hat{\mathbf{\Phi}}^{\triangleright}(k-1) + \frac{\eta^{\triangleright}(\Delta\mathbf{y}(k) - \hat{\mathbf{\Phi}}^{\triangleright}(k-1)\Delta\mathbf{h}(k-1))\Delta\mathbf{h}^{\mathrm{T}}(k-1)}{\mu^{\triangleright} + ||\Delta\mathbf{h}(k-1)||^2} \tag{10.62}$$

$$\begin{aligned} \hat{\phi}_{ij}^{\triangleright}(k) = \hat{\phi}_{ij}^{\triangleright}(0),\ if\ |\hat{\phi}_{ij}^{\triangleright}(k)| > \zeta \text{ or } sign(\hat{\phi}_{ij}^{\triangleright}(k)) \\ \neq sign(\hat{\phi}_{ij}^{\triangleright}(0)), i = 1, \ldots, N,\ j = 1, \ldots, M. \end{aligned} \tag{10.63}$$

where $\hat{\mathbf{\Phi}}^{\triangleright}(k)$ is the estimation value of $\mathbf{\Phi}^{\triangleright}(k)$, $\hat{\phi}_{ij}^{\triangleright}(0)$ is the initial value of $\hat{\phi}_{ij}^{\triangleright}(k), i = 1, \ldots, N, j = 1, \ldots, M$. $\eta^{\triangleright} \in (0, 2]$ is a step factor making the route guidance and perimeter control algorithm more flexible and more general. $\zeta^{\triangleright}$ is a positive constant.

2. PJM prediction
The CFDL data model (10.54) of the large-scale multi-region urban road network system can be rewritten as follow.

$$\mathbf{y}(k+1) = \mathbf{y}(k) + \mathbf{\Phi}^{\triangleright}(k)\Delta\mathbf{h}(k) \tag{10.64}$$

Let the prediction and control horizons, respectively are denoted by $N_p^{\triangleright}$ and $N_u^{\triangleright}$, then the following $N_p^{\triangleright}$-step ahead prediction equation of the large-scale multi-region urban traffic system can be easily obtained from (10.64):

$$\mathbf{Y}(k+1) = \mathbf{E}(k)\mathbf{y}(k) + \mathbf{A}^{\triangleright}(k)\Delta\mathbf{H}(k) \tag{10.65}$$

where the definition of the variables is introduced as follows:

$$\begin{cases} \mathbf{Y}(k+1) = [\mathbf{y}(k+1)^{\mathrm{T}}, \ldots, \mathbf{y}(k+N_p)^{\mathrm{T}}]^{\mathrm{T}} \in \Re^{NN_p^{\triangleright}} \\ \Delta\mathbf{H}(k) = [\Delta\mathbf{h}(k)^{\mathrm{T}}, \ldots, \Delta\mathbf{h}(k+N_u^{\triangleright}-1)^{\mathrm{T}}]^{\mathrm{T}} \in \Re^{MN_u^{\triangleright}} \\ \mathbf{E}(k) = [I_{N\times N}, I_{N\times N}, \ldots, I_{N\times N}]^{\mathrm{T}} \in \Re^{NN_p^{\triangleright}} \\ \mathbf{A}^{\triangleright}(k) = \begin{bmatrix} \mathbf{\Phi}^{\triangleright}(k) & 0 & 0 & 0 \\ \mathbf{\Phi}^{\triangleright}(k) & \mathbf{\Phi}^{\triangleright}(k+1) & 0 & 0 \\ \vdots & \vdots & \ddots & \vdots \\ \mathbf{\Phi}^{\triangleright}(k) & \mathbf{\Phi}^{\triangleright}(k+1) & \cdots & \mathbf{\Phi}^{\triangleright}(k+N_u^{\triangleright}-1) \\ \vdots & \vdots & \cdots & \vdots \\ \mathbf{\Phi}^{\triangleright}(k) & \mathbf{\Phi}^{\triangleright}(k+1) & \cdots & \mathbf{\Phi}^{\triangleright}(k+N_u^{\triangleright}-1) \end{bmatrix} \in \Re^{NN_p^{\triangleright}\times MN_u^{\triangleright}} \end{cases} \tag{10.66}$$

The following auto-regressive (AR) model is used to predicted the future PJM of the MRUTS:

$$\hat{\mathbf{\Phi}}^{\triangleright}(k+p) = \boldsymbol{\theta}_1^{\triangleright}(k)\hat{\mathbf{\Phi}}^{\triangleright}(k+p-1) + \boldsymbol{\theta}_2^{\triangleright}(k)\hat{\mathbf{\Phi}}^{\triangleright}(k+p-2) + \ldots$$
$$+\boldsymbol{\theta}_{n_p^{\triangleright}}^{\triangleright}(k)\hat{\mathbf{\Phi}}^{\triangleright}(k+p-n_p^{\triangleright}), p = 1, 2, \ldots, N_u^{\triangleright} - 1 \quad (10.67)$$

$$\hat{\phi}_{ij}^{\triangleright}(k+p) = \hat{\phi}_{ij}^{\triangleright}(0), \ if \ |\hat{\phi}_{ij}^{\triangleright}(k+p)| > \zeta \text{ or } sign(\hat{\phi}_{ij}^{\triangleright}(k+p))$$
$$\neq sign(\hat{\phi}_{ij}^{\triangleright}(0)), i = 1, \ldots, M, j = 1, \ldots, N. \quad (10.68)$$

where $\boldsymbol{\theta}_l^{\triangleright}(k) \in \Re^{N \times N}, l = 1, 2, , n_p^{\triangleright}$, is the coefficient matrix, and $n_p^{\triangleright}$ is a proper order of the AR model, which is usually chosen as 2–7 in(Han, 1984). Here it is selected as $n_p^{\triangleright}$=2.

Define $\hat{\mathbf{\Phi}}^{\triangleright}(k+p-1) = [\hat{\mathbf{\Phi}}^{\triangleright\mathrm{T}}(k+p-1), \ldots, \hat{\mathbf{\Phi}}^{\triangleright\mathrm{T}}(k+p-n_p^{\triangleright})]^{\mathrm{T}}$ and $\mathbf{\Theta}^{\triangleright}(k) = [\boldsymbol{\theta}_1^{\triangleright}(k), \ldots, \boldsymbol{\theta}_{n_p^{\triangleright}}^{\triangleright}(k)]$, then the PJM prediction model (10.67) can be transformed into the following vector form:

$$\hat{\mathbf{\Phi}}^{\triangleright}(k+p) = \mathbf{\Theta}^{\triangleright}(k)\hat{\mathbf{\Phi}}^{\triangleright}(k+p-1), p = 1, 2, \ldots, N_u^{\triangleright} - 1. \quad (10.69)$$

where the learning law of $\mathbf{\Theta}^{\triangleright}(k)$ can be calculated by the projection algorithm as follow:

$$\mathbf{\Theta}^{\triangleright}(k) = \mathbf{\Theta}^{\triangleright}(k-1) + \frac{(\hat{\mathbf{\Phi}}^{\triangleright}(k) - \mathbf{\Theta}^{\triangleright}(k-1)\hat{\mathbf{\Phi}}^{\triangleright}(k-1))}{\delta^{\triangleright} + ||\hat{\mathbf{\Phi}}^{\triangleright}(k-1)||^2}\hat{\mathbf{\Phi}}^{\triangleright\mathrm{T}}(k-1) \quad (10.70)$$

where $\delta^{\triangleright}$ is a positive factor, and it could be selected as $\delta^{\triangleright} \in (0, 1]$.

3. Computation of route guidance and perimeter control ratios

In order to design the route guidance and perimeter control strategy for the MRUTS, the following criterion function (10.71) is optimized to get the perimeter control ratios and route guidance proportions. The increment vectors of perimeter control ratios and route guidance proportions are, respectively, denoted as $\Delta\mathbf{u}(k) = \mathbf{u}(k) - \mathbf{u}(k-1)$ and $\Delta\mathbf{b}(k) = \mathbf{b}(k) - \mathbf{b}(k-1)$, and the following cost function is defined:

$$J(\Delta\mathbf{H}(k)) = \sum_{i=1}^{N_p^{\triangleright}} \left\|\mathbf{y}_d^{\triangleright}(k+i) - \mathbf{y}(k+i)\right\|^2 + \xi_u^{\triangleright}\lambda_u^{\triangleright}\sum_{j=0}^{N_u^{\triangleright}-1} \|\Delta\mathbf{u}(k+j)\|^2$$
$$+\xi_b^{\triangleright}\lambda_b^{\triangleright}\sum_{j=0}^{N_u^{\triangleright}-1} \|\Delta\mathbf{b}(k+j)\|^2 \quad (10.71)$$

where $\mathbf{y}_d^{\triangleright}(k+i) \in \Re^N$ is the desired number of vehicles in each region at time interval $k+i$, $\xi_u^{\triangleright}$ and $\xi_b^{\triangleright}$ are two positive factors balancing the order of magnitude of each item on the right side of the equal sign in (10.71), and $\lambda_u^{\triangleright}, \lambda_b^{\triangleright} > 0$ are two weighting factors making the change of route guidance and perimeter control ratios less drastic.

For the MRUTS, both of the inputs $\mathbf{u}(k)$, $\mathbf{b}(k)$ and the outputs $\mathbf{y}(k)$ should be physically subject to the following constraints.

For the system output, the number of vehicles in each region cannot be negative or greater than n_i^{jam}, i.e.,

$$\mathbf{0} \leq \mathbf{y}(k+i) \leq \mathbf{y}_{\text{jam}},\ i = 1, \ldots, N_p^{\triangleright} \tag{10.72}$$

where $\mathbf{y}_{\text{jam}} = [y_1^{\text{jam}}, \ldots, y_N^{\text{jam}}]^T \in \Re^N$ and $\mathbf{0} = [0, \ldots, 0]^T \in \Re^N$.

For the control input, the route guidance and perimeter control proportions are subject to the following constraints:

$$\mathbf{u}_{\min} \leq \mathbf{u}(k) \leq \mathbf{u}_{\max}(k) \tag{10.73}$$

$$\mathbf{0} \leq \mathbf{b}(k) \leq \mathbf{1}_b \tag{10.74}$$

$$\sum_{\substack{j \in \mathcal{N} \setminus N_i \\ m \in V_{ij}}} b_{imj}^{\triangleright}(k) = 1 \tag{10.75}$$

where $\mathbf{u}_{\min}$ and $\mathbf{u}_{\max}(k)$ are consist of all the minimum and maximum perimeter control ratios $u_{ij,\min}$ and $u_{ij,\max}(k)$, respectively, $i, j \in \mathcal{N}, j \in N_i$. That is, the meaning of (10.73) is the same with (10.52). $\mathbf{1}_b = [1, 1, \ldots, 1]^{\mathrm{T}} \in \Re^{M_b^{\triangleright}}$ is a constant vector, and it represents the upper bound of the route guidance proportions.

Combining (10.54) and (10.72)–(10.74), the inequality constraints imposed on $\Delta\mathbf{u}(k)$ and $\Delta\mathbf{b}(k)$ can be arranged into the following concise form:

$$\begin{bmatrix} \mathbf{A}^{\triangleright}(k) & & & & & \\ & -\mathbf{A}^{\triangleright}(k) & & & & \\ & & \boldsymbol{I}_U & & & \\ & & & -\boldsymbol{I}_U & & \\ & & & & \boldsymbol{I}_B & \\ & & & & & -\boldsymbol{I}_B \end{bmatrix} \begin{bmatrix} \Delta\mathbf{H}(k) \\ \Delta\mathbf{H}(k) \\ \Delta\mathbf{H}(k) \\ \Delta\mathbf{H}(k) \\ \Delta\mathbf{B}(k) \\ \Delta\mathbf{B}(k) \end{bmatrix} \leq \begin{bmatrix} \mathbf{N}_{\text{jam}} - \mathbf{E}(k)y(k) \\ \mathbf{E}(k)\mathbf{y}(k) \\ \mathbf{U}_{\max}(k) - \mathbf{U}(k-1) \\ \mathbf{U}(k-1) - \mathbf{U}_{\min} \\ \mathbf{1}_B - \mathbf{B}(k-1) \\ \mathbf{B}(k-1) \end{bmatrix} \tag{10.76}$$

where $\mathbf{U}(k-1) = [\mathbf{u}(k-1)^{\mathrm{T}}, \ldots, \mathbf{u}(k+N_u^{\triangleright}-2)^{\mathrm{T}}]^{\mathrm{T}}$, $\mathbf{U}_{\min} = [\mathbf{u}_{\min}^{\mathrm{T}}, \ldots, \mathbf{u}_{\min}^{\mathrm{T}}]^{\mathrm{T}} \in \Re^{M_u^{\triangleright} N_u^{\triangleright}}$, $\mathbf{B}(k-1) = [\mathbf{b}(k-1)^{\mathrm{T}}, \ldots, \mathbf{b}(k+N_u^{\triangleright}-2)^{\mathrm{T}}]^{\mathrm{T}}$, $\mathbf{Y}_{\mathrm{jam}} = [\mathbf{y}_{\mathrm{jam}}^{\mathrm{T}}, \ldots, \mathbf{y}_{\mathrm{jam}}^{\mathrm{T}}]^{\mathrm{T}} \in \Re^{NN_p^{\triangleright}}$, $\mathbf{U}_{\max}(k) = [\mathbf{u}_{\max}(k)^{\mathrm{T}}, \ldots, \mathbf{u}_{\max}(k+N_u^{\triangleright}-1)^{\mathrm{T}}]^{\mathrm{T}}$, $\Delta\mathbf{U}(k) = \mathbf{U}(k) - \mathbf{U}(k-1)$, $\Delta\mathbf{B}(k) = \mathbf{B}(k) - \mathbf{B}(k-1)$, $\mathbf{1}_B = [1, 1, \ldots, 1]^{\mathrm{T}} \in \Re^{M_b^{\triangleright} N_u^{\triangleright}}$. I_U and I_B are unit matrices with proper dimensions.

It is obvious that the optimization problem of minimizing (10.71) with respect to $\Delta\mathbf{u}(k)$ and $\Delta\mathbf{b}(k)$ subject to the constraints (10.75)–(10.76) is a convex quadratic programming problem, and it could be easily solved through many optimization methods. In this work, it is addressed by fmincon toolbox of Matlab. After $\Delta\mathbf{u}(k)$ and $\Delta\mathbf{b}(k)$ are obtained from (10.75)–(10.76), we have $\Delta\mathbf{u}(k) = \mathbf{g}_u^{\mathrm{T}}\Delta\mathbf{U}(k)$ and $\Delta\mathbf{b}(k) = \mathbf{g}_b^{\mathrm{T}}\Delta\mathbf{B}(k)$, where $\mathbf{g}_u = [1_u^{\mathrm{T}}, 0_u^{\mathrm{T}}]^{\mathrm{T}}$, $\mathbf{g}_b = [1_b^{\mathrm{T}}, 0_b^{\mathrm{T}}]^{\mathrm{T}}$, $\mathbf{1}_u = [1, 1, \ldots, 1]^{\mathrm{T}} \in \Re^{M_u^{\triangleright}}$, $\mathbf{1}_b = [1, 1, \ldots, 1]^{\mathrm{T}} \in \Re^{M_b^{\triangleright}}$, $\mathbf{0}_u = [0, 0, \ldots, 0]^{\mathrm{T}} \in \Re^{(N_u^{\triangleright}-1)M_u^{\triangleright}}$, and $\mathbf{0}_b = [0, 0, \ldots, 0]^{\mathrm{T}} \in \Re^{(N_u^{\triangleright}-1)M_b^{\triangleright}}$. Consequently, the route guidance and perimeter control ratios at time step k are obtained as below:

$$\mathbf{u}(k) = \mathbf{u}(k-1) + \mathbf{g}_u^{\mathrm{T}}\Delta\mathbf{U}(k) \tag{10.77}$$

$$\mathbf{b}(k) = \mathbf{b}(k-1) + \mathbf{g}_b^{\mathrm{T}}\Delta\mathbf{B}(k) \tag{10.78}$$

The multi-step cMFAPLC strategy consists of the PJM estimation algorithm (10.62)–(10.63), PJM prediction algorithm (10.67)–(10.70), and the route guidance and perimeter control algorithms (10.71)–(10.78). The reset mechanism (10.63) and (10.68) are designed to make the PJM estimation algorithm (10.62) and prediction algorithm (10.67) have a stronger ability to track time-varying parameters.

The main features of the multi-step cMFAPLC route guidance and perimeter control method proposed here are summarized as below:

(1) The multi-step cMFAPLC scheme is a data-driven route guidance and perimeter control method, that is, the structure, order, and parameters of the mathematical model, and even MFD of the large-scale urban road network system are all no longer necessary. Instead, only the measured route guidance and perimeter control input and the system output data are utilized to design the predictive route guidance and perimeter control approach. The input/output data of the urban traffic system can be easily measured by the various traffic detectors, such as navigation softwares, video detectors, and geomagnetic coil. The phenomenon of inaccurate prediction and discounted control effect caused by model mismatch will no longer appear. Besides, the in the multi-step cMFAPLC strategy, the MFD is no longer need in the designing process of the route guidance and perimeter control strategy.
(2) The multi-step cMFAPLC strategy can deal with the nonsquared MIMO nonlinear route guidance and perimeter control problems with actual traffic constraints in the large-scale multi-region urban traffic system, and it is the most outstanding merit of this scheme.

(3) The proposed multi-step cMFAPLC includes the prediction mechanism and the rolling optimization control technique, and the consideration to deal with practical constraints, which highlights the main differences.
(4) The dynamic linearization data modeling method which is method used here is quite different from other existing linearization methods, especially that used in the study of urban traffic systems. Here are some examples of other dynamic linearization approaches. In piecewise linearization technique (Geroliminis et al., 2013), the switching instant and residence time of the piecewise linearized dynamics and some other more model information are all necessary. In Taylors linearization method (Aboudolas and Geroliminis, 2013), it is an approximated description because the high-order terms in the original dynamics of the system are ignored. Further, it is difficult for them to deal with the parameter uncertainties or disturbances in urban traffic demands, the number of vehicles traveling in the urban road networks, etc. Different from the above linearization approaches, an equivalent description of the original MRUTS is provided through CFDL data modeling method, and there is no need to know or establish the mathematical model of urban transportation system. Besides, the dynamic linearization data modeling technique also has the following other outstanding advantages. First, this method is a data-driven method for unknown dynamic nonlinear systems, which can be used without obtaining the mathematical model information of urban transportation system. Second, it is simple and can be easily updated via the route guidance and perimeter control input and the system output data of the closed-loop urban traffic system. More discussion on the advantages and differences of this method compared with other existing linearization methods can be seen in (Hou and Xiong, 2019).
(5) There is not any unmodeled dynamics in the multi-step cMFAPLC scheme, since all the information on the large-scale multi-region urban road network system dynamics should be included in the input/output data due to there is no data dropout from the theorem proof. Further, the PJM of the urban road network system can be easily estimated and predicted via the input/output data of the MRUTS, and the proposed multi-step cMFAPLC-based route guidance and perimeter control scheme is a convex quadratic programming problem. It reveals the fact that there is no heavy computational burden in the proposed strategy.

10.3.3 Numerical Simulation Results

Some simulation experiments are presented here to test the performance of the multi-step cMFAPLC route guidance and perimeter control strategy proposed in the previous subsection for the MRUTS. In addition, some other commonly used route guidance and perimeter control approaches are also compared to highlight the superiority of the multi-step cMFAPLC strategy.

A. MRUTS Road Network Description and Parameter Settings

An MRUTS consists of four regions is presented here, see Fig. 10.20. Regions 1 and 3 are adjacent to all other areas in the road network, while regions 2 and 4 are not neighboring each other.

The MFD of each region of the MRUTS can be seen in Fig. 10.21. Let the basis MFD denoted by $\bar{G}^{\triangleright}$ represent the one in the downtown of Yokohama (Sirmatel and Geroliminis 2016), whose parameters are $\bar{A}^{\triangleright} = 4.133 \times 10^{-11}$, $\bar{B}^{\triangleright} = -8.282 \times 10^{-7}$, and $\bar{C}^{\triangleright} = 0.0042$, respectively. In the simulation, the MFDs of the 4 regions are: $G_1^{\triangleright} = 1.5\bar{G}^{\triangleright}$, $G_2^{\triangleright} = 1.2\bar{G}^{\triangleright}$, $G_3^{\triangleright} = 1.1\bar{G}^{\triangleright}$, and $G_5^{\triangleright} = 1.3\bar{G}^{\triangleright}$, respectively. The unit boundary capacity is $\bar{C}_{\max}^{\triangleright} = 3.2$(veh/s), and that of each region scales in proportion to $G^{\triangleright}$. $\alpha^{\triangleright}$ is equal to 0.64 for all the regions. The other parameters of the urban road network are: $n_i^{\text{cr}} = 3400$ (veh), $n_i^{\text{jam}} = 10,000$ (veh), and the unit maximum and jammed trip completion flow are $\bar{G}_i^{\triangleright}(n_i^{\text{cr}}) = 6.3$ (veh/s), $\bar{G}_i^{\triangleright}(n_i^{\text{jam}}) = 0.43$ (veh/s) $i = 1, 2, 3, 4$, respectively. It is noticeable that the proposed cMFAPLC strategy can be designed as long as the desired reference vehicles number value of the MRUTS is determined. The desired vehicle number is chosen as $n_i^* = 0.97 \cdot n_i^{\text{cr}}$, in order to make more vehicles reach the destination as far as possible without causing traffic congestion in the regions. The mathematical model of urban transportation system model (10.39)–(10.51) given above, the MFD models and some other parameters settings, are only used to simulate the data of urban transportation system, rather than design the multi-step cMFAPLC route guidance and perimeter control strategy.

In order to test and verify the effectiveness of the multi-step cMFAPLC route guidance and perimeter control strategy proposed in this section. At the initial stage of the simulation, it is assumed that all areas are in a congested state. The number of vehicles in the regions at the beginning are: $n_1(0) = 5000$(veh), and $n_i(0) = 4000$(veh), i=2,3,4,5, which are all greater than n_i^{cr}. The parameters of green signal ratio at boundary intersections are: $g_0 = 0.5$, $g_{\max} = 0.9$ and $g_{\min} = 0.1$, respectively, so that $u_{ij,\max}^g = 1.8$ and $u_{ij,\min}^g = 0.2$ according to the definition of perimeter control ratio.

The simulation time setting is the same as that in the previous sections. It is also set as four hours to simulate the process of morning peak from formation to dissipation. The traffic signal cycle time of all the boundary intersections is 120s, and the sampling time and the route guidance and perimeter control cycle length are both set as T=240s. The entire simulation time is divided into 60 time intervals. The predictive and the control horizons of the proposed multi-step cMFAPLC route guidance and perimeter control strategy are set as $N_p^{\triangleright}$=3 and $N_c^{\triangleright}$=2, respectively. Other parameters in the cMFAPLC strategy are: $\mu^{\triangleright} = \eta^{\triangleright} = \delta^{\triangleright} = 1$, $\zeta^{\triangleright} = 100$, $\xi_u^{\triangleright} = \xi_b^{\triangleright} = 1000$, $\lambda_u^{\triangleright} = 20$, and $\lambda_b^{\triangleright} = 10$. The time-varying traffic demands with noises are described in Fig. 10.22. The simulation is done with MATLAB.

In order to test the effectiveness and robustness of the multi-step cMFAPLC route guidance and perimeter control method, it is assumed that there exist some measurement noises in MFDs, traffic demands, and number of vehicles. The above noises, respectively, obey normal and uniform distributions, which are depicted as below:

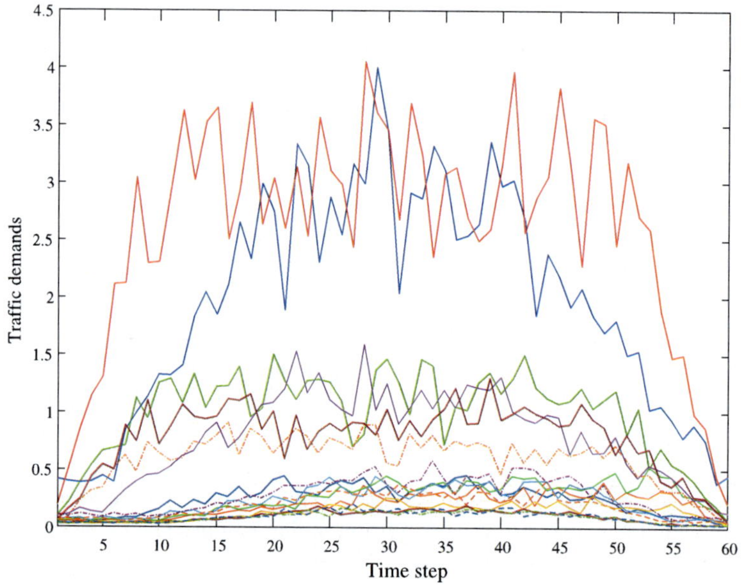

Fig. 10.22 Traffic demands among the four regions, in which the gray blue solid line denotes $q_{11}^{\triangleright}$, the orange solid line denotes $q_{12}^{\triangleright}$, the yellow solid line denotes $q_{13}^{\triangleright}$, the green solid line denotes $q_{14}^{\triangleright}$, the dark green solid line denotes $q_{21}^{\triangleright}$, the gray blue solid line denotes $q_{22}^{\triangleright}$, the brown solid line denotes $q_{23}^{\triangleright}$, the blue solid line denotes $q_{24}^{\triangleright}$, the brown chain line denotes $q_{31}^{\triangleright}$, the purple solid line denotes $q_{32}^{\triangleright}$, the dark purple chain line denotes $q_{33}^{\triangleright}$, the dark green chain line denotes $q_{34}^{\triangleright}$, the dark brown solid line denotes $q_{41}^{\triangleright}$, the red solid line denotes $q_{42}^{\triangleright}$, the gray blue dashed line denotes $q_{43}^{\triangleright}$, and the orange dashed line denotes $q_{44}^{\triangleright}$

$$\tilde{q}_{ij}^{\triangleright}(k) = q_{ij}^{\triangleright}(k)(1 + (0, \sigma_q^{\triangleright 2})) \tag{10.79}$$

$$\tilde{n}_i(k) = n_i(k)(1 + (0, \sigma_n^{\triangleright 2})) \tag{10.80}$$

$$\tilde{G}_i^{\triangleright}(n_i(k)) = G_i^{\triangleright}(n_i(k))(1 + (-\alpha_{G^{\triangleright}}^{\triangleright},\ \alpha_{G^{\triangleright}}^{\triangleright})) \tag{10.81}$$

where $\sigma_q^{\triangleright 2}$, $\sigma_n^{\triangleright 2}$, and $\alpha_{G^{\triangleright}}^{\triangleright}$ are all equal to 0.2.

Consider the fact that drivers often do not completely follow the route guidance in the urban traffic practice, it is assumed that under the case that without route guidance, the route choice proportions of the drivers are obtained through the logit model (Ramezani et al. 2015; Sirmatel and Geroliminis 2016). Thus, the realized route choice proportions under multi-step cMFAPLC method proposed in this section are calculated below:

$$b_{imj}^{\text{real}}(k) = \beta b_{imj}^{\text{cMFAPLC}}(k) + (1-\beta) b_{imj}^{\text{logit}}(k) \tag{10.82}$$

where $b_{imj}^{\triangleright\text{real}}(k)$, $b_{imj}^{\triangleright\text{cMFAPLC}}(k)$, and $b_{imj}^{\triangleright\text{logit}}(k)$ represent the realized route choice ratios, route guidance proportions calculated by cMFAPLC and logit model, respectively. $\beta^{\triangleright}$ is the route guidance compliance rate of the drivers. In this simulation study, $\beta^{\triangleright} = 0.7$.

B. Benchmark route guidance and perimeter control Methods

1. NC

As mentioned in Sect. 10.2, all the perimeter control ratios u_{ij} are equal to 1 from first to last under NC strategy (Kouvelas et al., 2017). The drivers route choice proportions are obtained by the logit model mentioned in the previous subsection.

2. BBC

BBC is another commonly used perimeter control strategy. As shown in (Daganzo, 2007; Aboudolas and Geroliminis, 2013), the main idea of BBC is that each perimeter control ratio switches back and forth between its maximum and minimum values. The details of BBC perimeter control method can be found in Chap. 9. The route choice proportions under BBC approach are obtained by the logit model.

3. Greedy Control (GC)

The control ideas of GC and BBC are similar. Different from BBC, the relative traffic situation of the adjacent regions is taken more account in the switching condition of perimeter control ratios in GC. In GC strategy, the perimeter control ratios from the less congested regions to the more congested regions are taken as the minimum value, while that from the more congested regions to the less congested regions are taken as the maximum value at each control cycle. The details of GC strategy for perimeter control can be found in (Geroliminis et al., 2013). In GC scheme, the acquisition of route choice proportions is the same with that in NC and BBC strategies.

4. MPC

MPC is a frequently used model-based method for route guidance and perimeter control. The objective of MPC route guidance and perimeter control strategy is to minimum the TTS within a finite horizon for the large-scale multi-region urban road network (Sirmatel and Geroliminis, 2016). The model of the MRUTS (10.39)–(10.51) is utilized to predict the future traffic states of the MRUTS for the designing of the route guidance and perimeter control strategy. The parameters of MPC are also set as $N_p^{\triangleright}$=3 and $N_c^{\triangleright}$=2, which is the same with that of multi-step cMFAPLC, and the realized route choice ratios are acquired similar with (10.82), only replacing $b_{imj}^{\triangleright\text{cMFAPLC}}(k)$ with $b_{imj}^{\triangleright\text{MPC}}(k)$, where the latter represents the route guidance proportions calculated by MPC scheme.

Among the above route guidance and perimeter control methods, MPC is based on the urban traffic model, while cMFAPLC, BBC, and GC are based on the traffic data. All the aforementioned route guidance and perimeter control methods are compared under the same simulation environment and initial conditions.

C. Simulation Analysis

The simulation results of different strategies are illustrated in Figs. 10.23, 10.24, 10.25, 10.26, 10.27, 10.28, 10.29, 10.30, 10.31 and 10.32.

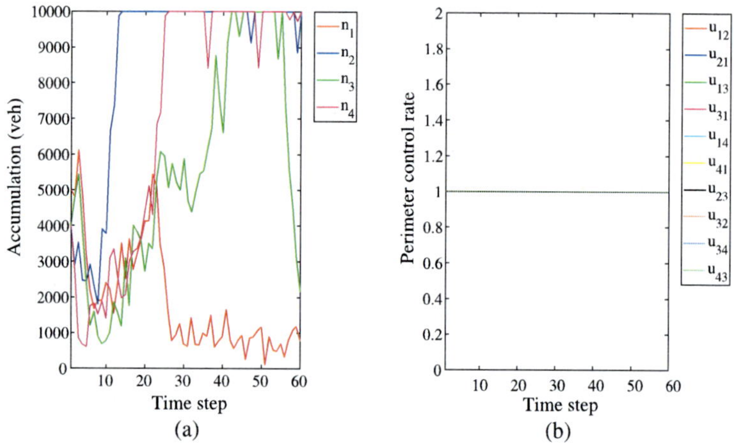

Fig. 10.23 Result of NC strategy (part one): (**a**) evolution of the number of vehicles traveling in each region; (**b**) perimeter control ratios

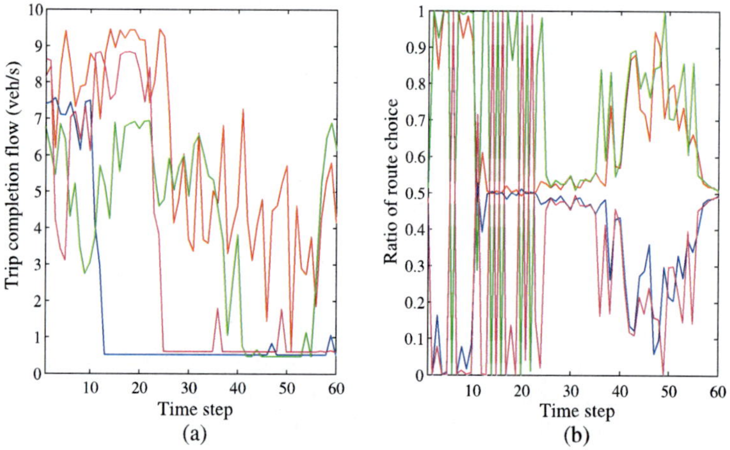

Fig. 10.24 Result of NC strategy (part two): (**a**) trip completion flow for each region; (**b**) ratio of route choice. Where in sub-figure (**a**), the red solid line denotes $G_1^{\triangleright}$, the blue solid line denotes $G_2^{\triangleright}$, the green solid line denotes $G_3^{\triangleright}$, the magenta solid line denotes $G_4^{\triangleright}$. In sub-figure (**b**), the red solid line denotes $b_{214}^{\triangleright}$, the blue solid line denotes $b_{234}^{\triangleright}$, the green solid line denotes $b_{412}^{\triangleright}$, and the magenta solid line denotes $b_{432}^{\triangleright}$

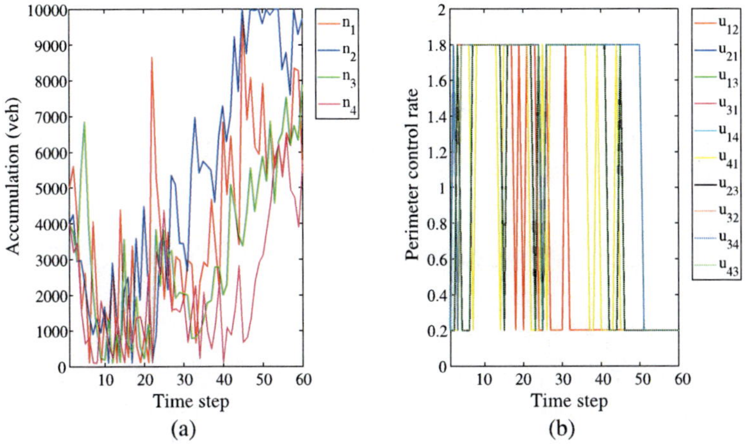

Fig. 10.25 Result of BBC strategy (part one): (**a**) evolution of the number of vehicles traveling in each region; (**b**) perimeter control ratios

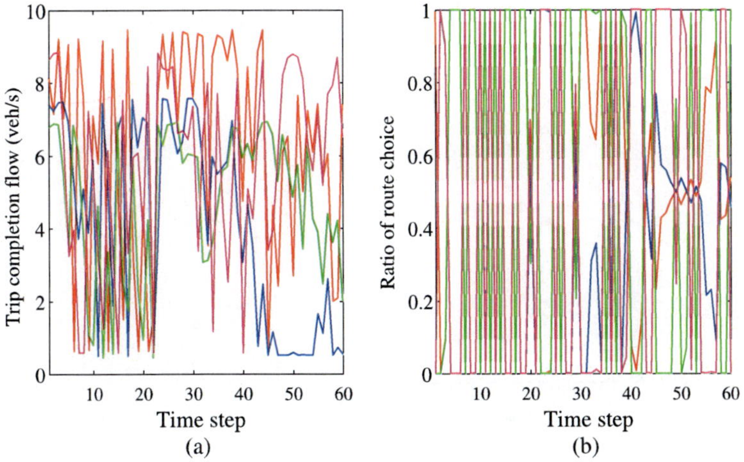

Fig. 10.26 Result of BBC strategy (part two): (**a**) trip completion flow for each region; (**b**) ratio of route choice. Where in sub-figure (**a**), the red solid line denotes $G_1^{\triangleright}$, the blue solid line denotes $G_2^{\triangleright}$, the green solid line denotes $G_3^{\triangleright}$, the magenta solid line denotes $G_4^{\triangleright}$. In sub-figure (**b**), the red solid line denotes $b_{214}^{\triangleright}$, the blue solid line denotes $b_{234}^{\triangleright}$, the green solid line denotes $b_{412}^{\triangleright}$, and the magenta solid line denotes $b_{432}^{\triangleright}$

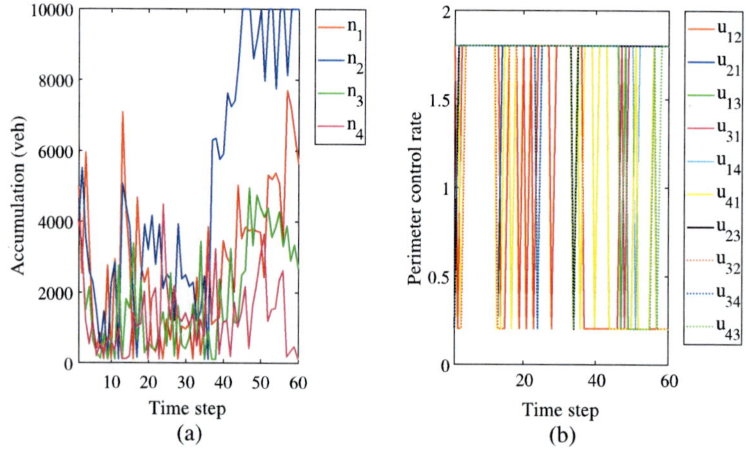

Fig. 10.27 Result of GC strategy (part one): (**a**) evolution of the number of vehicles traveling in each region; (**b**) perimeter control ratios

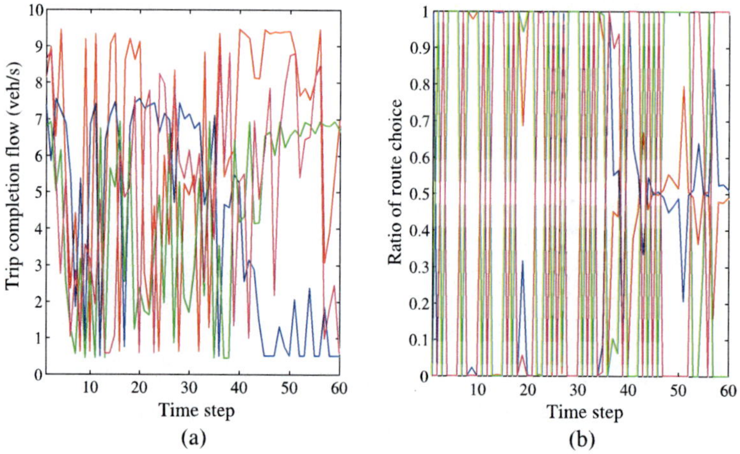

Fig. 10.28 Result of GC strategy (part two): (**a**) trip completion flow for each region; (**b**) ratio of route choice. Where in sub-figure (**a**), the red solid line denotes $G_1^{\triangleright}$, the blue solid line denotes $G_2^{\triangleright}$, the green solid line denotes $G_3^{\triangleright}$, the magenta solid line denotes $G_4^{\triangleright}$. In sub-figure (**b**), the red solid line denotes $b_{214}^{\triangleright}$, the blue solid line denotes $b_{234}^{\triangleright}$, the green solid line denotes $b_{412}^{\triangleright}$, and the magenta solid line denotes $b_{432}^{\triangleright}$

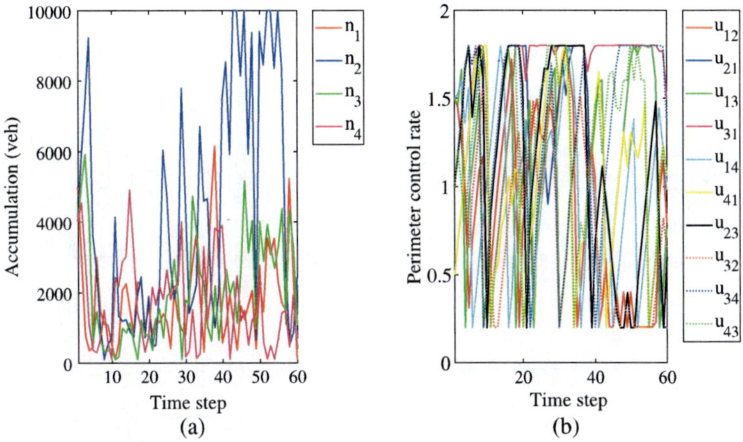

Fig. 10.29 Result of MPC strategy (part one): (**a**) evolution of the number of vehicles traveling in each region; (**b**) perimeter control ratios

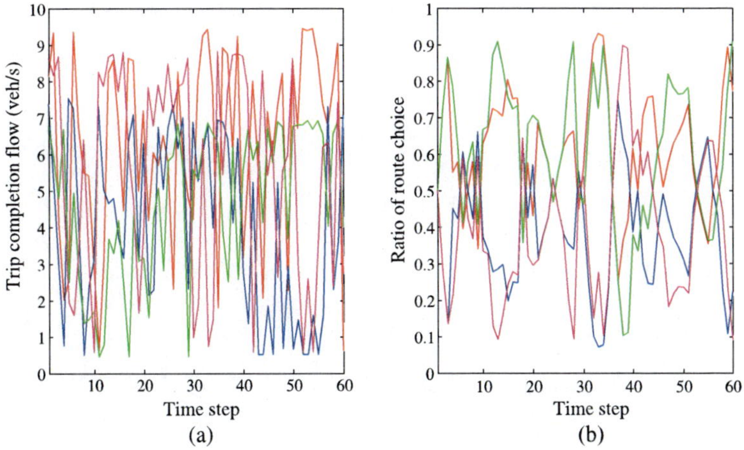

Fig. 10.30 Result of MPC strategy (part two): (**a**) trip completion flow for each region; (**b**) ratio of route choice. Where in sub-figure (**a**), the red solid line denotes $G_1^{\triangleright}$, the blue solid line denotes $G_2^{\triangleright}$, the green solid line denotes $G_3^{\triangleright}$, the magenta solid line denotes $G_4^{\triangleright}$. In sub-figure (**b**), the red solid line denotes $b_{214}^{\triangleright}$, the blue solid line denotes $b_{234}^{\triangleright}$, the green solid line denotes $b_{412}^{\triangleright}$, and the magenta solid line denotes $b_{432}^{\triangleright}$

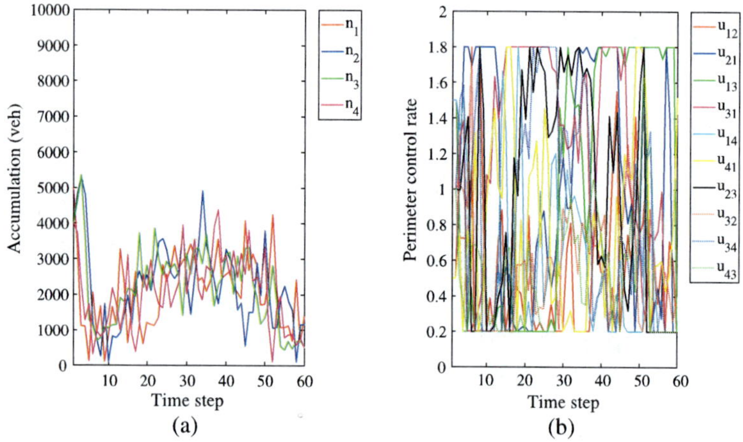

Fig. 10.31 Result of multi-step cMFAPLC strategy (part one): (**a**) evolution of the number of vehicles traveling in each region; (**b**) perimeter control ratios

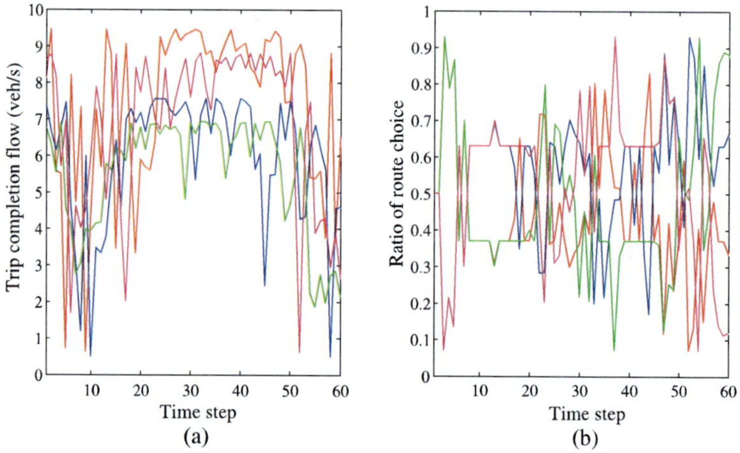

Fig. 10.32 Result of multi-step cMFAPLC strategy (part two): (**a**) trip completion flow for each region; (**b**) ratio of route choice. Where in sub-figure (**a**), the red solid line denotes $G_1^{\triangleright}$, the blue solid line denotes $G_2^{\triangleright}$, the green solid line denotes $G_3^{\triangleright}$, the magenta solid line denotes $G_4^{\triangleright}$. In sub-figure (**b**), the red solid line denotes $b_{214}^{\triangleright}$, the blue solid line denotes $b_{234}^{\triangleright}$, the green solid line denotes $b_{412}^{\triangleright}$, and the magenta solid line denotes $b_{432}^{\triangleright}$

The result of NC is shown in Figs. 10.23 and 10.24. It can be seen from Fig. 10.23a that after a short period of unblocked traffic state, the number of in three of the four regions increase to the maximum values, then traffic jams occur in the three regions, and the jammed state continue until the end of the simulation in regions 2 and 4. Accordingly, the trip completion flows in the corresponding regions also drop rapidly to the minimum value, and maintain to the end of the peak hour, see Fig. 10.24a, which is quite bad for the dissipation of traffic jam.

Figures 10.25, 10.26, 10.27 and 10.28 describe the results of BBC and GC. As depicted in Figs. 10.25a and 10.27a, congestion and traffic jam also occur in some regions under both of the two strategies. Fortunately, the durations of traffic jam under the two methods are shorter than that under NC strategy. The perimeter control ratios of the two strategies are shown in Figs. 10.25b and 10.27b, and it can be seen from the two sub figures that the perimeter control ratios switch back and forth between the maximum and minimum values. Figures 10.26a and 10.28a show that the trip completion flow in each region fluctuates severely under the two perimeter control methods. The route choice proportions of the drivers are shown in Figs. 10.26b and 10.28b.

The results of MPC and multi-step cMFAPLC are compared in Figs. 10.29, 10.30, 10.31 and 10.32, respectively. It can be seen from Figs. 10.29 and 10.30 that under MPC method, there is still a traffic jam in region 2, while all the other regions work well. On the contrary, Figs. 10.31 and 10.32 illustrate that all the four regions are unblocked during the whole simulation process under multi-step cMFAPLC scheme, and neither congestion nor traffic jam appears in each region during the simulation. Accordingly, the trip completion flows of the regions keep at a high level under multi-step cMFAPLC strategy, while that under MPC scheme is relatively low in the jammed region. The perimeter control ratios and route guidance proportions of the two strategies are presented in Figs. 10.29b, Fig. 10.30b, 10.31b and 10.32b, respectively.

As in Sect. 10.2, TTS, TNT, ATT, and TCT are also utilized here to quantitatively evaluate the performances of the above route guidance and perimeter control methods. Under the same simulation settings, if a strategy has the more TNT and the less TTS, ATT, TCT, then it has the better performance.

Among the above four evaluation indicators, TNT and TTS are most common and important. The evolution process of TNT and TTS of the MRUTS under the five route guidance and perimeter control strategies is shown in Figs. 10.33 and 10.34, respectively. As illustrated in the figures, multi-step cMFAPLC has the least TTS. MPC has the most TNT, and TNT of cMFAPLC is similar to MPC. It means that multi-step cMFAPLC and MPC worked better than the other strategies.

Furthermore, the performances of different route guidance and perimeter control methods are compared in Table 10.4. It can be seen from the table that cMFAPLC and MPC are both superior to NC, BBC, and GC. Besides, BBC and GC both worked better than NC. The MRUTS has similar TNT under multi-step cMFAPLC and MPC strategies, but the TTS, ATT, and TCT under multi-step cMFAPLC strategy are less than that under MPC approach. It implies that multi-step cMFAPLC worked better than MPC. Overall, multi-step cMFAPLC is superior to all the other route

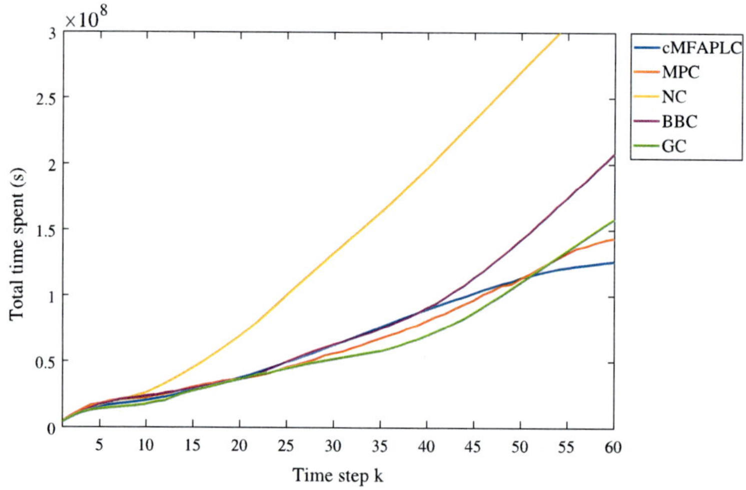

Fig. 10.33 Total time spent of the urban traffic system under the six strategies

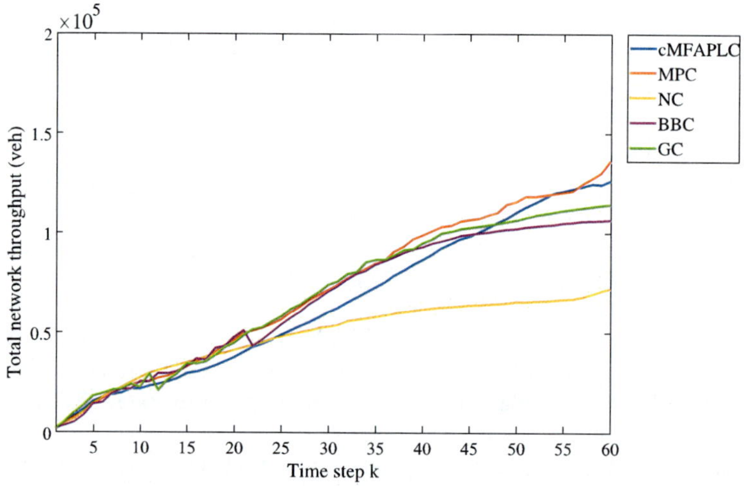

Fig. 10.34 Total network throughput of the urban traffic system under the six strategies

Table 10.4 Evaluation results of different strategies

Strategy	TNT (veh)	TTS (s)	ATT (s)	TCT (s)
NC	7.2205×10^4	3.3483×10^8	4.6372×10^3	0.21
BBC	1.0651×10^5	2.0659×10^8	1.9396×10^3	2.19
GC	1.1428×10^5	1.5667×10^8	1.3709×10^3	4.03
MPC	1.3644×10^5	1.4188×10^8	1.0399×10^3	169.98
cMFAPLC	1.2629×10^5	1.2465×10^8	986.99	32.01

guidance and perimeter control methods. It is worth noting that the TCT of multi-step cMFAPLC is only 32.01 s, so that at each control cycle, the computation time of solving route guidance and perimeter control ratios is much less than the traffic signal cycle length 2 min. Thus, the proposed multi-step cMFAPLC method could be real-time applied.

10.4 Conclusion

Two new data-driven control method called one-step DED-MFAPLC and multi-step cMFAPLC for route guidance and perimeter control of MRUTS are proposed in this chapter. A prominent advantage of the two strategies is that only the measured input/output data of the MRUTS is needed throughout the process of the route guidance and perimeter control strategies design. The effectiveness and superiority of the proposed strategy are verified via simulation and comparison results with some other commonly used route guidance and perimeter control approaches.

References

Aboudolas K, Geroliminis N (2013) Perimeter and boundary flow control in multi-reservoir heterogeneous networks. Transp Res Part B: Methodol 55(9):265–281

Ben-Akiva M, Bierlaire M (1999) Discrete choice methods and their applications to short term travel decisions. In: Handbook of transportation science. Springer New York, NY, USA, pp 5–33

Daganzo CF (2007) Urban gridlock: macroscopic modeling and mitigation approaches. Trans Res Part B: Method 41(1):49–62

Dijkstra EW (1959) A note on two problems in connexion with graphs. Numer Mathe 1(1):269–271

Geroliminis N, Haddad J, Ramezani M (2013) Optimal perimeter control for two urban regions with macroscopic fundamental diagrams: a model predictive approach. IEEE Trans Intell Transp Syst 14(1):348–359

Haddad J, Geroliminis N (2012) On the stability of traffic perimeter control in two-region urban cities. Trans Res Part B: Methodol 46(9):1159–1176

Han Z (1984) On the identification of time-varying parameters in dynamic systems. Acta Automat Sinica 10:330–337

Hou ZS, Jin ST (2011) Data-driven model-free adaptive control for a class of MIMO nonlinear discrete-time systems. IEEE Trans Neural Netw Learn Syst 22(12):2173–2188

Hou ZS, Jin ST (2013) Model free adaptive control: theory and applications. CRC Press, Florida

Hou ZS, Xiong SS (2019) On model-free adaptive control and its stability analysis. IEEE Trans Autom Contr 64(11):4555–4569

Ji Y, Geroliminis N (2012) On the spatial partitioning of urban transportation networks. Trans Res Part B: Methodol 46(10):1639–1656

Kouvelas A, Saeedmanesh M, Geroliminis N (2017) Enhancing model-based feedback perimeter control with data-driven online adaptive optimization. Trans Res Part B: Methodol 96:26–45

Ramezani M, Haddad J, Geroliminis N (2015) Dynamics of heterogeneity in urban networks: aggregated traffic modeling and hierarchical control. Trans Res Part B: Method 74:1–19

Sirmatel II, Geroliminis N (2016) Economic model predictive control of large-scale urban road networks via perimeter control and regional route guidance. IEEE Trans Intell Transp Syst 19(4):1112–1121